Introduction to
Chemical
Engineering
Analysis

*Get the habit of analysis—analysis
will in time enable synthesis to become
your habit of mind.*

FRANK LLOYD WRIGHT

Introduction to
Chemical Engineering Analysis

T. W. F. RUSSELL
M. M. DENN
Professors of Chemical Engineering
University of Delaware

John Wiley & Sons, Inc., New York . London . Sydney . Toronto

Library of Congress Cataloging in Publication Data:

Russell, T   W   F   1934–
    Introduction to chemical engineering analysis.

    Includes bibliographical references.
    1. Chemical engineering. I. Denn, Morton M.,
1939–   joint author. II. Title.
TP155.R88   660.2   72-2173
ISBN 0-471-74545-6

Printed in the United States of America.

10 9 8 7 6

Preface

Chemical engineering education has changed significantly during the past two decades. This textbook is intended for use in the introductory course of a modern chemical engineering curriculum. Such a curriculum should be based on engineering science principles; but it should be oriented toward practical engineering application. A firm foundation in what we term *analysis* is essential if a principle-based curriculum is to be oriented successfully toward practical application.

There has been considerable discussion within the profession about the proper content of the introductory course. We feel that it would be useful to recount the evolution of our thinking as the present course has developed. We are motivated by the widespread criticism of the usual sophomore course in industrial stoichiometry. This course has changed little during the era of major modification of upper-class undergraduate and graduate curricula. Although "modernized" at times by introduction of the digital computer to assist in the solution of major problems, the course content remains basically the solution of steady state mass and energy balances for existing processes. Skills are haphazardly developed through many example problems and little attention is paid to the development of a consistent logical approach to problem solution. Recent developments that have greatly improved and expanded senior-level high-school and university freshman courses are ignored. In an attempt to "simplify" problems for the sophomore level many concepts, particularly in basic thermodynamics, are introduced incorrectly and a relearning must take place in the courses that follow. *Considerations of design are never included because the concept of a rate is not introduced.* The inadequacy of the traditional sophomore course is compounded by the lack of continuity between it and the engineering science courses that immediately follow. There is a difference in approach, type of problem, and analytical level. The transition is a difficult one for many students.

This textbook was developed to remedy these problems. It is a result of six years' experience in teaching the subject to sophomore students at the University of Delaware in both the regular and extension programs. Some portions of the book have also been used with graduate engineers who participated in AIChE continuing education courses at both the national (Today Series) and local levels and as the basis of an undergraduate course for nonmajors. Both of us have been involved in all aspects of the teaching and for three intermediate years taught the sophomore course together, each with a section of about 25 students. The material covered in each section was coordinated on a lecture-to-lecture basis, and its impact on students was evaluated after each lecture and again after each major topic was covered. Substantial student feedback has been received, and several undergraduates have worked with us for extended periods in evaluation and revision. This method of teaching has allowed us to experiment with various ways of organizing and presenting the material.

During the first year of development we taught a slightly modified industrial stoichiometry course, followed by an applied mathematics course that concentrated on solutions of various types of differential equations. It quickly became apparent that the major problem the student faced was developing the equations that described a particular situation. Since the student did not feel adequately trained in this skill, there was a strong tendency to separate the mathematical description and its behavior from the situation that it described. This had two equally undesirable effects. Some students concentrated on the mathematical manipulation and thought little about the relationship to the physical situation, while some students became confused as to the role of mathematics and tended to dismiss the material as being an academic exercise unrelated to physical reality. This is not unlike what has happened outside of the university in professional practice. (It has been our experience in continuing education activities that the greatest number of engineers who profess a need for "more mathematics" are really in need of a better understanding of model development.)

In an attempt to overcome these serious problems and also to revise what we considered to be an incorrect or inadequate presentation in the standard stoichiometry course, we set out to restructure the introductory sequence to emphasize the relationship between mathematical description and physical reality. We began by developing mathematical models for a series of increasingly complex physical situations. For each of the next three years we made changes, in every case increasing emphasis on the complementary roles of conservation equations and constitutive relations, on the significant role of experiment in model development, and on the use of models in design. We reduced the mathematical content each year, finding that much of the material that we were inclined to teach was actually irrelevant to the physical problems of interest and that it could be easily learned subsequently by the student who was motivated from physical considerations. With the content and structure of the course finally established after four years, we then spent the remaining time polishing the presentation and the examples.

The text has the following ultimate objectives.

1. To develop the basic skills of analysis to provide a sound foundation for subsequent courses.
2. To reinforce, amplify, and apply the material covered in basic chemistry, physics, and mathematics.
3. To develop an early appreciation for design by involving the student in simple but significant chemical engineering design problems.

The first part of the text (Chapters 1 to 3) develops the basic logic of engineering analysis, emphasizing the distinction between conservation principles and constitutive relations. The essential role of experiment is stressed. Part Two, on isothermal liquid systems, deals with the principle of conservation of mass. *There is no easier problem in chemical engineering than the analysis of isothermal liquid phase reactors.* By focusing on flow and batch isothermal liquid systems it is possible, in a short time, to discuss meaningful design problems such as the sizing of a chemical reactor. The constitutive relations developed here deal mostly with rates of reaction and mass transfer, and considerable emphasis is again placed on experiment.

In Part Three (beginning with Chapter 10) the concept of internal energy is introduced and applied to nonreacting and reacting liquid systems. The restriction to liquid systems makes it possible to deal completely with the nonisothermal aspects of the problems studied in Part Two without the complication of thermodynamic pressure and volume effects. The mathematical models developed here are slightly more complex than in Part Two, and Chapter 13 motivates a need for the ability to solve linear differential equations. Chapter 14, on gases, shows how the same principles apply to nonliquid processes.

The mathematical topics needed in the text are reviewed in Part Four, "Mathematical Review Notes," and are referred to in the text as needed. The material in this part is frequently covered in mathematics courses, although in our own course at Delaware we make extensive use of Chapter 17 in conjunction with Chapter 13. This is the way in which our students learn to solve differential equations.

Problems are included after each chapter. The text is written with the expectation that problems will be used to amplify and extend the material. It has been our teaching experience that problems are an essential part of the learning process.

We have not included physical property data beyond the minimal needs of a self-contained treatment. Access to tables such as those in Perry's *Chemical Engineers' Handbook* is a necessity for engineering work, and we assume that students will familiarize themselves with these sources at an early stage.

We recognize that this is a challenging text for the beginning engineering student. Although the concepts are not difficult, in analysis we utilize the shorthand language of mathematics, and the reading of mathematics for information instead of manipulation is a new experience for students. It is necessary to read slowly with a pencil, to reproduce results, and to make notes in the margin.

In preparing the book we have received a great deal of help. Our wives and children have quietly tolerated our absences. Our chairmen, R. L. Pigford, A. B. Metzner, and the late J. A. Gerster, showed understanding in allowing us the freedom to experiment. J. J. Carberry, T. R. Keane, C. A. Petty, and S. I. Sandler used portions of the text, and their comments and those of their students were very helpful. Undergraduates J. F. Garner and T. B. Willing and graduate students J. J. Ostermaier and R. Rothenberger read large portions of the text for clarity and accuracy and helped in problem and example development. Undergraduate G. E. Braden and graduate students J. R. Black, J. R. Hepola, P. T. Cichy, and T. L. Holmes helped with problem development while assisting as graders. Six classes of Delaware undergraduates have allowed us to experiment with them—to them we are particularly grateful.

T. W. F. RUSSELL
M. M. DENN

Contents

Notation

We have generally adopted standard notation for physical quantities. In a few cases this has necessitated using the same symbol more than once, but these instances are rare and widely separated in the text. Symbols used only once in a narrow context are not listed. Dimensions F (force), L (length), M (mass), T (temperature), θ (time), and μ (moles) are shown for each symbol. For a discussion of dimensions see Section 3.5.

Notational conventions not in common use are as follows:

( )	"is a function of"; i.e , $f(t)$: f is a function of t. A parenthesis *never* denotes multiplication
[ ], { }, ⟨ ⟩, etc.	always denotes multiplication
_ (underbar)	per unit mass; $\underline{H}$ = enthalpy per unit mass
~ (undertilde)	per mole; $\underset{\sim}{H}$ = enthalpy per mole
~ (overtilde)	partial molar quantity; $\tilde{H}_A$ = partial molar enthalpy of A

The remaining notation is as follows:

a	interfacial area, L^2
a_0, a_1	constants defined by Equations 13.18
A	cross-sectional area, L^2
A_0	area of orifice, L^2
BP	boiling point, T
c	concentration, ML^{-3} or μL^{-3}. Usually subscripted to denote a species or stream, as c_A (concentration of A) or c_1 (concentration of solute in stream 1)
$\underline{c}_p, \underset{\sim}{c}_p$	heat capacity per unit mass and mole at constant pressure, $FLM^{-1}T^{-1}$, $FL\mu^{-1}T^{-1}$
$\underline{c}_V, \underset{\sim}{c}_V$	heat capacity per unit mass and mole at constant volume, $FLM^{-1}T^{-1}$, $FL\mu^{-1}T^{-1}$
C_p, C_V	total heat capacity at constant pressure and volume, FLT^{-1}
°C	degrees Centigrade

D	diameter, L
$\mathscr{D}$	number of dimensions
e	$2.718\ldots$
E	activation energy, $FL\mu^{-1}$; mass flow rate of extract in Chapter 9, $M\theta^{-1}$
$\mathscr{E}$	stage efficiency
F	force, F; mass flow rate of feed in Chapter 9, $M\theta^{-1}$
$°F$	degrees Fahrenheit
g	acceleration due to gravity, $L\theta^{-2}$
g_c	conversion factor for systems using mass and force, $MLF^{-1}\theta^{-2}$
h	height, L; heat transfer coefficient in Chapters 11 to 13, $FL^{-1}T^{-1}\theta^{-1}$
H	enthalpy, FL
$\Delta \tilde{H}_F$	heat of formation, $FL\mu^{-1}$
$\Delta \tilde{H}_R$	heat of reaction, $FL\mu^{-1}$
ΔH_s	heat of solution, FL
i	$\sqrt{-1}$
J	$-\Delta \tilde{H}_R/\rho \varsigma_p$, $L^3 T \mu^{-1}$
k	reaction rate constant, $L^{3n-3}\mu^{1-n}\theta^{-1}$ for reaction of nth order. Usually subscripted, as k_1 for reaction Number 1.
k_0	temperature independent part of k, $L^{3n-3}\mu^{1-n}\theta^{-1}$
$k^{\mathrm{I}}, k^{\mathrm{II}}$	mass transfer coefficients for phases I and II, $M\theta^{-1}$
K	proportional controller gain, Chapters 4 and 13; distribution coefficient based on mass fraction, Chapter 9; $ha/\rho_c q_c \varsigma_{pc}$, Chapters 11 to 13.
$°K$	degrees Kelvin
K_e	equilibrium constant
KE	kinetic energy, FL
L	length, L
m	mass, M
M	distribution coefficient based on concentration
M_w	molecular weight
n	number of moles, μ. Usually subscripted to denote a species
N_{Fr}	Froude Number u^2/gD
N_{Re}	Reynolds Number $Dv\rho/\mu$
p	pressure, FL^{-2}
P	power, $FL\theta^{-1}$
PE	potential energy, FL
q	volumetric flow rate, $L^3\theta^{-1}$. Often subscripted to denote a stream
$\mathbf{q}$	heat added to system, FL
Q	rate at which heat is added, $FL\theta^{-1}$
r	intrinsic rate of reaction, $\mu L^{-3}\theta^{-1}$

$\mathbf{r}$	rate of mass transfer, $ML^{-2}\theta^{-1}$. Sometimes subscripted to denote a species
r_{i+}, r_{i-}	rate of formation, disappearance of species i by reaction, $\mu L^{-3}\theta^{-1}$
$\mathbf{r}_{i+}, \mathbf{r}_{i-}$	rate of accumulation, depletion of species i by mass transfer, $ML^{-2}\theta^{-1}$
R	gas constant, $FL\mu^{-1}T^{-1}$; mass flow rate of raffinate in Chapter 9, $M\theta^{-1}$
$°R$	degrees Rankine
s_N, s_{cN}	separation ratio for N-stage, N-stage countercurrent system
S	mass flow rate of solvent, $M\theta^{-1}$
t	time, θ
T	temperature, T
u	bubble rise velocity in Chapter 3, $L\theta^{-1}$; $\rho_c q_c \mathcal{c}_{pc} K/[1 + K]\rho q \mathcal{c}_p$ in Chapters 12, 13
U	internal energy, FL
v	velocity, $L\theta^{-1}$
V	volume, L^3
$\mathcal{V}$	number of variables
w	work done by system, FL
W	rate of doing work, $FL\theta^{-1}$
W_s	rate of doing shaft work, $FL\theta^{-1}$
x_i	c_i/c_{if} in Chapter 7; mass fraction of species i in phase II in Chapter 9
y_i	mass fraction of species i in phase I
z	axial position, L; pV/nRT in Chapter 14
α_i	stoichiometric coefficient for reactant i
γ	$\mathcal{c}_p/\mathcal{c}_V$
δ_i	stoichiometric coefficient for product i
$\Delta t, \Delta z, \ldots$	small change in $t, z, \ldots$
ε	$[q_1 + q_2 - q_3]/[q_1 + q_2]$ in Chapter 4; power per unit volume in Chapter 8, FL^{-2}
ε_i^k	stoichiometric coefficient for element k in the reaction to form species i
θ	holding or residence time, characteristic time, θ
Θ	holding time for tubular reactor, θ
λ	q^{II}/q^{I}, m_A/m, m_1/m, $\rho_1 q_1/\rho q$, depending on context
Λ	F/S
μ	viscosity, $ML^{-1}\theta^{-1}$; Λ/K in Chapter 9
ν	n_{A1}/n_A, $q_1 c_{A1}/q c_A$ depending on context
π	$3.14159\ldots$
ρ	density, ML^{-3}

subscripts

c	coolant; critical in Chapter 14
e	equilibrium
f	feed
g	gas
I	imaginary part
l	stream leaving
L	liquid
0	initial value
R	real part; reduced in Chapter 14
s	saturation; steady state in Chapter 13

superscripts

I, II	phase I, II

Introduction to Chemical Engineering Analysis

PART I

Introduction

TO THE STUDENT

We have taught introductory courses in chemical engineering analysis for six years. Formal student evaluations and our own personal evaluations are in agreement that these courses have been successful. During this period, by trial and error and with the help of considerable faculty-student interplay, we have developed the approach taken in this book. To the best of our ability the text is a faithful transformation of what we know to be an effective learning experience. You will find that this approach to learning is probably new to you. You will have to draw on your previous studies in physics, chemistry, and mathematics, and combine these with your observational experience of the physical world. You will find it necessary to use mathematics as a descriptive language to provide information about engineering problems. This is common in engineering practice, but represents a point of view which differs from that which you have probably associated with your mathematics courses to date.

The essence of analysis is the ability to convert an engineering problem into a mathematical statement, and then to use the mathematical statement to obtain useful engineering information. This means that you must learn to *read* mathematics. Mathematics is a particularly convenient language because its symbolic structure and logical rules enable you to proceed through a problem in a concise and systematic manner. When you have become sufficiently familiar with this approach you will find it a pleasure to use. As with any new language, however, the initial learning stages may seem difficult. Even by the experienced the language of mathematics must be read slowly, using pencil and paper, with frequent stops to perform and verify the indicated manipulations. There is no other way to develop a true command of the language as a working tool.

The material in the text can be learned and applied by a student beginning a chemical engineering program. We have demonstrated this fact successfully in the six years of classroom development and testing that preceded the publication of the text. We have even used this material in an introductory course for nonmajors, including students from chemistry, secondary education, biology, and electrical engineering. The text will force you to think about the relation between laboratory experiment, mathematical manipulation, and engineering design. It will require you to review material from chemistry and physics. It will make you *use* the calculus that you have studied, and you should find yourself referring often to your calculus text or the Mathematical Review Notes, Chapters 15 to 17.

This text is not a summary of all the important areas of chemical engineering, and you should not fall into the too-common trap of believing that analysis and engineering are synonymous. Analysis can be applied to many classes of problems that are not covered in this introductory text, and skill in analysis is a necessary foundation for the study of engineering. The art of engineering goes far beyond analysis, however.

CHAPTER 1

The Role of Analysis

1.1 CHEMICAL ENGINEERING

Chemical engineering is the profession concerned with the creative application of the scientific principles underlying the transport of mass, energy, and momentum and the physical and chemical change of matter. The broad implications of this definition have been justified over the past few decades by the kinds of problems that chemical engineers have solved, though the profession has devoted its attention in the main to the chemical process industries. As a result chemical engineers have been defined more traditionally as those applied scientists trained to deal with the research, development, design, and operation problems of the chemical, petroleum, and related industries. Experience has shown that the principles required to meet the needs of the process industries are applicable to a significantly wider class of problems, and the modern chemical engineer is bringing his established tools to bear on such new areas as the environmental and life sciences.

Chemical engineering developed as a distinct discipline during the twentieth century in answer to the needs of a chemical industry no longer able to operate efficiently with manufacturing processes which in many cases were simply larger scale versions of laboratory equipment. Thus, the primary emphasis in the profession was initially devoted to the general subject of how to use the results of laboratory experiments to design process equipment capable of meeting industrial production rates. This led naturally to the characterization of design procedures in terms of the *unit operations*, those elements common to many different processes. The basic unit operations include fluid flow, heat exchange, distillation, extraction, etc. A typical manufacturing process will be made up of combinations of the unit operations. Hence, skill in designing each of the units at a production scale would provide the means of designing the entire process.

The unit operations concept dominated chemical engineering education and practice until the mid-1950s, when a movement away from this equipment-oriented philosophy toward an *engineering science* approach began. This approach holds that the unifying concept is not specific processing operations, but rather the understanding of the fundamental phenomena of mass, energy, and momentum transport that are common to all of the unit operations, and it is argued that concentration on unit operations obscures the similarity of many operations at a fundamental level.

Although there is no real conflict between the goals of the unit operations and engineering science approaches, the latter has tended to emphasize mathematical skills and to deemphasize the design aspects of engineering education. Such a conflict need not exist, and recent educational effort has been directed toward the development of the skills that will enable the creative engineering use of the fundamentals, or a synthesis of the engineering science and unit operations approaches. One essential skill in reaching this goal is the ability to express engineering problems meaningfully in precise quantitative terms. Only in this way can the chemical engineer correctly formulate, interpret, and use fundamental experiments and physical principles in real-world applications outside of the laboratory. This skill, which is distinct from ability in mathematics, we call *analysis*.

This book is a basic text in analysis for chemical engineers. Despite the widened scope of the modern profession we have oriented it toward the process applications. This is not because we are old-fashioned, but because we firmly believe that the traditional applications represent the best learning mechanism for those principles of transport and change that are more widely applicable. The student who may be inclined toward one of the newer applications, such as the life sciences, will find his needs equally met when taken together with studies in the area toward which he is oriented.

1.2   CHEMICAL ENGINEERING PROBLEMS

By discussing the three problems posed in this section we will attempt to demonstrate in a general way how the chemical engineer attacks a problem, the typical kinds of equipment he is concerned with, what features are common to the problems, and how analysis is involved in each.

Example 1.1

Find a way to produce 100 million pounds per year of monoethylene glycol

$$\begin{array}{cc} H & H \\ | & | \\ HO-C-C-OH \\ | & | \\ H & H \end{array}$$

·Ethylene glycol is a colorless, odorless liquid which resembles water but is denser and more viscous. A constituent of antifreeze, it is one of the major organic chemicals produced industrially, with sales of about 1.5 billion pounds per year. The entire glycol family, organic molecules with two hydroxyl (—OH) groups per molecule, is of chemical importance, for glycols are good solvents for many organic compounds. They are themselves water soluble, and they can undergo chemical reaction of one or both hydroxyl groups to form other compounds. Since monoethylene glycol does not exist in natural form it must be manufactured.

Laboratory experiments have shown that ethylene glycol can be produced by a number of different chemical paths:

1. Hydration of ethylene oxide

$$H_2C\!\!-\!\!-\!\!CH_2 + HOH \rightarrow CH_2OHCH_2OH$$
$$\diagdown\!\!O\!\!\diagup$$

2. Hydrolysis of ethylene chlorohydrin

$$CH_3OCH_2Cl + NaHCO_3 \rightarrow CH_2OHCH_2OH + NaCl + CO_2$$

3. From Formaldehyde and Carbon Monoxide

$$CH_2O + CO + H_2O \rightarrow CH_2OHCOOH$$

$$CH_2OHCOOH + ROH \rightarrow CH_2OHCOOR + H_2O$$

$$CH_2OHCOOR + 2H_2 \rightarrow CH_2OHCH_2OH + ROH$$

We will simplify our considerations here greatly by focusing only on the first path, hydration of ethylene oxide. Ethylene oxide can be obtained by oxidation of ethylene and is readily available, with an annual U.S. production in excess of 15 billion pounds per year.

Now, clearly, the first essential feature of a process to manufacture ethylene glycol is a vessel of some sort—a *reactor*—in which ethylene oxide and water can be brought together to react. Schematically, we might show this step as in Figure 1.1. The *product* stream will, of course, contain ethylene

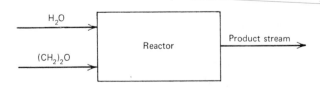

FIGURE 1.1   Reactor in which to form ethylene glycol from water and ethylene oxide.

glycol, but it might contain other material. There may be unreacted water and/or ethylene oxide. The hydroxyl group is very reactive and we must also allow for at least the following reactions:

1. Formation of diethylene glycol from ethylene glycol

$$CH_2OHCH_2OH + (CH_2)_2O \rightarrow CH_2OHCH_2OCH_2CH_2OH$$

2. Formation of triethylene glycol from diethylene glycol

$$CH_2OHCH_2OCH_2CH_2OH + (CH_2)_2O \rightarrow$$

$$CH_2OHCH_2OCH_2CH_2OCH_2CH_2OH$$

Thus, following the reactor in the process we must allow for the possibility that there will need to be a unit that removes water, one that removes unreacted ethylene oxide, and one or more units that will separate the mono-ethylene glycol from the di- and triglycols. Such a process is sketched schematically in Figure 1.2.

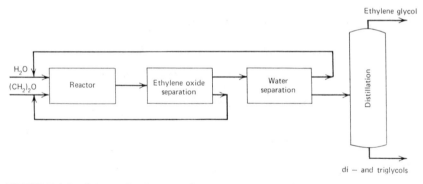

FIGURE 1.2   Schematic diagram of a process to manufacture ethylene glycol.

A schematic process block diagram is only a first step in the solution of the problem posed here. The following questions must be answered to establish even a preliminary process design.

1. What factors control the amounts of mono-, di-, and triglycol in the reactor effluent?
2. How large a reactor is required to produce the 100 million tons per year of monoethylene glycol, and what shape should it have?
3. What kind of equipment is needed to separate the water and ethylene oxide from the reactor effluent, and how much of each must be removed?
4. What kind of unit is needed to separate the glycols from each other?

The search for an answer to the first two questions begins in the chemistry laboratory with a series of experiments. In the case of a mature chemical like ethylene glycol, which has been manufactured for many years, the results of such experiments are well known. Laboratory experiments are usually carried out in small *batch reactors*, vessels in which the raw materials are placed and the progress of the reaction observed over a period of time. For the glycols these experiments show that the relative amounts of mono-, di-, and triglycol depend on the ratio of ethylene oxide to water in the reactor at the start of the experiment. Experiments at different temperatures show that the relative amounts of the three glycols are unaffected by temperature, but that the rate at which they are formed is highly temperature dependent.

The planning, interpretation, and use of these experimentsr equires an application of analysis. It is necessary to derive a set of equations that constitute a *mathematical model* describing the laboratory experiment. The test of such a model is its ability to predict laboratory results for changing conditions, and the model is used to ensure that the entire range of importance has been examined. A model of the reactor for the continuous processing unit must also be constructed. As we shall see in Chapters 5 and 7, the same information about the rate at which the chemical reaction occurs is required in both models. Thus, information gained at the laboratory scale can be applied to a different type of reactor at the production level. The mathematical model of the plant unit can then be examined to determine the composition of the effluent under various operating conditions and an optimum size, configuration, and throughput rate selected. We shall study this specific problem in detail in Section 7.6.

The approach to separation of species is a similar one. Laboratory experiments or data recorded in handbooks need to be used together with the construction of mathematical models for ultimate design. Separation of water from the glycols, for example, might be carried out by evaporating the water and collecting and condensing the vapor. (Water boils at 100°C, all of the glycols at temperatures greater than 197°C.) The continuously flowing mixture must be heated in some device such as the *heat exchanger* shown schematically in Figure 1.3. Here the model must incorporate the rate at which heat can be transferred between streams in order to enable the designer to choose the temperature and flow rate of the stream, the area of

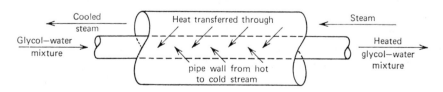

FIGURE 1.3   Schematic diagram of a heat exchanger.

contact between the hot steam line and the glycol-water line, and so on. Heat transfer problems of this type are discussed in Chapter 11.

Example 1.2

Design a secondary treatment unit that will process 60 million gallons per day of domestic sewage in such a way that it can be discharged into natural water without harmful effect.

In secondary treatment of sewage dissolved organic material is removed by oxidation to form carbon dioxide and water. The reactor for this process is usually a large open tank which has air sparged through pipes on the bottom. The basic design questions are the relation of effluent purity to tank size, sparger design, and air feed rate.

Application of analysis to this process is significantly more complicated than in the glycol problem. The understanding of the oxidation reaction is incomplete, since a large number of organic species of unknown composition is usually present and the reactions must be catalyzed by enzymes produced by living microorganisms present in the sewage. Thus, the rate of oxidation of sewage is intimately related to the birth, growth, and death rates of the organisms. Furthermore, the system is a three-phase one, consisting of water, air, and porous solid flocs of microorganisms. The rate of the overall process depends in part on the rate at which oxygen dissolves in the liquid, the rate at which oxygen and organic move through the floc to the individual cells and through the cell walls, etc. That is, the chemical reaction between organic and oxygen is governed in part by a nonchemical phenomenon, the transfer rate of mass.

It is possible to isolate many of these phenomena experimentally and to model them. In particular, we discuss rates of mass transfer in Chapter 8 and rates at which bubbles rise in a liquid in Section 3.5.3. Present design methods for treatment facilities do not attempt to characterize the detailed chemistry, but are based on an overall estimate of organic content called biological oxygen demand (BOD). BOD is a measure of the amount of oxygen consumed by a sample of sewage seeded with the microorganism in a batch laboratory reactor over a five-day period. The total mass of organism can be monitored as a function of time during the BOD experiment to estimate the biological growth rate. BOD can be thought of as the equivalent of a chemical species that must be reacted, much like the ethylene oxide in the preceding example. In principle, then, the problem of design is similar to that for the glycol, except now the model for the full-scale process must adequately take into account not only the rate of chemical reaction but also the rates of the bacterial growth process and the mass transfer processes. In Section 8.5 we will consider a simplified case of a chemical reaction with associated mass transfer, illustrating the basic approach, but a problem of the full scope of the sewage treatment is too complex for an introductory text.

Example 1.3

Develop a device that can be employed to take over the function of the human kidney.

It is often commented that the human body is simply a compact and sophisticated chemical plant. This is, of course, a gross and misleading simplification. Nonetheless, certain physiological functions do bear a striking resemblance to chemical processing units. The kidney, which removes metabolic waste products and some poisons from the blood, regulates the body's acid-base balance by ion transfer between the blood and other fluids, and participates in other processes such as formation of red blood cells and regulation of blood pressure. The first two of these can be represented schematically as in Figure 1.4. The low molecular weight materials—metabolic

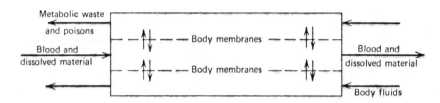

FIGURE 1.4   Schematic diagram of the kidney.

products (e.g., urea), poisons (e.g., barbituates), Na^+ and H^+ ions, etc.— move through membranes whose passages are too small to allow movement of important high molecular weight proteins such as hemoglobin.

The device currently in clinical use to replace certain of the kidney's functions is called a *hemodialyser*, or, more commonly, an artificial kidney. Dialysis is a process in which a membrane is placed between two liquids and works like a microscopic screen in allowing passage of small molecules and holding back large ones. It is employed commercially in the purification of sugar containing small amounts of high molecular weight substances. The hemodialyser is shown schematically in Figure 1.5. Blood flows from an artery, usually in the wrist, past a cellophane membrane whose pores are too

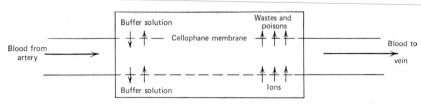

FIGURE 1.5   Schematic diagram of a hemodialyser.

small to allow movement of protein. On the other side of the membrane is a bath of a solution of NaCl, NaCOCH$_3$, KCl, CaCl$_2$, MgCl$_2$, and dextrose. Urea and other wastes and poisons such as barbituates pass through the membrane from the blood (as do some essential amino acids and vitamins not removed by the natural kidney), while ions from the bath pass in the opposite direction. The blood is then returned to a vein, also in the wrist. Not shown are the treatments needed to prevent coagulation.

There are a number of questions that need to be answered for efficient design and operation of a hemodialyser. The most obvious is, of course, development of efficient, biologically neutral membranes with the proper pore properties. This is, at present, an empirical science with little application for analysis. For a given membrane, however, it is necessary to determine transfer rates for various species over a range of flow conditions in order to incorporate them into a design model. Relevant quantities to be calculated then are spacing between membranes, contact area, and number of parallel flow routes.

Alternatives to dialysis have been proposed. As one example, activated charcoal is often used in purification processes because it adsorbs material on its surface. A column filled with charcoal granules through which the blood passes has been studied as a possible means of removing unwanted material from the blood. The obvious questions are how large should the granules be and how long a column?

1.3   CONCLUDING REMARKS

The processes discussed in the preceding examples share some common features, despite the variety of the applications. They involve chemical reaction, change of state of matter, and the transfer of material, energy, and momentum, though perhaps the momentum transfer was not so obvious as the other two. These change and transport processes take place within units such as the reactor and heat exchanger, and it is the relationship between the design and operating variables of these unit operations and the composition and properties of the output streams that we need in order to evaluate, design, and operate the overall process.

Chemical engineering analysis is concerned with developing an understanding of the rate processes of transport and chemical and physical change and in expressing these processes quantitatively. Such quantitative descriptions are then applied to the units of interest in order to obtain *mathematical models* relating process behavior and performance to the variables that are at the engineer's disposal. A mathematical description is important not only because it is a concise and effective shorthand for communication of information, but even more so because a mathematical description of a phenomenon like chemical reactor behavior can be manipulated under the prescribed rules of mathematics to yield essential information concerning expected performance and feasibility. A thorough understanding of the use of mathematical models is essential if one is to properly define, organize, and understand chemical engineering problems.

More specifically, we mean here by analysis the systematic techniques for carrying out the following operations:

1. Description of a physical situation in mathematical symbols.
2. Manipulation of the mathematical model to determine expected behavior of the physical situation.
3. Comparison of the model with the true physical situation.
4. Careful study of the limitations of the mathematical model.
5. Use of the model for equipment design and prediction of performance.

In the following chapters we shall first examine the foundations underlying analysis and then apply these principles to a variety of situations of practical chemical engineering interest.

At this point you may wish to supplement the brief treatment of chemical engineering contained in this chapter. A discussion of the traditional aspects can be found in a small book available in paperbound edition,

1.1 Killeffer, D. H., *Chemical Engineering*, Doubleday, New York, 1967.

Killeffer focuses on the development of several specific processes which nicely demonstrate chemical engineering practice, the evolution of technology, and the respective roles of chemist and chemical engineer. A fine detailed source is

1.2 *Kirk-Othmer Encyclopedia of Chemical Technology*, 2nd Ed., Interscience, New York, 1963.

Each section in Kirk-Othmer deals with the history, chemistry, and manufacturing operations of the subject. It is well worth browsing at this stage of study. In particular, see such sections as ammonia, glycols, penicillins, petroleum refining processes, polyamides (nylon), and sulfuric acid. Penicillin is discussed in greater detail in

1.3 "The History of Penicillin Production," *Chemical Engineering Progress Symposium Series No. 100*, Vol. 66 (1970).

The *Symposium Series* is a periodical in which each number is devoted to a specific topic. You may also wish to consult two texts which contain a large number of process descriptions and can, to a large degree, be understood with only the background in basic chemistry expected of readers of this book:

1.4 P. H. Groggins, Ed., *Unit Processes in Organic Synthesis*, McGraw-Hill, New York, 1958.

1.5 R. N. Shreve, *The Chemical Process Industries*, 2nd ed., McGraw-Hill, New York, 1956.

A readable introduction to polymer processing can be found in the first, fourth, and fifth sections of

1.6 F. W. Billmeyer, Jr., *Textbook of Polymer Science*, Interscience, New York, 1962.

and Chapters 5 and 12 of

1.7 F. Rodriguez, *Principles of Polymer Systems*, McGraw-Hill, New York, 1970.

The engineering aspects of waste water treatment are summarized in a paper in the journal *Power*,

1.8 R. H. Marks, "Waste-Water Treatment," *Power*, June 1967.

A survey of some of the physiologically oriented activities of chemical engineers, though rather incomplete and thus, to some extent, misleading in emphasis, can be found in three collections:

1.9 D. Hershey, Ed., *Chemical Engineering in Medicine and Biology*, Plenum, New York, 1966.

1.10 "Chemical Engineering in Medicine," *Chemical Engineering Progress Symposium Series No. 66*, Vol. 62 (1966).

1.11 "The Artificial Kidney," *Chemical Engineering Progress Symposium Series No. 84*, Vol. 64 (1968).

Most libraries will also have the proceedings of the Annual Conference on Engineering in Medicine and Biology.

CHAPTER 2

Basic Concepts of Analysis

2.1 INTRODUCTION

The successful solution of chemical engineering problems requires that we have the ability to quantitatively describe, or *model*, the behavior of the elements of a process. We do this by using the principles of chemistry, physics, and mathematics to obtain equations. These equations can then be manipulated to predict what will happen under given circumstances. Thus, we will know the effect on the end product of changing the temperature at which a reactor operates, or the size of a pipe in a heat exchanger. The analysis process is straightforward and systematic. In this chapter we will examine the approach, see how a useful model of a simple process unit can be obtained, and get somewhat of a preview of the things to look for in more complex situations.

2.2 THE ANALYSIS PROCESS

The specific goals of analysis were outlined at the end of Chapter 1: description of a physical situation, prediction of behavior, comparison with true behavior, evaluation of the limitations of the model, and prediction and design. The logical sequence of the process is shown as a block flow diagram in Figure 2.1. Our aim is to become proficient in detail at each step of the process and to understand thoroughly the interactions between the various steps. We will begin by briefly discussing each step as shown in the figure.

We have seen the variety of physical situations that are of interest to chemical engineers. These include process equipment such as reactors, heat exchangers, and distillation columns, as well as small-scale laboratory equipment for determining basic design data or for investigating physical principles. We might need a mathematical description of the properties of a

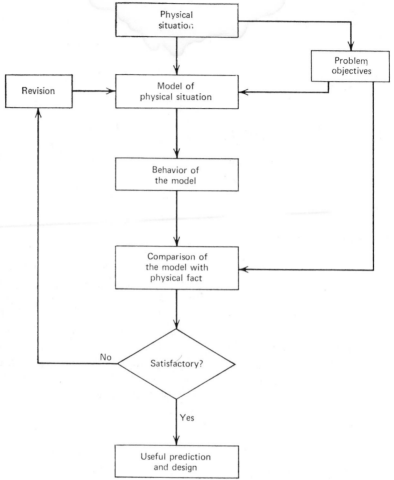

FIGURE 2.1   The analysis process.

material, say a gas in terms of its molecular configuration or a porous membrane in terms of its composition and preparation. Whether we are trying to describe the behavior of a piece of equipment, a part of the human circulatory system, or any other physicochemical phenomenon, the development of a mathematical model proceeds in the same manner.

The basic sources of any mathematical description are the conservation principles for mass, energy, and momentum. Taken together with the other fundamental principles of physics, such as gravitational attraction, it should be possible *in principle* to obtain a mathematical description of any phenomenon. That this is an unreasonable expectation in fact is obvious at once, for the science of physics is devoted to precisely the end of obtaining a complete description of nature from fundamental principles, and physics is still a very active science. Just recall from the basic physics course the complexity of describing the state of a pure, nonreacting gas in terms of the behavior of the millions of interacting molecules. Thus, we may expect that there will be many situations of engineering interest which are too complex for the laws of physics to be applied in their most fundamental form. We therefore need a secondary source from which we may draw to develop mathematical models. This nonfundamental source, so essential to engineering analysis, produces what we have chosen to designate as a *constitutive relationship*. These relationships are generally developed from careful and clever experimentation for specific situations of interest. Development of a systematic approach to mathematical description using the conservation laws and constitutive relationships is a major concern of this text and much of what follows is devoted to meeting this goal.

Most mathematical descriptions will represent an essential compromise between the complexity required for description of a physical situation true in every detail and the simplicity required so that the model may be compared with experiment and then used for design. The degree of compromise depends on the specific problem objectives and enters into the effort that we devote to obtaining a model.

Given a mathematical description, it is necessary to verify its correctness before using it for an engineering purpose. This requires solving the equations to predict the behavior of the mathematical model under conditions where a direct comparison can be made with the behavior of the real physical situation. Algebra and differential and integral calculus are usually required to determine the behavior of a model, and in many cases an appreciation of numerical techniques and the use of automatic computing devices is needed. These applied mathematics are an essential tool in the analysis process, although our major emphasis here is on the other parts of analysis.

We can manipulate the mathematical model in order to see how the variables of the problem interact and what the effect of parameters of the model is on model behavior. This information may be used to plan appropriate experiments for testing the range of validity of the mathematical model, as well as to interpret experimental results. The comparison step is an essential feature of the analysis of engineering processes, for it is here that the engineer forms his value judgments about the usefulness and reliability of a model for subsequent design and prediction. If, for a given set of objectives, the comparison between model and physical reality is adequate, then we may proceed to use the model; if not, we must consider why the comparison is inadequate, make appropriate modifications, and compare again.

2.3 A SIMPLE EXAMPLE

Prior to the detailed study of the analysis process it is helpful to illustrate some of the concepts briefly described in the previous section by a particularly simple example. The physical situation is shown in Figure 2.2. A tank of constant cross section, initially filled with some liquid to a height h_0, is emptied by a flow through a small hole, or *orifice*, in the base. Our objectives,

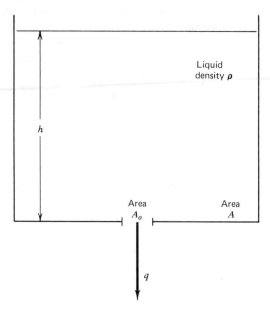

FIGURE 2.2   Tank draining through an orifice in the bottom.

which must always be defined, might be the answer to all or some of the following questions:

1. How long will it take the tank to drain?
2. How does the height of liquid vary with time?
3. How does the flow rate through the orifice vary with the depth of liquid?

This tank draining is a physical situation with which most people have had direct experience, and if not, experience can quickly be obtained by punching a hole in a large can, filling it with water, and then observing the behavior as the liquid flows out. Observation shows us that the level in the tank decreases with time, the process takes place at constant temperature, and that the flow rate of liquid through the orifice varies with the height of the liquid and with the size of the exit orifice. Little more can be said with verbal statements and it is necessary to symbolize the problem and make use of a mathematical description if we are to obtain quantitative answers to the questions posed. We first define a set of symbols that seems convenient:

q volumetric flow rate from tank, measured in cubic feet per second
A cross-sectional area of tank, in square feet
A_0 cross-sectional area of orifice, in square feet
h height of liquid at any time, in feet
ρ density of liquid, in pounds per cubic foot
t time, in seconds

The precise application of conservation principles to the description of this process will be illustrated in later chapters. The simplicity of the problem allows us to generate a model on a more intuitive basis at this point. *The conservation of mass clearly requires that the rate of outflow of mass through the orifice be equal to the rate at which the mass changes within the tank.* The mass in the tank is equal to density times volume of liquid, ρAh. The meaning of the time derivative as a rate of change then indicates that the rate of change of mass in the tank is $d[\rho Ah]/dt$. The rate at which mass flows out is equal to the product of volumetric flow rate, q, and density, or mass per volume, ρ. Thus

$$\frac{d[\rho Ah]}{dt} = -\rho q \tag{2.1}$$

where the minus sign indicates a flow out and a resulting decrease in mass. Since ρ and A are constants, Equation 2.1 can be written

$$\rho A \frac{dh}{dt} = -\rho q$$

or

$$\frac{dh}{dt} = -\frac{q}{A} \tag{2.2}$$

Equation 2.2 is a single equation involving two quantities that we do not know, the height of liquid, h, and the rate at which it is flowing from the tank, q. Since we have two unknowns and only one equation we must seek a second relationship. In Chapter 3 we shall see how a consideration of dimensions partly establishes a relation between the variables in this problem, q and h, and in Chapter 10 we shall exploit the principle of conservation of energy to this end. For the present, however, we shall presume that we are faced with the not uncommon situation of being unable or unwilling to apply further conservation principles at a convenient level of complexity. Our additional relationship between q and h, the *constitutive relationship*, must then be obtained by intuition and/or experiment. We would anticipate that a relationship obtained in this way will be rather less general than one based on fundamental conservation principles, and we must use great care in applying the results to situations that differ very much from the conditions of any experiments that we have performed.

Now, we know that the flow occurs through the orifice because the pressure in the liquid at the base of the tank is greater than the pressure of the atmosphere, thus forcing the liquid out, and that the greater the pressure difference the greater the flow. We can express the general relationship as follows:

$$q = q(\Delta p)$$

by which we mean that q, the flow rate, is a function of Δp, the pressure change across the orifice. If the top of the tank is open, then the pressure there, too, is atmospheric. The pressure in the liquid at the bottom of the tank exceeds the pressure of the atmosphere by the weight of the liquid column, which is proportional to the height of liquid. The pressure difference, or "driving force" for flow, is therefore proportional to h. It is the functional relationship of q to h, denoted as $q(h)$, which we seek as our second relation to be combined with Equation 2.2.

The approach that we shall take is to postulate the form of the dependence of q on h (our constitutive relation), solve the model Equation 2.2 for h, and then check the prediction of the model with the experimental data. If the model and data do not agree, we will use the way in which they disagree as an aid in postulating a new dependence. The simplest dependence of q on h that we can imagine is no dependence at all, or q equal to a constant, $q = c$. This relation cannot be accurate for all values of h, for we know that as h goes to zero q must also go to zero—there is no flow from an empty tank. Nevertheless, it is instructive to consider this simplest case, for it will both direct our further study and prove to be of some inherent interest in itself.

For $q = c$, Equation 2.2 becomes

$$\frac{dh}{dt} = -\frac{c}{A} \tag{2.3}$$

This equation involves the derivative of h. But we want h itself, so an integration must be performed. The answer, of course, is obvious, for only a linear function of time has a constant derivative, but we shall still go through all the logical steps. If Equation 2.3 is true at all times, then we still have an equality if we add together the equation at two different times, or, in fact, at all times. Addition over all times is simply integration, so it must be true that the integral of the left-hand side between any two times must be equal to the integral of the right between the same times. That is, for any times t_1 and t_2

$$\int_{t_1}^{t_2} \frac{dh}{dt}\, dt = \int_{t_1}^{t_2} -\frac{c}{A}\, dt = -\frac{c}{A}[t_2 - t_1] \tag{2.4}$$

The integral of a derivative is simply the function evaluated at the endpoints, so Equation 2.4 is

$$h(t_2) - h(t_1) = -\frac{c}{A}[t_2 - t_1]$$

It is convenient to take t_1 as zero and to denote the initial height as h_0, while letting t_2 be any time, denoted simply by t. Thus at any time

$$h(t) = h_0 - \frac{ct}{A} = h_0\left[1 - \frac{ct}{Ah_0}\right] \tag{2.5}$$

According to our model, then, the height should be a linear function of time with intercept h_0 and slope $-c/A$. Table 2.1 shows some data of liquid height

TABLE 2.1   Liquid height versus time for the tank-emptying experiment (tank diameter = 10.75 in., tank height = 12 in., orifice diameter = 0.609 in.)

Height of Liquid (inches)	Time (seconds)	Height of liquid (inches)	Time (seconds)
12	0	6	36.4
	0		35.8
	0		36.4
11	5.8	5	43.8
	6.1		42.8
	5.9		43.8
10	10.9	4	51.0
	11.5		50.5
	11.6		51.6
9	16.6	3	60.2
	17.8		59.2
	17.2		60.6
8	23.0	2	71.0
	23.5		69.8
	23.0		71.4
7	30.0	1	85.0
	29.2		84.0
	29.8		85.2

versus time for three experimental runs in a draining tank, and the data are plotted in Figure 2.3 as h versus t. The small solid circles in Figure 2.3 represent the data from Table 2.1. In most cases one circle represents the three experimentally determined times. Examining the plot it is immediately evident that any line which passes through all the data will be curved and not the straight line predicted by Equation 2.5. If we look at the points more

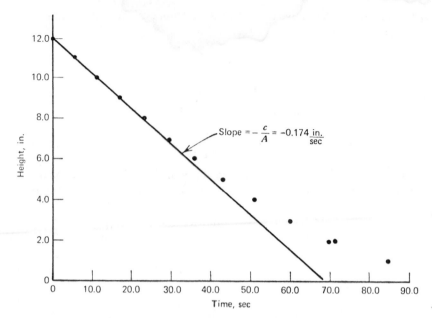

FIGURE 2.3   Liquid height versus time. The straight line is Equation 2.5.

closely, however, we see that *h is* linear with *t* for small values of *t*. In fact, we can draw a straight line that, as closely as we can visually estimate, passes through the first four or five data points. The solid line on Figure 2.3 is the best estimate of this line that we can sketch by eye and has a slope $-c/A = -0.174$ in./sec. (We will have more to say about the "best fit" in Chapter 6.) Our model represented graphically by this line predicts $h(t)$ quite well at values of *t* less than about 25 sec, but at higher values of *t* the model predicts heights much lower than the experimental data.

In retrospect, this behavior is what we might have anticipated from the assumption of a constant flow rate. During the early stages the change in height as a fraction of the total height is small, so it does not matter how *q* depends on *h* because during these early stages *q* will be nearly constant. At longer times, however, when the height is getting small and the flow rate is decreasing, the assumption of constant flow rate means that the model predicts too fast a flow and hence too low a level. Thus, if we plan to use the model for times longer than about thirty seconds we cannot avoid accounting for the fact that *q* goes to zero as *h* goes to zero.

The simplest form that will account for the vanishing of the flow from an empty tank is a linear one

$$q = bh$$

We would logically try this correction next, for we wish the *simplest* representation that adequately describes the emptying over a wide time span. Equation 2.2 then becomes

$$\frac{dh}{dt} = -\frac{bh}{A} \qquad (2.6)$$

For this second model the logical sequence of performing the integration to obtain h as a function of t is somewhat different. It is certainly true that the integral of the left-hand side equals the integral of the right

$$h(t_2) - h(t_1) = \int_{t_1}^{t_2} -\frac{bh}{A} \, dt$$

but the result is of little use, for in order to perform the integration on the right we would need to know how $h(t)$ depends on t, which is the very answer that we seek.

The integration is performed by using the shortcut "separation of variables" method described in Section 15.9 to rewrite Equation 2.6 symbolically as

$$\frac{dh}{h} = -\frac{b}{A} \, dt \qquad (2.7)$$

or, integrating between times t_1 and t_2 and the height from its value at t_1 to its value at t_2 we have

$$\int_{h(t_1)}^{h(t_2)} \frac{dh}{h} = -\frac{b}{A} \int_{t_1}^{t_2} dt$$

or

$$\ln \frac{h(t_2)}{h(t_1)} = -\frac{b}{A} [t_2 - t_1] \qquad (2.8)$$

Setting t_1 to zero and letting t_2 denote any time, t,

$$\ln \frac{h(t)}{h_0} = -\frac{bt}{A} \qquad (2.9)$$

and the exponential of both sides yields

$$h(t) = h_0 e^{-bt/A} \qquad (2.10)$$

Figure 2.4 is again a plot of the data from Table 2.1. Our second model predicts an exponential dependence of h on t, with the particular curve defined by the quantity b/A. We could pick various values of b/A and construct predicted curves until we had one that we considered as the best fit to the data. Without the benefit of some mathematical guidance this would have to be visually determined, a difficult task with curved lines. We avoid

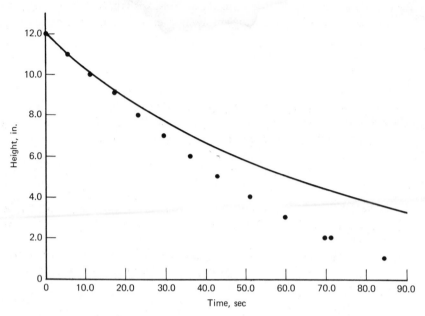

FIGURE 2.4   Liquid height versus time. The line is Equation 2.10.

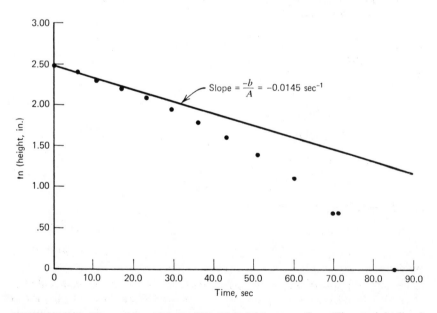

FIGURE 2.5   Natural logarithm of liquid height versus time. The straight line is Equation 2.9.

the problem by examining Equation 2.9 and observing that $\ln h$ is linear with t. A plot of $\ln h$ versus t is shown as Figure 2.5. Fitting a straight line through the first few data points in the same manner as we did for the previous model (by eye) we find that $-b/A = -0.0145/\text{sec}$. This determines the solid line in Figure 2.4. We have, by letting q vary with h, shifted the long time behavior in the desired direction, but in fact too far. Complete emptying would require an infinite time.

A reasonable generalization of the approach that we have been taking is to consider the function relating q and h to be of the form

$$q = kh^n$$

$n = 1$ is the case just considered, $n = 0$ is the first case, and n will lie somewhere between zero and unity, $0 < n < 1$, since any larger value of n will require an infinite time to empty. Equation 2.2 then becomes

$$\frac{dh}{dt} = -\frac{kh^n}{A}$$

or

$$\frac{1}{h^n}\frac{dh}{dt} = -\frac{k}{A}$$

Using the shortcut notation we can separate the variables in the equation to

$$\frac{dh}{h^n} = -\frac{k}{A}dt$$

or, integrating the right side from time zero to the present time and the left side with respect to h from h_0 at time zero to h at the present we obtain

$$\int_{h_0}^{h(t)} \frac{dh}{h^n} = -\frac{k}{A}\int_0^t d\tau$$

Carrying out the integrations leads to

$$\frac{h^{1-n}}{1-n} - \frac{h_0^{1-n}}{1-n} = -\frac{kt}{A} \tag{2.11}$$

and solving the algebraic equation for h we obtain

$$h(t) = h_0\left[1 - \frac{k[1-n]}{Ah_0^{1-n}}t\right]^{1/[1-n]} \tag{2.12}$$

The system empties in a finite time only for $n < 1$, so it is clear that no model using $n > 1$ can be valid over very long time intervals.

Now this slight generalization has, in fact, greatly increased the complexity of the analysis, for there are now *two* parameters, k and n, which must be determined from the experiment. A rational approach might be to choose a value for n, plot h^{1-n} versus t as motivated by Equation 2.11, and check for linearity, choosing a new value of n according to how the data deviate from linearity. Ultimately we will arrive at the "best" value of n to represent the data and find the corresponding value of k. We shall show in the next section and again in the next chapter that there is good reason to believe, however, that $n = 1/2$ is the proper value, in which case Equation 2.11 becomes

$$h^{1/2} = h_0^{1/2} - \frac{kt}{2A}$$

$$h = h_0\left[1 - \frac{kt}{2Ah_0^{1/2}}\right]^2$$

Figure 2.6 shows a plot of $h^{1/2}$ versus t, and although there is some scatter to the data the best line through all points has a slope $-k/2A = -0.0287$ in.$^{1/2}$/sec. We conclude, then, that the flow rate, q, in this experiment is well represented by the relation

$$q = 5.195h^{1/2}$$

since we know the value of A from Table 2.1. (We can also fit a line to the first four data points on Figure 2.6. If we do this we find that the slope of

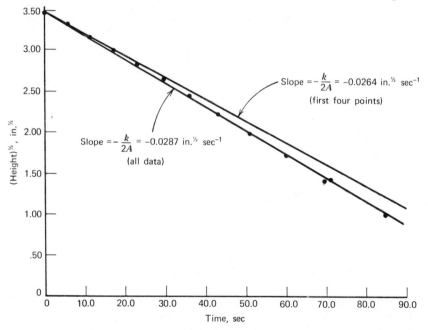

FIGURE 2.6 Square root of liquid height versus time. The straight lines are Equation 2.11 for $n = 1/2$.

such a line is $-k/2A = -0.0264$, slightly different from the value obtained using all data points. This is due to scatter in the data.)

The analysis process has now been illustrated to the point where the particular physical situation has been adequately modeled. We still need to consider how reliable the mathematical description may be when used to predict behavior in similar systems for purposes of design. We know nothing for example, about the nature of the parameter k beyond this experiment, particularly how its value changes with different size tanks and holes. We are not sure that $n = 1/2$ will work for all cases where we have flow from a vessel caused by the liquid head. The further information that we need can be obtained empirically if a series of experiments can be designed to test all parameters that may be of importance. Some further analysis, undertaken in the next chapter, can greatly simplify and direct this experimental design.

In our study of the comparison between the proposed models and actual experimental results we have observed a phenomenon that is of great significance and bears further investigation. The three equations that we have obtained, Equations 2.5, 2.10, and 2.12 are rewritten below.

$$q = c \qquad h(t) = h_0 \left[1 - \frac{ct}{Ah_0} \right] \qquad (2.5)$$

$$q = bh \qquad h(t) = h_0 e^{-bt/A} \qquad (2.10)$$

$$q = kh^n \qquad h(t) = h_0 \left[1 - \frac{k[1-n]t}{Ah_0^{1-n}} \right]^{1/[1-n]} \qquad (2.12)$$

We have seen that any of these models can represent the data near $t = 0$. The use of Taylor's theorem, Section 15.8, leads to an explanation. Consider, first, Equation 2.10, which contains an exponential. We notice that the exponential, e^x, has a series representation

$$e^x = 1 + x + \frac{x^2}{2!} + \frac{x^3}{3!} + \cdots$$

Denoting $(-bt/A)$ by x in Equation 2.10 we can then write for the linear model

$$h(t) = h_0 \left\{ 1 - \frac{bt}{A} + \frac{1}{2} \left[\frac{bt}{A} \right]^2 - \cdots \right\}$$

When t is small the quadratic, cubic, and higher terms will be much smaller than the first two (when $x = 0.1$, $x^2 = 0.01$, $x^3 = 0.001$, etc.), so that the second model predicts small time behavior closely approximated by

$$h(t) \simeq h_0 \left[1 - \frac{bt}{A} \right] \qquad (2.13)$$

Equation 2.13 and 2.5 are identical except for the coefficient of t, and the values of $b/A = 0.0145$ and $c/Ah_0 = 0.0145$ are the same, as they should be when t is small.

Similarly, we may use the series expansion

$$[1 + x]^\alpha = 1 + \alpha x + \frac{\alpha[\alpha - 1]}{2} x^2 + \cdots$$

to write Equation 2.12, with $x = -k[1 - n]t/Ah_0^{1-n}$ and $\alpha = 1/[1 - n]$, as

$$h(t) = h_0 \left\{ 1 - \frac{kt}{Ah_0^{1-n}} + \frac{n}{2} \left[\frac{kt}{Ah_0^{1-n}} \right]^2 - \cdots \right\}$$

For small values of time the behavior is closely approximated by

$$h(t) \simeq h_0 \left[1 - \frac{kt}{Ah_0^{1-n}} \right] \tag{2.14}$$

The experimentally observed value of $k/Ah_0^{1/2}$ can be simply computed from the slope shown in Figure 2.6. Its value for small values of t is 0.0145, an identical result to those obtained from the first two models.

Two conclusions stand out. All models investigated give an identical (linear) functional dependence between h and t for small time. Hence, *a poorly designed experiment which took insufficient data would show no differences between models* and might be used to justify the first model investigated, although the use of that model for predictive purposes might lead to gross errors in a problem where the level changes are a significant fraction of the total height. On the other hand, we might conceive of applications where only the behavior over a short time is required. In such a situation the simplest model suffices and no further sophistication is required. This demonstrates rather forcefully the role that problem objective plays in model formulation and comparison, as shown in Figure 2.1.

One further observation is in order. We have been discussing "small time" and "long time" as though time, as measured by a clock, were the pertinent variable. This is not the case, a fact which has important physical implications. Examining Equations 2.5, 2.13, and 2.14 we observe that $h(t)$ is a function only of the quantity ct/Ah_0 for the first model, bt/A for the second, and $kt/Ah_0^{1/2}$ for the particular case of the third when $n = 1/2$. It is these quantities that must be "small" if the approximations used in deriving Equations 2.13 and 2.14 are to be valid. Just what we mean by "small" depends on the particular situation. In the example we have been discussing, a value of c/Ah_0 (or b/A or $k/Ah_0^{1/2}$) equal to 0.0145/sec was small enough to produce exact agreement between all three equations at times up to 17 sec. If we could be satisfied with agreement on the order of 10 percent, the terms could be larger and still fall into the "small" category. We thus see that it is not time that must be small but a ratio t/θ, where we will call θ a *charac-teristic time*. The characteristic time for each model is, respectively, Ah_0/c, A/b, and $Ah_0^{1/2}/k$. That θ truly has dimensions of time is easily verified. For the first case c is the flow rate, with dimensions of volume per time, while Ah_0 is the initial volume. A volume divided by a volume per time has dimensions of time.

Realizing that t/θ is the key grouping of parameters in this problem enables us to decide on model complexity required in any given physical situation. If, for instance, we were designing a control system to maintain a constant level in a tank when flows to and from the tank change from their desired values over a time duration t_D, then we could use the simple first model if t_D/θ for the tank in question were small enough that there would be negligible error in using the simple model in place of the correct one. On the other hand, if the deviations between the predictions of the two models should be signifi-cant, then we would have to make a decision either to obtain accurate pre-diction with a complex model or to compromise with less accurate prediction and a simpler model. Since both models are really fairly simple in this case the decision is not crucial. In more complicated situations the best level of com-promise is often difficult to reach. A careful analysis of how the model equation fits into the overall problem must always be carried out. If it is a part of a more complex mathematical description simplicity is obviously important, while if the particular model is to be used alone simplicity may not be as desirable as accurate prediction.

2.4 ESTIMATING AN ORDER

In the preceding discussion it was clear that in the two-parameter con-stitutive relation

$$q = kh^n$$

knowing the proper value of n greatly simplifies the analysis. In a power relation of this type the value of n is known as the *order* of the process. With data that are reasonably free of experimental scatter it is often possible to estimate the order within reasonably narrow bounds and hence reduce or eliminate the trial-and-error nature of the computations.

The basic model equation and the flow rate equation, 2.2, combine to give

$$-\frac{dh}{dt} = \frac{k}{A} h^n$$

If we take logarithms of both sides we obtain

$$\ln - \left[\frac{dh}{dt}\right] = \ln \frac{k}{A} + n \ln h \tag{2.15}$$

Thus, a plot of ln $[-dh/dt]$ versus ln h will have slope n. However, we do not have dh/dt available. If the data are reasonably free of error and closely spaced over regions where h is changing rapidly, we can approximate $-dh/dt$ as

$$-\frac{dh}{dt} \simeq -\frac{\Delta h}{\Delta t}$$

where Δh refers to the change in height over a corresponding change Δt in time. Then, approximately,

$$\ln\left[-\frac{\Delta h}{\Delta t}\right] \simeq \ln\frac{k}{A} + n\ln h$$

The height-time data from Table 2.1 are reproduced in Table 2.2 using average times from the three experimental runs and plotted as logarithm of $-\Delta h/\Delta t$ versus logarithm of h in Figure 2.7. Since the derivative is estimated

TABLE 2.2   Liquid height versus average time for the tank-emptying experiment in order to obtain ln $(-\Delta h/\Delta t)$ versus ln h

h	t	$-\Delta h$	Δt	$\ln\left(\dfrac{-\Delta h}{\Delta t}\right)$	$\ln h$
12	0				2.5
		1	5.9	−1.78	
11	5.9				2.4
		1	5.4	−1.69	
10	11.3				2.3
		1	5.9	−1.78	
9	17.2				2.2
		1	6.0	−1.79	
8	23.2				2.1
		1	6.5	−1.87	
7	29.7				1.95
		1	6.5	−1.87	
6	36.2				1.8
		1	7.3	−1.99	
5	43.5				1.6
		1	7.5	−2.02	
4	51.0				1.4
		1	9.0	−2.20	
3	60.0				1.1
		1	10.7	−2.37	
2	70.7				0.69
		1	14.0	−2.64	
1	84.2				0

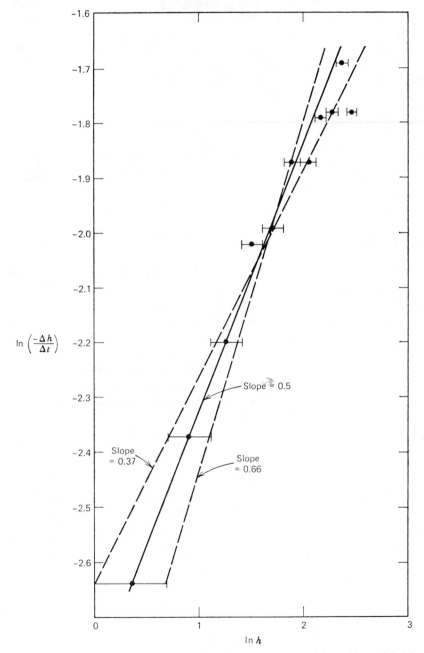

FIGURE 2.7 Natural logarithm of $-\Delta h/\Delta t$ versus natural logarithm of height. The band represents the range of heights over which the slope was calculated.

over a range of heights the data are plotted as the mean height with a band to represent the range. Although no definitive statement can be made because of the spread in the data, it is clear that the mean values in the region of least scatter are best fit with a line of slope $n = 0.5$. When h is greater than 9 in. (ln $h > 2.1$) the height changes too rapidly and there is too much scatter in these estimates of derivatives to be of any use, and any slope will fit the data equally well. This is equivalent to our previous observation that all models are equivalent at short times.

It is essential to emphasize that although approximating derivatives is a helpful and sensitive way to estimate order (power) when the data are good and closely spaced, any attempt to estimate k by means of the intercept will give extremely poor accuracy. For the data shown here errors in k of the order of 20 percent can be expected because of the spread about the mean at each point. The system equation must be integrated to obtain the parameter k.

2.5 CONCLUDING REMARKS

This simple example of an emptying tank has been introduced in order to demonstrate some of the difficulties encountered in the analysis process for a situation with which we are reasonably familiar, as well as to point up some of the insight into the physical problem that can be obtained. In more complex situations the same difficulties arise, although they may not be as easy to pinpoint. Careful analysis is always essential for understanding the process. In the chapters which follow we will expand our discussion on analysis by means of chemical engineering examples of increasing complexity so that the general concepts and philosophy may become practical operating tools.

2.6 PROBLEMS

2.1 In the draining tank problem we found that it was necessary to find a relationship between q and h before a mathematical description of the physical situation could be derived. Consider the following relationships and comment on their suitability for the tank-draining problem.

(a) $q = \dfrac{k}{h}$

(b) $q = M \sin h$

(c) $q = kh^3$

(d) $q = C_1 + C_2h + C_3h^2$

(e) $q = \dfrac{k_1}{1 + k_2h}$

2.2   Using parts (a) to (e) from Problem 2.1, find h as a function of t by substituting into Equation 2.2 and integrating. (A table of integrals may be helpful.)

2.3   Find q as a function of t for each model in the chapter. [Hint: you already know $q(h)$ and $h(t)$.]

2.4   The following data were obtained in a cylindrical vessel of diameter 1.0 in. which was allowed to drain through an orifice of diameter 0.043 in.

Height (inches)	Time (seconds)
15	0
14	6.0
13	12.2
12	18.7
11	25.5
10	32.7
9	40.3
8	48.3
7	56.7
6	66.1
5	76.2
4	87.6
3	101.0
2	117.5
1	140.7

Plot these data as h versus t and find the slope of the straight line through the first six data points. Using Equation 2.5, write down the expression for h in terms of t using the result from your graph. What is the error between predicted and experimental height values at $t = 10.0, 6.0, 4.0,$ and 2.0 sec? At $t = 100$ sec?

2.5   Check the data in Problem 2.4 with the mathematical description obtained when $q = bh$. Compute b/A by drawing a straight line through the first six points on a $\ln h$ versus t plot. What is the error between predicted and experimental values at $t = 10.0, 6.0, 4.0,$ and 2.0 sec? At $t = 100$ sec?

2.6   Using the expression $q = kh^{1/2}$, find the best value of k for the data of Problem 2.4.

2.7   Consider the tank-draining problem and develop the mathematical description, h as $h(t)$, if q is assumed equal to kh^2. Check your answer with Equation 2.12. Show the derivation of Equation 2.12 from Equation 2.11 as part of your answer.

2.8 A chemical operator is draining a tank originally full of crude oil. The tank measures 3 ft in diameter and is 10 ft high. The operator opens the valve (having a 1.25 in. diameter) in the tank base. He notices the change of fluid level from initial height down to 6 ft and records the following.

Height (feet)	Time (seconds)
10.0	0
9.0	54.0
7.5	142.9
7.0	172.5
6.0	240.2

He considers the outflow rate to be constant and on this basis develops a model equation describing the liquid height as a function of time. He plots the data.

(a) What is the slope of the straight line drawn through the data? What is its physical significance?

(b) For this specific mathematical model, how long will it take for the tank to drain?

(c) Compute the heights corresponding to the clock times 20.0, 60.0, 100.0, and 150.0 sec.

2.9 The chemical operator of Problem 2.8 does not know, as you do, that $q(h) = kh^{1/2}$ for flow driven by a liquid head.

(a) There exists the relation (which you will study more thoroughly in Chapters 3 and 10 when we deal with conservation of energy) $q = c_0 A_0 \sqrt{2gh}$ where c_0 is a dimensionless number known as the orifice coefficient. The value of c_0 can be taken as 0.61, and $g = 32.2$ ft/sec² What is the value of k?

(b) For this particular tank and functional form of q, derive the mathematical model describing the situation and compute the heights corresponding to $t = 20.0, 60.0, 100, 150, 250, 450, 850$, and 1000 sec.

(c) Assuming $q = kh^{1/2}$ is correct, compute the error in h for the times in part (b) if the assumption $q = c$ were used.

(d) On the basis of your results in part (c), is $q = $ constant a fairly good approximation for "short time," say, at least the first 2-1/2 min?

(e) Compare the times required to completely drain the tank for both models.

2.10 Draw several lines of slope equal to 0.5 which you feel fit the data in Figure 2.7 and determine a value of k from each. Is this an accurate way to measure k?

CHAPTER 3

Source of the Model Equations

3.1 INTRODUCTION

The analysis of a physical phenomenon begins with the development of a suitable mathematical description. That this step lies at the heart of the analysis process is obvious, for all further steps outlined in Chapter 2 presume that a mathematical description is available. We shall place considerable emphasis, therefore, in this chapter and the several that follow, on the development of a systematic procedure for expressing the description of a physical situation in precise mathematical terms.

A procedure for constructing a mathematical model for an extremely simple physical situation is shown in Figure 3.1. Let us presume for definiteness that we are seeking to describe the behavior of a piece of process equipment, such as the tank of the preceding chapter. The first step is the selection of what we shall call the *fundamental dependent variables*. This is the collection of quantities whose value at any time contains all of the information about the process necessary for investigation of any phenomena. There are only three such fundamental variables in most problems of interest to us: mass, energy, and momentum. In any given situation all or only some may be required. The description of an emptying tank in Chapter 2, for example, dealt only with the mass and did not appear to require consideration of momentum or energy.

In many instances the fundamental dependent variables are not conveniently measured. In the tank-emptying problem for example, though mass of liquid is our fundamental variable, it is the density, ρ, and the height of liquid, h, that we actually measure. These latter two quantities, taken together with the area of the tank, A, characterize the mass. Thus, the next step in the process is the selection of the *characterizing dependent variables*, those variables that can be measured conveniently and, properly grouped, determine the value of the fundamental variables. Mass, energy, and momentum might be characterized in a process by density, temperature, pressure,

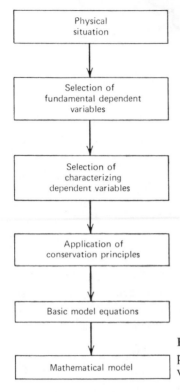

FIGURE 3.1 Model development for simple physical situations involving one dependent variable.

flow rate, etc. Generally, more than one characterizing variable is needed for each fundamental variable. The value of all of the characterizing variables at any time and at any point in space defines the *state* of the system, and characterizing variables are sometimes called state variables.

The basic model equation, a relationship between the dependent variables of the problem and the independent variables, is derived by applying the conservation laws. There are four independent variables of concern to us in engineering problems: time, t, and the coordinates necessary to establish position in space, x, y, and z. A major task in modeling a physical situation is selection of the pertinent dependent and independent variables. We can establish a systematic procedure for this by examining the way in which the conservation laws are employed.

3.2 CONSERVATION EQUATIONS

The basic quantitative step in model development is the application of the fundamental conservation principles of physics, conservation of mass, energy, and momentum. This step, which we shall discuss in some detail, leads to what we shall designate as the basic model equations. Our use of the conservation laws is essentially a bookkeeping one, in which we maintain a balance sheet for the mass, energy, or momentum in our process.

Consider any region of space enclosed for the purpose of "accounting" by a (fictitious) surface, which we shall denote as the "control surface," and the volume within the surface is referred to as the "control volume." Call the quantity to be conserved X, which may be mass, momentum, or energy. (See Figure 3.2.) The conservation law may be stated as follows.

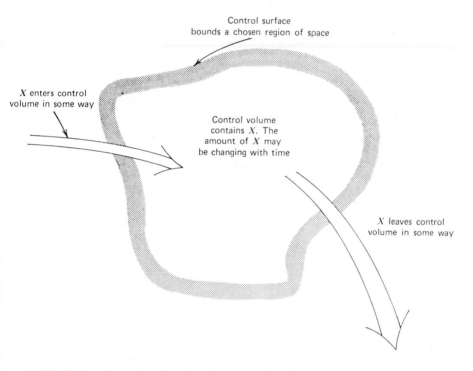

FIGURE 3.2 Control surface bounds a control volume to which a conservation law is applied.

The total amount of X contained within the control volume at time t_2 must be equal to the total amount of X contained within the control volume at time t_1, plus the total amount of X which has appeared in the control volume in the time interval t_1 to t_2 by all processes, less the total amount of X which has disappeared from the control volume in the time interval t_1 to t_2 by all processes. We express this verbal statement partly in symbols as follows.

$$X|_{t_2} = X|_{t_1} + \text{amount of } X$$
$$\text{entering during } (t_1, t_2)$$
$$- \text{amount of } X$$
$$\text{leaving during } (t_1, t_2)$$

Notice that mass and energy are scalar quantities while momentum has direction associated with it, and thus momentum conservation must be considered in each coordinate direction separately.

Consider, for example, the trivial problem of filling a tank with a stream of *constant* flow rate q, ft³/sec. The fundamental dependent variable is mass, expressed by the characterizing variables ρ, A, and h. We are interested in the accumulation of mass within the tank, so our control volume is simply the volume of the tank. The control surface, shown in Figure 3.3, is partly made up of the tank walls and bottom and partly made up with a fictitious

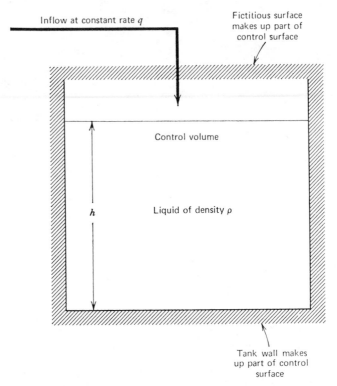

Inflow at constant rate q

Fictitious surface makes up part of control surface

Control volume

Liquid of density ρ

h

Tank wall makes up part of control surface

FIGURE 3.3   Control volume for tank filling with liquid of constant density.

surface drawn across the top of the tank. Since we are interested only in the total mass, rather than something happening at particular points in space, the only independent variable in the problem is time. The fundamental dependent variable of importance is mass, and we apply the law of conservation of mass as follows.

The total amount of mass ($\rho A h$) contained within the control volume (the tank) at time t_2 is equal to the total amount of mass ($\rho A h$) contained within the control volume at time t_1, plus the total amount of mass that has entered the control volume during the time interval t_1 to t_2 ($\rho q[t_2 - t_1]$), less the amount of mass that has left the control volume (none). Expressed in symbols we have

$$\rho Ah|_{t_2} = \rho Ah|_{t_1} + \rho q[t_2 - t_1] - 0[t_2 - t_1]$$

This statement is reexpressed in terms of the characterizing dependent variables and time in a manner to be discussed in the next chapter. The resulting mathematical model is

$$\frac{d\rho Ah}{dt} = \rho q$$

$$\frac{dh}{dt} = \frac{q}{A}$$

(3.1)

Since q and A are constants, Equation 3.1 can be readily solved to yield h as a function of t:

$$h(t) = h_0 + \frac{q}{A} t$$

It is immediately clear from our familiarity with the tank-draining problem that the procedure outlined in Figure 3.1 only works for very simple situations. Let us reconsider the tank-draining problem in the present context. The draining problem is similar to the filling problem in that the same control volume and characterizing variables are used (Figure 3.4). The word statement of the law of conservation of mass is written as follows.

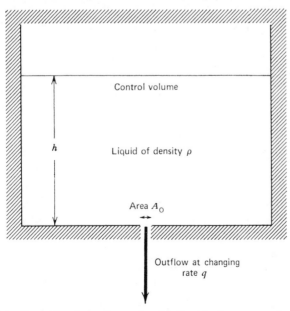

FIGURE 3.4 Control volume for tank with liquid of constant density draining through an orifice in the bottom.

The total amount of mass (ρAh) contained within the control volume (the tank) at time t_2 is equal to the total amount of mass (ρAh) contained within the control volume at time t_1, plus the total amount of mass that has entered the control volume during the time interval t_1 to t_2 (none), less the total amount of mass that has left the control volume by flow through the hole during the time interval t_1 to t_2.

Because the exit flow rate is not constant, the amount that leaves between t_1 and t_2 cannot be written down quite as intuitively as the amount that entered in the filling problem, so we will leave the symbolic representation to the next chapter. The final equation which results is

$$\frac{dh}{dt} = -\frac{q}{A} \tag{3.2}$$

In Equation 3.2 there are two unknowns, q and h (q was a constant in the tank-filling problem). Since we only have one equation in our model at this

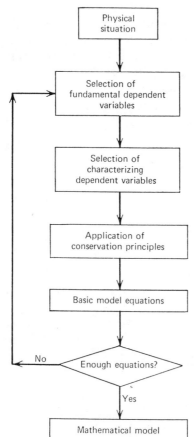

FIGURE 3.5 Model development for physical situations requiring more than one dependent variable.

point it is obvious that an additional relationship is needed. A modification of the general procedure illustrated in Figure 3.1 is required which allows for a check to determine if there are sufficient equations to describe the state of the system. Figure 3.5 includes a loop which allows for an iteration of this type and inclusion of additional fundamental variables, if necessary. In the tank-emptying problem we shall show (Chapter 10) that mass alone is not sufficient, and that consideration of the law of conservation of energy adds to our understanding of the problem. With the correct functional relationship between q and h obtained using conservation of energy Equation 3.2 may be integrated.

3.3 CONSTITUTIVE EQUATIONS

In principle, the sequence of steps outlined in Figure 3.5 showing how the conservation laws are applied would appear to be all that is necessary to obtain a complete mathematical model. However, in many engineering situations it is either not necessary, not reasonably convenient, or, in truly complex situations, not humanly possible to obtain all the information we need by direct application of conservation laws and the other fundamental principles of physics. (Just consider trying to describe the complex interactions of the millions of gas molecules contained in a vessel by direct application of the principle of gravitational attraction and the laws of conservation!) The procedure therefore needs to be modified further. A new type of relationship called a *constitutive relation* is defined in an operational manner by Figure 3.6. A search for this constitutive relationship is initiated after one has decided that additional relations are necessary which are not possible to obtain by applying the conservation laws at a reasonable level of complexity. A constitutive equation may be derived entirely from experiment. It may have a form suggested by a theory but some parameters evaluated by experiment. It may be obtained entirely on theoretical grounds, say by applying the conservation laws at a molecular level and making use of statistical and quantum mechanics. Or, it may be obtained by some reasonable combination of any of the above approaches. The search for the necessary constitutive relation or relations for any particular problem requires a thorough understanding of the physical situation and, indeed, may often not be successful unless experiments are carried out as part of the model development program. The tank-draining problem is an example of a case where the constitutive relation $q = 5.195h^{1/2}$ was derived entirely by experiment.

It is evident that by defining a constitutive equation as the relation needed to complete a physical description after the basic principles have been fully used we can draw the implication that a constitutive equation is *specific*. It is an equation that applies to a particular material, or to a particular situation, but does not have general applicability. The relation between pressure,

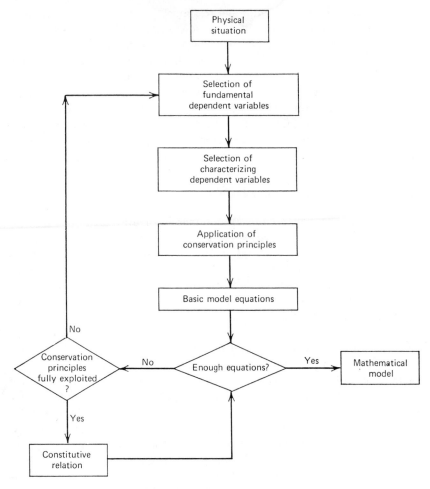

FIGURE 3.6   Model development using a constitutive relation.

volume, temperature, and number of moles of a gas known as the "ideal gas law" is a good example. This expression

$$pV = nRT$$

was first proposed to approximately represent some experiments on gases at low density. It does not represent the behavior of any real gas at high density, and is clearly not a law at all, but simply a convenient representation of a rather specific case. The "law" is often derived in basic physics or chemistry courses by application of conservation principles after making a large number of assumptions about the nature of a gas, assumptions that will be valid only at low density. In many cases uses of the ideal gas equation will be inappropriate, and a different, more accurate, constitutive equation must be used.

There are clearly many familiar relations that apply to specific situations. The density of an aqueous sodium chloride solution has a different dependence on solute concentration than does an aqueous sugar solution. The expansion of iron per degree of heating is different from the expansion of nylon. A common feature of the relationships which come easily to mind is that they describe what are essentially molecular phenomena, phenomena which, like an eighteenth century natural philosopher, one might hope to predict from an understanding of the forces between molecules and the physics of systems of large numbers of particles. There is another category, which includes situations like the tank-draining problem. These are engineering situations in which we choose to obtain an equation by experiment because it is easier to do than to engage in a detailed application of the conservation laws. In complex physical situations the model development process will nearly always include a well-planned experimental program or in some way utilize previous experimental results. It is difficult to make general statements about this latter type of constitutive relationship since each problem is unique, but the approach will be demonstrated frequently.

3.4 CONTROL VOLUMES

The word statement of the conservation law supplies us with a means of obtaining relationships between the dependent and independent variables of a problem. The precise manner in which we proceed from the word statement to symbols will be discussed in the next chapter. In this section we consider one final feature in the logic of the modeling process in order that the spatial dependence of the dependent variable may be taken into account. Time is always considered to be an independent variable of concern in the way in which the conservation laws are stated. Spatial dependence is closely associated with the way in which a control volume is defined for a particular problem.

In the problems and the procedure that we have discussed so far the piece of process equipment has been considered as the control volume. In many cases a satisfactory model can be derived on this basis, which clearly assumes that the dependent variables have no spatial variation. In that case the variables can be assigned a specific value at any instant of time which is independent of spatial position. In the tank filling and draining problems ρ, A, and h meet this qualification, so the model equation only includes time as the independent variable. Had we been dealing with a suspension of solid particles in the liquid, ρ at any instant of time might vary with spatial position because of the settling effect due to gravity, as shown in Figure 3.7. If this physical situation were to be modeled, a decision would have to be made

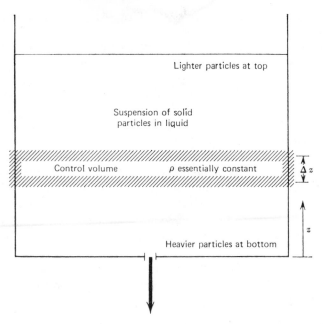

FIGURE 3.7   Control volume for a draining tank in which liquid density depends on position.

either to ignore the effect and continue with the simple formulation using time as the only independent variable, or to select a new control volume and consider the spatial variation in the direction perpendicular to the earth's surface. The new control volume, if we should choose to follow the latter course, might be selected as a partial volume of the tank with cross-sectional area A and small height Δz, as shown in the figure, in which ρ is essentially unchanged. (Notice the similarity to procedures in calculus in which we treat a continually varying quantity as constant for all practical purposes over a very small interval, such as in Section 15.6.) With this control volume we can then apply the conservation laws and continue, but now the final equations will depend on the location, z, as well as on t. The most general case will have dependence on x, y, z, and t.

   The final version of the logical flow chart is shown in Figure 3.8, where we have added a step to account for the necessity of choosing a control volume in which the characterizing variables can be specified by a unique value. When we can neglect the variation of a dependent variable in some spatial direction we speak of the mathematical model as a *lumped* model, while we call a model *distributed* if spatial variation is explicitly accounted for. This is rather imprecise terminology, for clearly a model which accounts for variations in the z direction, but not x and y, is lumped with respect to the x and y directions and distributed with respect to z. In this text we will never deal

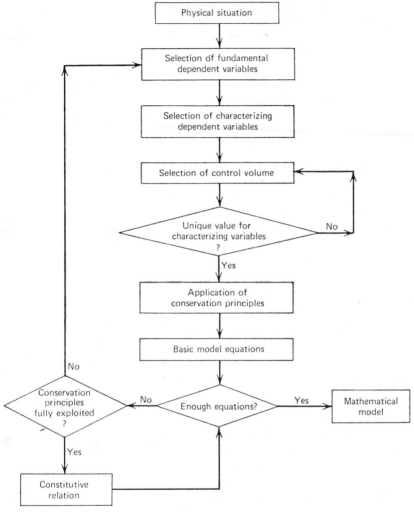

FIGURE 3.8 Model development for any physical situation.

with spatial variation in more than one direction, so when we refer to a lumped model it is one in which time is the only independent variable. We shall initially direct our attention to lumped models and the assumptions about a physical situation that lead to such a model. In Chapter 7 we will turn to the consideration of distributed models as well.

3.5 DIMENSIONAL ANALYSIS

When we seek relationships among the variables necessary to describe any physical phenomenon, each symbol employed represents a quantity that has certain definite *dimensions*, such as length or mass per volume. These quantities are measured in *units*, which can be chosen arbitrarily. For example, length might be measured in centimeters or inches, and it is a fundamental principle of physics that the behavior of a system cannot depend on the arbitrary choice of measuring units that we choose to employ.

The equation or equations representing the physical situation are relations among the symbols representing the pertinent variables. Each term in any one of these equations must have the same dimensions if the equation is to be mathematically consistent with the physical situation. Recognition of this simple and basic property of equations describing physical situations gives us a means of checking initial formulations and any subsequent algebraic manipulations. Although it might seem too elementary to be of practical significance, a check for dimensional consistency often turns up errors in formulation and manipulation and is always carried out by experienced engineers. Furthermore, if we recognize that dimensional consistency must always hold, even when it is not possible to actually derive the equations for a particular situation, it is sometimes possible to employ a process called dimensional analysis to obtain a clue as to the form of a constitutive relationship.

Our goal in this section, then, is to explore these two ideas of dimensional consistency and dimensional analysis in more detail. It is necessary first to review the systems of units that are commonly employed in scientific and engineering problems.

3.5.1 Systems of Units

There are five systems of units that are commonly used in English-speaking countries, two "metric" and three "English," and it is necessary to be familiar with each. The quantities we generally measure are length (L), time (θ), mass (M), force (F), and temperature (T). The fundamental unit of time is always seconds, but the units of one or more of the other quantities vary from system to system. A basic relation necessary to our discussion is that *between* force and mass, *Newton's principle of acceleration* (*Newton's Second Principle*), which states that the force required to uniformly accelerate a given mass is equal to the product of mass and acceleration

$$f = ma \tag{3.3}$$

The dimensions of acceleration are length per time per time or, symbolically,

$$a[=]L\theta^{-2} \tag{3.4}$$

The symbol $[=]$ means "has dimensions of." Thus, the dimensions of force are defined in terms of M, L, and θ

$$F[=]ML\theta^{-2} \tag{3.5}$$

The system of units most commonly used in scientific work, and hence probably the most familiar at this point from basic courses in physics and chemistry, is the cgs, or *centimeter-gram-second*, system. Here length is measured in centimeters (cm), mass in grams (g), time in seconds (sec), and temperature in degrees centigrade (°C). From Equation 3.5, then,

$$\text{force } [=] \text{ g cm sec}^{-2}$$

The unit of force is called the dyne (d), so we have the definition

$$1 \text{ d} = 1 \text{ g cm sec}^{-2}$$

Now there is an important, but subtle, point here. We have introduced the units centimeter, gram, second, and dyne, *four* quantities of which only *three* are independent. If we wish to measure force in dynes, mass in grams, length in centimeters, and time in seconds, then Equation 3.3 does not have consistent units, and we need to rewrite it to include a conversion factor, commonly called g_c:

$$
\underset{\left[1\dfrac{\text{g cm}}{\text{d sec}^2}\right]}{g_c} \; \underset{[\text{d}]}{f} = \underset{[\text{g}]}{m} \; \underset{\left[\dfrac{\text{cm}}{\text{sec}^2}\right]}{a}
\tag{3.6}
$$

The conversion factor is frequently omitted in the cgs system because its numerical value is unity. Such is not the case in all systems, however, and it is good practice to *include g_c whenever force and mass both appear*. Finally, let us recall that the *weight* of an object, w, is the force caused by the acceleration due to gravity,

$$g_c w = mg \tag{3.7}$$

The acceleration due to gravity is approximately 980 cm/sec², so the weight of a 1 g mass is 980 d. There is an unfortunate confusion between mass and weight among laymen. The mass of an object is the same anywhere in the universe, although the weight depends on the gravitational acceleration and varies slightly from place to place even on earth. The cgs system is summarized in Table 3.1.

TABLE 3.1 Units of the cgs (centimeter-gram-second) system

Length	[=] centimeter, cm
Mass	[=] gram, g
Time	[=] second, sec
Force	[=] dyne, d
Temperature	[=] degrees centigrade, °C
g_c	= 1 g cm/d sec²

A closely related system is the S.I., or *International System* (Système International d'Unités), in which the units of length, mass, time, and temperature are, respectively, meter (m), kilogram (kg), second, and degree centigrade. The unit of force, called a Newton (n), is numerically equal to one kilogram meter per second per second, so that g_c, again, has a numerical value of unity. The system is summarized in Table 3.2. The S.I. has been adopted by a large number of countries and is now the required system of units for *AIChE Journal*, published by the American Institute of Chemical Engineers. Engineering usage is not yet common, however.

TABLE 3.2   Units of the S.I. (International System)

Length	[=] meter, m; 1 m = 100 cm
Mass	[=] kilogram, kg; 1 kg = 1000 g
Time	[=] second, sec
Force	[=] Newton, n; 1 n = 10^5d
Temperature	[=] degrees centigrade, °C
g_c	= 1 kg m/n sec^2

The basic English system is the fps, or *foot-pound-second* system. Here the units of length, mass, time, and temperature are, respectively, the foot (ft), pound-mass (lb_m), second, and degree Fahrenheit (°F). In the fps system the unit of force is defined in such a way as to keep g_c numerically equal to unity. This force unit, the poundal (lbl), is defined as

$$1 \text{ lbl} = 1 \text{ lb}_m \text{ ft sec}^{-2}$$

The system is summarized in Table 3.3. The fps is a logical system, but it is

TABLE 3.3   Units of the fps (foot-pound-second) system

Length	[=] foot, ft; 1 ft = 30.48 cm
Mass	[=] pound-mass; 1 lb_m = 453.59 g
Time	[=] second, sec
Force	[=] poundal, lbl; 1 lbl = 13,826 d
Temperature	[=] degrees Fahrenheit, °F; 1°F = 5/9°C
g_c	= 1 lb_m ft/lbl sec^2

seldom used, because of the unfortunate general confusion between mass and weight. Although incorrect, it has become common to talk of both mass and force, particularly the force, weight, in pounds, as though the two were the same. The engineering system has been developed in an attempt to avoid confusion. It is the most common system in the English speaking world and the one which we shall generally use throughout this book.

TABLE 3.4   Units of the engineering system

Length	[=] foot, ft; 1 ft = 30.48 cm
Mass	[=] pound-mass, lb_m; 1 lb_m = 453.59 g
Time	[=] second, sec
Force	[=] pound-force, lb_f; 1 lb_f = 4.45 × 10⁵ d
Temperature	[=] degrees Fahrenheit, °F; 1°F = 5/9°C
g_c	= 32.174 lb_m ft/lb_f sec²

In the engineering system, summarized in Table 3.4, the units of length, mass, time, and temperature remain the foot, pound-mass, second, and degree Fahrenheit. A new unit of force is defined, however, the *pound-force* (lb_f), with

$$1 \ lb_f = 32.174 \ lbl$$

or, from Equation 3.6,

$$1 \ lb_f = 32.174 \ lb_m \ ft/sec^2$$
$$g_c = 32.174 \ lb_m \ ft/lb_f \ sec^2$$

The choice of 32.174 is, of course, not arbitrary. The acceleration due to gravity in English units is 32.174 ft/sec² at sea level and 45° latitude, so that when we express the *weight* from Equation 3.7 as

$$w = \frac{g}{g_c} m$$

we find that the weight of one pound-mass is numerically equal to one pound-force. This agreement with common usage is gained at the expense of a conversion factor which is not unity. Furthermore, it is important to note that the engineering system is an earthbound system. g_c is a fixed number, while g varies slightly from point to point on earth, but to sufficient accuracy the numerical ratio g/g_c is always unity. In a markedly different gravitational field, however, the ratio g/g_c will differ greatly from unity and the sole advantage of the engineering system, the numerical equality of mass and weight, is lost.

Finally, we note in passing that there is another English system that is used on occasion, the *gravitational system*. The units of length, mass, time, force, and temperature are, respectively, foot, slug, second, pound-force, and degree Fahrenheit. The slug is defined

$$1 \ slug = 32.174 \ lb_m$$

$$g_c = 1 \ slug \ ft/lb_f \ sec^2$$

The gravitational system is summarized in Table 3.5. There are also two metric systems using a gram-force and kilogram-force, but these are not used in English-speaking countries.

TABLE 3.5   Units of the gravitational system

Length	[=] foot, ft; 1 ft = 30.48 cm
Mass	[=] slug; 1 slug = 14,592 g
Time	[=] second, sec
Force	[=] pound force, lb_f; 1 lb_f = 4.45 × 10^5 d
Temperature	[=] degrees Fahrenheit, °F; 1°F = 5/9°C
g_c	= 1 slug ft/lb_f sec^2

A few further words are in order about temperature. The zero points on the centigrade and Fahrenheit scales are completely arbitrary (0°C = freezing point of water, 0°F = freezing point of water less 32 degrees!) and the conversion factor 1°F = 5/9°C refers only to temperature *differences*. More natural scales, though less commonly used by laymen, are the absolute Rankine (°R) and Kelvin (°K) scales, where the zero corresponds to the point at which molecular motion in an ideal gas ceases. The relations are

$$T(°R) = T(°F) + 459.58$$

$$T(°K) = T(°C) + 273.15$$

It is necessary to distinguish carefully between the common and absolute scales. *Most fundamental relations in physics and chemistry require absolute temperatures.*

We shall deal with energy in detail in subsequent chapters. It is helpful here, though, to draw on the previous experience of elementary physics that energy and work correspond to a force times distance. The natural units of energy are then as shown in Table 3.6.

TABLE 3.6   Units of energy

cgs:	1 dynecm ≡ 1 erg
S.I.:	1 Newtonm ≡ 1 joule; 1 joule = 10^7 erg
engineering:	1 foot lb_f; 1 ft lb_f = 1.356 × 10^7 erg

Conversion between systems of units is a trivial operation, as demonstrated by the following examples. It is frequently necessary, however, since various laboratory data for a given problem may have been recorded in several different systems.

Example 3.1

Density, ρ, is frequently measured in the laboratory in grams per cubic centimeter. If the density of a certain sample of water is 1.00 g/cm^3, what is the density in pound-mass per cubic foot?

$$\rho = \left[1.00 \, \frac{g}{cm^3}\right]\left[\frac{1 \, lb_m}{453.6 \, g}\right]\left[\frac{(2.54)^3 \, cm^3}{1 \, in^3}\right]\left[\frac{12^3 \, in^3}{1 \, ft^3}\right] = 62.4 \, lb_m/ft^3$$

Example 3.2

Viscosity, μ, a measure of the resistance of a fluid to motion, is commonly measured in poise (p), where

$$1 \, p = 1 \, g/cmsec$$

The viscosity of water is approximately 10^{-2} p at room temperature. Express in engineering units.

$$\mu = \left[10^{-2} \, \frac{g}{cmsec}\right]\left[\frac{1 \, lb_m}{453.6 \, g}\right]\left[\frac{1 \, cm}{0.0328 \, ft}\right] = 6.72 \times 10^{-4} \, \frac{lb_m}{ftsec}$$

Example 3.3

When a fluid moves in a pipe the ratio

$$N_{Re} = \frac{Dv\rho}{\mu}$$

is often required, where D is the pipe diameter, v the fluid velocity, ρ and μ the density and viscosity, respectively, and N_{Re} the *Reynolds Number*, named for the nineteenth century English scientist Osborne Reynolds. What are the dimensions of N_{Re}?

$$D[=]L, \quad v[=]L\theta^{-1}, \quad \rho[=]ML^{-3}, \quad \mu[=]ML^{-1}\theta^{-1}$$

Thus

$$N_{Re}[=]\frac{[L][L\theta^{-1}][ML^{-3}]}{ML^{-1}\theta^{-1}} = 1$$

That is, N_{Re} is *dimensionless*, and, as long as we always measure all quantities in the same set of units, the numerical value will be independent of the particular system used. To emphasize this point, take water with $D = 1$ in., $v = 10$ ft/sec, $\rho = 62.4 \, lb_m/ft^3$, and $\mu = 6.72 \times 10^{-4} \, lb_m/ftsec$. Then

$$N_{Re} = \frac{[1 \, in][1 \, ft/12 \, in][10 \, ft/sec][62.4 \, lb_m/ft^3]}{6.72 \times 10^{-4} \, lb_m/ftsec} = 7.74 \times 10^4$$

In cgs units we get $D = 2.54$ cm, $v = 304.8$ cm/sec, $\rho = 1 \, g/cm^3$, $\mu = 10^{-2}$ g/cmsec.

$$N_{Re} = \frac{[2.54 \, cm][304.8 \, cm/sec][1 \, g/cm^3]}{10^{-2} \, g/cmsec} = 7.74 \times 10^4$$

3.5.2 Dimensional Consistency in Mathematical Descriptions

In describing a physical situation each symbol which we employ in the equations has an inseparable part of its meaning associated with its dimensions. Any time we formulate a problem using one of the conservation laws the equations should be checked to make sure that each term has the dimension of mass, energy, or momentum, as dictated by the conservation law employed. This almost trivial check can be a real aid to the student becoming familiar with model formulation and should *always* be done after any equation is initially derived. As an example we refer again to the tank-draining problem in Chapter 2.

After employing the law of conservation of mass we have the following relation, Equation 2.1,

$$\frac{d}{dt}[\rho A h] = -\rho q \tag{2.1}$$

All the symbols have been defined in Chapter 2 in terms of their units in the engineering system with the exception of d, which is an operator like $+$ or $-$ and has no physical dimensions. There are two terms in the equation, and each should have units representing mass/time, say lb_m/sec. We check for dimensional consistency by writing the units of each term as it appears in the equation:

$$\frac{d}{dt} \quad \rho \quad A \quad h = -\rho \quad q$$

$$\frac{1}{\text{sec}} \frac{lb_m}{ft^3} ft^2\, ft \qquad \frac{lb_m}{ft^3} \frac{ft^3}{\text{sec}}$$

Since the units ft/ft cancel we obtain

$$\frac{lb_m}{\text{sec}} = \frac{lb_m}{\text{sec}}$$

We can also use the concept of dimensional consistency to check on any algebraic manipulations of an equation. We perform the algebra on the symbols appearing in the equation, and the algebra might be quite complex. The units associated with a symbol are also operated on by the same algebraic manipulations, and a partial verification of the algebra can be made by checking dimensional consistency. As a trivial example, dividing Equation 2.1 by ρA gives Equation 2.2,

$$\frac{dh}{dt} = -\frac{q}{A} \tag{2.2}$$

$$\frac{ft}{\text{sec}} = \frac{ft^3/\text{sec}}{ft^2} = \frac{ft}{\text{sec}}$$

The equation is at least dimensionally consistent.

A more complicated example to check for consistency is Equation 3.8 below, which will be derived in Chapter 5 to describe a chemically reacting liquid system in a tank with continuous flow in and out.

$$\frac{d}{dt}[Vc_A] = q_f c_{Af} - qc_A - Vkc_A \tag{3.8}$$

V liquid volume, ft³

q_f, q flow rates in and out, respectively, ft³/min

c_{Af}, c_A concentration of material A in streams entering and leaving the tank, respectively, lb_m/ft³

k parameter in a constitutive relation, the *specific reaction rate constant*, 1/min

Replacing the symbols with their units we see that each term has units of lb_m/min.

$$\frac{d}{dt} \quad V \ c_A = q_f \ c_{Af} - \ q \ c_A - V \ k \ c_A$$

$$\frac{1}{min} \ ft^3 \frac{lb_m}{ft^3} \quad \frac{ft^3}{min} \frac{lb_m}{ft^3} \quad \frac{ft^3}{min} \frac{lb_m}{ft^3} \quad ft^3 \frac{1}{min} \frac{lb_m}{ft^3}$$

$$\frac{lb_m}{min} \qquad \frac{lb_m}{min} \qquad \frac{lb_m}{min} \qquad \frac{lb_m}{min}$$

These checks for dimensional consistency in relations must become habit if one is to gain the necessary proficiency in model development. They are trivial to perform, but any critical reader of engineering and scientific material can frequently find errors in dimensionality of equations, a comment on a lesson not adequately learned.

We can always find a grouping of the variables such that each term in any equation is dimensionless. For example, Equations 2.2 and 3.8 can be written

$$\frac{A}{q} \quad \frac{dh}{dt} = -1$$

$$\frac{ft^2}{ft^3/min} \frac{ft}{min} = 1, \quad \text{dimensionless}$$

$$\frac{1}{q_f c_{Af}} \frac{d}{dt} \ [V \ c_A] = 1 \ - \ \frac{q}{q_f} \ \frac{c_A}{c_{Af}} - \frac{Vk}{q_f} \ \frac{c_A}{c_{Af}} \tag{3.9}$$

$$\frac{1}{\frac{ft^3}{min} \frac{lb_m}{ft_3}} \ ft^3 \frac{lb_m}{ft^3} \quad 1 \quad \frac{\frac{ft^3}{min} \frac{lb_m}{ft^3} \ ft^3 \frac{1}{min} \frac{lb_m}{ft^3}}{\frac{ft^3}{min} \frac{lb_m}{ft^3} \ \frac{ft^3}{min} \ \frac{lb_m}{ft^3}}$$

At first glance this rearrangement does not seem to provide any advantage over the original formulation, but let us pursue the idea a bit further. We focus on Equation 3.9, the reacting system. Under certain conditions, to be discussed in subsequent chapters, it is found that

$$q_f = q = \text{constant}$$

$$c_{Af} = \text{constant}$$

A direct consequence of the first restriction is that the volume of liquid, V, is a constant. Making use of the fact that constants can be moved in and out of the differentiation operator, d/dt, we rewrite Equation 3.9 as

$$\frac{V}{q} \frac{d}{dt} \frac{c_A}{c_{Af}} = 1 - \frac{c_A}{c_{Af}} - \frac{V}{q} k \frac{c_A}{c_{Af}}$$

The ratio c_A/c_{Af} is dimensionless, and it is convenient to give it a name,

$$\frac{c_A}{c_{Af}} \equiv x_A$$

Furthermore, V/q has dimensions of time. We call it the *residence time* and give it a name

$$\frac{V}{q} \equiv \theta$$

Thus we have the equation

$$\theta \frac{dx_A}{dt} = 1 - x_A - [k\theta] x_A$$

We note further that the ratio t/θ is dimensionless. Let

$$\tau \equiv \frac{t}{\theta}$$

Then

$$dt = \theta \, d\tau$$

and we obtain, finally,

$$\frac{dx_A}{d\tau} = 1 - x_A - [k\theta] x_A \tag{3.10}$$

Equation 3.10 relates three dimensionless groups of variables, x_A, $k\theta$, and τ. Since it involves a derivative we also need the value of x_A at time zero, x_{A0}. Thus, solution gives x as a function of three variables, the two constants, x_{A0} and $k\theta$, and the reduced time, τ. Contrast this with the solution of the original equation (3.8), which gives c_A as a function of *five* constants, q, V, k, c_{Af}, and c_{A0} (the initial value of c_A), and time, t. The dimensionless representation shows that the time behavior depends only on certain *groupings* of constants. The complexity is therefore significantly reduced, making the equation easier to write and look at, with less chance for error in manipulation. If we have three constants in the dimensionless equation and wish to study the behavior for, say, ten values of each, then we must study $10 \times 10 \times 10 = 1000$ cases. To study ten values each of the five constants in the original formulation requires studying $10 \times 10 \times 10 \times 10 \times 10 = 100,000$ cases!

3.5.3 Dimensional Analysis and Constitutive Relationships

A constitutive relationship is sought at the point in the model development where further information cannot be conveniently obtained from an application of the conservation laws. One or more equations relating two or more of the dependent variables are required, and they are often obtained by performing experiments or by utilizing the experimental information of others. In many situations of engineering interest the physical situation may be so complicated that it is not easy to plan a reasonable experimental program or to interpret the results without a lot of trial and error manipulations and guesswork. In the very simple example of Chapter 2 the experiments were easy to plan, but seeking a functional relationship between q and h was done by a combination of guesswork and successive testing of more complex models after the simple ones failed to fit the data adequately.

In situations where we need a constitutive relationship one or more equations are not conveniently derivable. Even when we do not know the functional form for an equation we recognize that dimensional consistency must exist; that is, that each term in the equation must have the same units as every other term and *that we can rearrange the variables so that each term in the equation is dimensionless*. This simple fact can be exploited for further insight into the construction of constitutive equations following a limited discussion of the form of equations.

Many of the constitutive equations with which we deal are algebraic (e.g., $q = kh^{1/2}$, $pV = nRT$, etc.). If the physical situation involves two variables, x and y, each with dimensions, we can always express the relationship between x and y in the form

$$\alpha x^a y^b = \beta x^c y^d + \gamma x^e y^f + \delta x^g y^h + \cdots \tag{3.11}$$

where $\alpha, \beta, \gamma, \delta, \ldots, a, b, c, d, \ldots$ are real numbers without dimensions. Table 3.7 shows examples of how various equations can be put into the form of Equation 3.11. By multiplication and division, as shown previously, we can always make the terms $x^a y^b$, $x^c y^d$, etc. in Equation 3.11 dimensionless. In a similar way, an algebraic equation involving three variables x, y, and z can be put in the dimensionless form

$$\alpha x^a y^b z^c = \beta x^d y^e z^f + \gamma x^g y^h z^i + \cdots$$

TABLE 3.7 Various equations in the form of Equation 3.11

Equation	Form of Equation 3.11
$x = 5y^2$	$x = 5y^2$
$x = 5e^y$	$x = 5 + 5y + \frac{5}{2}y^2 + \frac{5}{6}y^3 + \cdots$
$3e^{2x} = \sin y$	$3 + 6x + 6x^2 + \cdots = y - \frac{1}{6}y^3 + \frac{1}{120}y^5 - \cdots$
$2xy^2 = e^{2x}\sin y$	$2xy^2 = [1 + 2x + 2x^2 + \cdots]$
	$\times [y - \frac{1}{6}y^3 + \cdots]$
	$= y + 2xy + 2x^2y - \frac{1}{6}y^3 - \frac{1}{3}xy^3$
	$- \frac{1}{3}x^2y^3 + \cdots$

and, for n variables $x_1, x_2, \ldots, x_n$,

$$\alpha x_1^{a_1} x_2^{a_2} x_3^{a_3} \cdots x_n^{a_n} = \beta x_1^{b_1} x_2^{b_2} x_3^{b_3} \cdots x_n^{b_n} + \gamma x_1^{c_1} x_2^{c_2} x_3^{c_3} \cdots x_n^{c_n} + \cdots$$

where $\alpha, \beta, \gamma, \ldots, a_1, a_2, \ldots, a_n, b_1, b_2, \ldots, b_n, c_1, c_2, \ldots, c_n, \ldots$ are numbers.

For purposes of introducing dimensional analysis we shall deal only with algebraic relationships between the dependent variables. As we shall see throughout the book, it is rare that a constitutive relationship is not expressible as an algebraic equation, so the procedures that we shall develop are of quite general applicability. To illustrate, suppose that the dependent variables of the problem are x, y, and z. The *dimensionless* equation relating them can always be arranged to have the form

$$\alpha x^a y^b z^c = \beta x^d y^e z^f + \gamma x^g y^h z^i + \delta x^j y^k z^l + \cdots \tag{3.12}$$

where $\alpha, \beta, \gamma, \delta, \ldots, a, b, c, d, \ldots$ are numbers without dimensions. Now let the dimensions of the problem be L (length) and θ (time), and suppose that x is an area (L^2), y a time (θ), and z an acceleration ($L\theta^{-2}$). Then the dimensions of the term on the left can be written

$$\begin{bmatrix} \text{dimensions} \\ \text{of} \\ x \end{bmatrix}^a \begin{bmatrix} \text{dimensions} \\ \text{of} \\ y \end{bmatrix}^b \begin{bmatrix} \text{dimensions} \\ \text{of} \\ z \end{bmatrix}^c$$

$$[L^2]^a [\theta]^b [L\theta^{-2}]^c$$

or, combining terms by simple algebra

$$\frac{\text{dimensions of}}{\text{left-hand side}} = L^{2a+c} \theta^{b-2c}$$

Since each term is dimensionless, the exponents of L and θ must each be equal to zero. Thus we must have

$$2a + c = 0$$
$$b - 2c = 0 \tag{3.13}$$

Equations 3.13 are, of course, just two linear, homogeneous algebraic equations in the three unknowns, a, b, and c, and we can solve for any two in terms of the third. Solving in terms of c, for example, we obtain

$$b = 2c$$

$$a = -\tfrac{1}{2}c$$

so that the first term on the left-hand side of Equation 3.12 is

$$\alpha x^{-c/2} y^{2c} z^{c} = \alpha \left[\frac{y^2 z}{x^{1/2}}\right]^{c}$$

$$= \alpha \left[\frac{y^4 z^2}{x}\right]^{c/2}$$

Similarly, the term $\beta x^d y^e z^f$ in Equation 3.12 has dimensions

$$\left[\begin{array}{c} L^2 \\ \text{dimensions} \\ \text{of } x \end{array}\right]^d \left[\begin{array}{c} 0 \\ \text{dimensions} \\ \text{of } y \end{array}\right]^e \left[\begin{array}{c} L\theta^{-2} \\ \text{dimensions} \\ \text{of } z \end{array}\right]^f = L^{2d+f}\theta^{e-2f}$$

so, to be dimensionless the exponents of L and θ must be zero:

$$2d + f = 0$$

$$e - 2f = 0$$

Solving in terms of, say, e

$$f = \tfrac{1}{2}e$$

$$d = -\tfrac{1}{4}e$$

Thus

$$\beta x^d y^e z^f = \beta x^{-e/4} y^e z^{e/2} = \beta \left[\frac{y z^{1/2}}{x^{1/4}}\right]^e = \beta \left[\frac{y^4 z^2}{x}\right]^{e/4}$$

Notice that exactly the same grouping of variables, $y^4 z^2/x$, appears! In retrospect, of course, this is obvious and will be true of all of the other terms as well. Equation 3.12 must then have the form

$$\alpha \left[\frac{y^4 z^2}{x}\right]^{c/2} = \beta \left[\frac{y^4 z^2}{x}\right]^{e/4} + \gamma \left[\frac{y^4 z^2}{x}\right]^{i/2} + \delta \left[\frac{y^4 z^2}{x}\right]^{-j} + \cdots$$

This is an equation in the single dimensionless quantity $y^4 z^2/x$, which we will denote by N. Thus

$$\alpha N^{c/2} = \beta N^{e/4} + \gamma N^{i/2} + \delta N^{-j} + \cdots \tag{3.14}$$

Now, if we knew the numbers $\alpha, \beta, \gamma, \delta, \ldots, c, e, i, j, \ldots$ we could solve this one algebraic equation in one unknown for the solution, a real number which we might call N_0. Since we do not know $\alpha, \beta, \gamma, \delta, \ldots, c, e, i, j, \ldots$ we cannot write down the value of N_0, but we do know that there is such a number, and although we cannot determine it from Equation 3.14 it should be possible to fix it by experiment. Thus

$$\frac{y^4 z^2}{x} = N_0$$

The relationship between the variables x, y, and z in the physical problem *must* be of the form

$$N_0 x = y^4 z^2$$

and all that remains to complete the constitutive equation is the experimental determination of the number N_0.

In any problem we will specify the relevant variables and their dimensions. Let the number of variables be $\mathscr{V}$ and the number of dimensions be $\mathscr{D}$. The difference, $G = \mathscr{V} - \mathscr{D}$, is an important quantity in dimensional analysis. In the preceding example and several to follow we have $G = 1$, which is a particularly simple case, for it will always lead to a single number to be determined experimentally. Frequently, G is greater than unity, in which case an entire function must be determined by a series of experiments that must be performed in a manner specified by the dimensional analysis. We will justify these observations and deal with the case $G > 1$ subsequently.

Example 3.4   The Orifice Equation

As a specific and practical example of the use of dimensional analysis in developing a constitutive equation we can usefully examine the tank-draining problem again. Recall that we needed a constitutive relationship between the characteristic dependent variables q and h. The volumetric flow rate, q, is related to Δp, the pressure drop from one side of the orifice to the other. [In our discussion in Chapter 2 we considered q as a function of height, h, a simplification that is sound only if the tank is open to the atmosphere. For reasons which we shall see subsequently, it turns out here to be convenient to treat the more general situation of $q = q(\Delta p)$ as initially discussed in setting up the problem.] We suspect that the liquid density, ρ, is important, and that A_0, the area of the exit orifice, must be considered. The variables, with their dimensions, are then

$$q \; [=] \; L^3 \theta^{-1}$$

$$\Delta p \; [=] \; FL^{-2}$$

$$\rho \; [=] \; ML^{-3}$$

$$A_0 \; [=] \; L^2$$

where L, M, F, and θ denote length, mass, force, and time, respectively. Any time we use an engineering system with force and mass we also must include g_c:

$$g_c \ [=] \ MF^{-1}L\theta^{-2}$$

We thus have five variables to relate to one another, $\mathscr{V} = 5$, and four dimensions, $\mathscr{D} = 4$. Then $G = \mathscr{V} - \mathscr{D} = 1$.

When the constitutive equation is written in dimensionless form each term must have the form

$$q^a \, \Delta p^b \rho^c A_0{}^d g_c{}^e$$

with dimensions

$$[L^3\theta^{-1}]^a [FL^{-2}]^b [ML^{-3}]^c [L^2]^d [MF^{-1}L\theta^{-2}]^e$$

$$= L^{3a-2b-3c+2d+e} M^{c+e} F^{b-e} \theta^{-a-2e}$$

Since the term is dimensionless the exponent of each dimension must be zero. Thus

$$3a - 2b - 3c + 2d + e = 0$$

$$c + e = 0$$

$$b - e = 0$$

$$-a - 2e = 0$$

These four equations in five unknowns can be solved in terms of any one unknown. Suppose we solve in terms of a. Then

$$b = e = -c = -a/2$$

$$d = -a$$

Each term in the equation then has the form

$$q^a \, \Delta p^{-a/2} \rho^{a/2} A_0^{-a} g_c^{-a/2} = \left\{ \frac{q}{A_0} \left[\frac{\rho}{g_c \, \Delta p} \right]^{1/2} \right\}^a$$

The constitutive equation is then an equation in terms of a single dimensionless grouping of the variables, $[q/A_0][\rho/g_c \, \Delta p]^{1/2}$. In principle that equation could be solved to obtain a value for the group. We will call that value C, in which case we obtain

$$\frac{q}{A_0} \left[\frac{\rho}{g_c \, \Delta p} \right]^{1/2} = C$$

$$q = CA_0 \left[\frac{g_c \, \Delta p}{\rho} \right]^{1/2}$$

We do not know the value of the constant, C, but we can check the *form* of the relation for the experimental situation studied in Chapter 2. In the open tank the pressure on the bottom is atmospheric pressure plus the additional pressure resulting from the weight of the liquid column, $\rho g h / g_c$, while the pressure outside is atmospheric. Thus

$$\Delta p = \frac{\rho g h}{g_c}$$

and, for that experiment,

$$q = [CA_0 g^{1/2}]h^{1/2}$$

In Chapter 2 we showed that the relationship $q = kh^{1/2}$ gave very good agreement between the model predictions and data. From the data in Chapter 2 we then obtain the value $C = 0.91$. *Notice that this result applies to all orifices, while the experiments in Chapter 2 appeared to apply to only a particular tank and orifice.* Thus, from dimensional analysis and an experiment we have the result that the flow rate through an orifice because of a pressure difference will be

$$q = 0.91 A_0 \left[\frac{g_c \Delta p}{\rho}\right]^{1/2}$$

In practice, for reasons that will be evident in Chapter 10 with application of the principle of conservation of energy, the result is generally written

$$q = C_0 A_0 \left[\frac{2 g_c \Delta p}{\rho}\right]^{1/2}$$

where C_0, the *orifice coefficient*, is a number consistently found to be close to 0.6. In the experiment reported here we found $C_0 = 0.65$.

This striking success of dimensional analysis in predicting the square root relation necessitates the insertion of an equally strong note of caution. *Only those variables that we include can appear in the result.* Thus, had we neglected to include A_0 as one of the variables that might influence q and had we inserted h instead, the constitutive equation derived would have been

$$q = Cg^{1/2}h^{5/2}$$

which is dimensionally consistent but not at all in agreement with the experiment. Careful experimental verification is essential in the search for a constitutive relationship, for the dimensional analysis can only guide our thinking.

Example 3.5 The Ideal Gas

As chemical engineers we frequently deal with processes in which some of the components are gases. We illustrate here another development of a constitutive equation from dimensional analysis by working with a simple gas problem which should be familiar from the basic courses in physics and chemistry, the relationship of the pressure of a gas to the other characterizing variables.

A simple model of a gas is one in which the molecules are assumed to be rigid spheres moving randomly with a mean speed u. The pressure, p, depends on u; the number of moles, n; the molecular weight, M_w; and the chamber volume, V. The variables and dimensions are then

$$p \;[=]\; FL^{-2}$$

$$u \;[=]\; L\theta^{-1}$$

$$n \;[=]\; \mu \;\text{(moles)}$$

$$M_w \;[=]\; M\,\mu^{-1}$$

$$V \;[=]\; L^3$$

$$g_c \;[=]\; MF^{-1}L\theta^{-2}$$

g_c is needed because both force and mass are used as dimensions. We have six variables and five dimensions, so that $G = 1$ and we anticipate there will be a single dimensionless group relating the variables through one experimental parameter.

Each term of the dimensionless constitutive equation can be written

$$p^a u^b n^c M_w{}^d V^e g_c{}^f$$

and the dimensions are then

$$[FL^{-2}]^a [L\theta^{-1}]^b [\mu]^c [M\,\mu^{-1}]^d [L^3]^e [MF^{-1}L\theta^{-2}]^f$$

$$= L^{-2a+b+3e+f} M^{d+f} \mu^{c-d} F^{a-f} \theta^{-b-2f}$$

Setting the exponent of each dimension to zero to obtain nondimensionality gives

$$-2a + b + 3e + f = 0$$

$$d + f = 0$$

$$c - d = 0$$

$$a - f = 0$$

$$-b - 2f = 0$$

and solving in terms of a

$$e = f = -d = -c = a$$

$$b = -2a$$

Thus each term in the equation has the form

$$p^a u^{-2a} n^{-a} M_w^{-a} V^a g_c^{\ a} = \left[\frac{g_c p V}{n M_w u^2} \right]^a$$

We have an equation in the single quantity $g_c p V / n M_w u^2$ which has a numerical value if the equation is solved. Since we do not know the equation at this point, however, we cannot evaluate the number. But we can now write

$$pV = Nn \frac{M_w u^2}{g_c}$$

where N is a constant to be evaluated experimentally.

We now proceed one step further by recalling from physics and chemistry that the average kinetic energy of this "billiard ball" model of a gas is directly measured by the absolute temperature. The kinetic energy per mole is

$$\mathrm{KE} = \frac{1}{2g_c} M_w u^2 \, [=] \, \mathrm{ftlb_f/mole}$$

Thus we set

$$\mathrm{KE} = CT$$

If T is measured in degrees Rankine, appropriate units of the dimensional constant C would be $\mathrm{ftlb_f/mole\ °R}$. Substituting

$$pV = (2NC)nT$$

or, combining constants, we obtain the *ideal gas equation*

$$pV = nRT$$

The *universal gas constant*, R, has an experimental value

$$R = 1545 \ \mathrm{ftlb_f/mole\ °R}$$

Of course the molecular model is at best a crude one, and real gases follow the equation only over a restricted range where intermolecular forces, molecular vibration, nonsphericity, and so on are not important. Figure 3.9 shows experimental data for several gases plotted as pV/nRT versus p at various temperatures. The constant value derived here is valid at low pressures

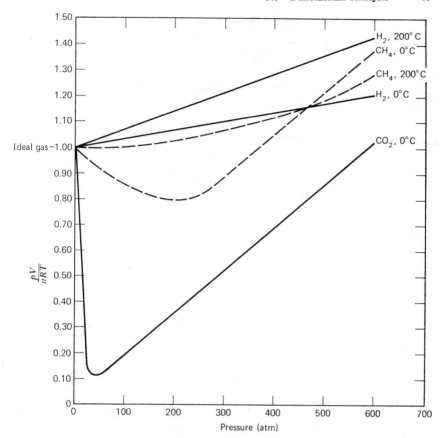

FIGURE 3.9   Deviation from ideal gas behavior, $pV/nRT = 1$, for several gases. From D. Himmelblau, *Basic Principles and Calculations in Chemical Engineering*, 2nd Ed., Prentice-Hall, Englewood Cliffs, N.J., 1967. Reproduced by permission.

and high temperatures, corresponding to low densities, where we would expect the molecular idealization to apply best.

When the temperature is involved in a dimensional analysis a convenient shortcut is usually taken. The mean particle speed is used only to obtain an energy so that temperature can be introduced. This is avoided by simply using the product RT as a variable, which has units of energy per mole, say ft lb_f/mole. For the simple gas model, then, we have variables p, n, V, and RT, with dimensions F, μ, and L, and the ideal gas equation follows immediately.

In most cases we will find that $G = \mathcal{V} - \mathcal{D}$ is two or more. Inspection of the procedure that we have been following in making the terms dimensionless indicates that we will always have $\mathcal{D}$ algebraic equations in $\mathcal{V}$ unknowns.

Thus, we can always obtain G different solutions to these algebraic equations, and we will therefore find in the general case G different dimensionless groupings of the problem variables. Suppose, for example, that $G = 2$. We will then have two groups, N_1 and N_2, and one dimensionless equation relating them. Then, in principle, we can solve for one in terms of the other to obtain the formal relationship

$$N_1 = f(N_2)$$

Now we need to find a *function* by experiment, rather than a number, so a sequence of experiments is required. If $G = 3$ we will have

$$N_1 = f(N_2, N_3)$$

and so on.

Before proceeding to further examples we can systematize the approach to dimensional analysis as follows:

1. Identify and tabulate all variables of concern, with their dimensions.
2. Determine the number of dimensionless groups by computing

$$G = \mathscr{V} - \mathscr{D}$$

The arrangement of variables within each dimensionless group is then found as follows:

3. Write the variables as a product of powers, $x^a y^b z^c, \ldots$
4. Raise the dimensions of each variable to the same power as the variable itself.
5. Develop $\mathscr{D}$ equations by writing one equation for each dimension so that the algebraic sum of its exponents is set equal to zero.
6. Solve the $\mathscr{D}$ linear algebraic equations for the exponents in terms of $\mathscr{V} - \mathscr{D}$ arbitrarily selected exponents.
7. Each exponent determines a different dimensionless group, $N_1, N_2, \ldots$. Express the constitutive equation in the form

$$N_1 = f(N_2, N_3, \ldots)$$

If there is only one group, $f =$ constant.

The procedure is applied in the following examples in which $G > 1$.

Example 3.6   The Orifice Equation

Consider the problem of a tank draining through an orifice now from a different point of view. We will focus on the tank open to the atmosphere and seek the dependence of q on h directly. We will still expect a dependence on A_0 and ρ. Since the liquid leaves the tank only because of the presence of a gravitational field the acceleration due to gravity, g, must also enter. The variables and dimensions are then

$$q \; [=] \; L^3 \theta^{-1}$$

$$h \; [=] \; L$$

$$A_0 \; [=] \; L^2$$

$$\rho \; [=] \; ML^{-3}$$

$$g \; [=] \; L\theta^{-2}$$

Only mass is used, not force, so g_c is not required. Here $\mathscr{V} = 5$, $\mathscr{D} = 3$, $G = 2$.

When the constitutive equation is written in dimensionless form we have for each term

$$q^a h^b A_0{}^c \rho^d g^e$$

and dimensions

$$[L^3\theta^{-1}]^a [L]^b [L^2]^c [ML^{-3}]^d [L\theta^{-2}]^e = L^{3a+b+2c-3d+e} M^d \theta^{-a-2e}$$

Independence of dimensions then requires that the exponent of L, M, and θ be set to zero:

$$3a + b + 2c - 3d + e = 0$$

$$d = 0$$

$$-a - 2e = 0$$

We see immediately since $d = 0$ that the density cannot be a factor in the volumetric flow. Elimination of e from the remaining two equations then leads to

$$\tfrac{5}{2}a + b + 2c = 0$$

This is one equation in *three* unknowns, so we can solve only for one in terms of the other two. Solving for b,

$$b = -2c - \tfrac{5}{2}a$$

and we obtain the term as

$$q^a h^{-2c-5a/2} A_0{}^c g^{-a/2} = \left[\frac{q}{g^{1/2}h^{5/2}} \right]^a \left[\frac{A_0}{h^2} \right]^c$$

Notice that we have *two* dimensionless groups

$$N_1 = \frac{q}{g^{1/2}h^{5/2}}, \qquad N_2 = \frac{A_0}{h^2}$$

Thus the constitutive equation is one equation relating the two quantities, N_1 and N_2. If we solve for N_1 we obtain it in terms of N_2

$$N_1 = f(N_2)$$

$$\frac{q}{g^{1/2}h^{5/2}} = f\left(\frac{A_0}{h^2}\right)$$

$$q = [gh^5]^{1/2}f\left(\frac{A_0}{h^2}\right) \tag{3.15}$$

The difference between this result and the one obtained in Example 3.4 lies in the value of G for the variables chosen. In the previous analysis the choice of variables led to $G = 1$, so that only a single number need be found experimentally. Here, wth $G = 2$, a function is needed.

Finally, before leaving the orifice problem, some further comments on experimental planning are useful. We note from Equation 3.15 that experiments are required to find the form of the function $f(A_0/h^2)$. We could, of course, fix A_0 and vary h as before, but the analysis could be rather tedious. It is much easier for the subsequent analysis to *fix h* (by adding liquid at the rate at which it leaves) and measure q for various values of A_0, though the actual experiment might be a trifle more tedious. What we find, of course, is that

$$q = \text{constant} \times A_0 \text{ at fixed } h$$

or

$$f\left(\frac{A_0}{h^2}\right) = C\frac{A_0}{h^2}$$

$$q = CA_0\sqrt{gh}$$

which is the result obtained previously.

Example 3.7   Bubble Rise Velocities

Numerous processes, such as the sewage treatment system in Example 1.2, require interaction between a gas and liquid phase. The interaction is often accomplished by bubbling the gas through the liquid. The contact then depends on the time required for a gas bubble to rise through the liquid, or, more simply, on the rise velocity of the gas. We can use dimensional analysis to provide a significant amount of information on rise velocity.

The rise velocity, u, can be expected to depend on the bubble size as measured by the diameter, D; the liquid and gas densities, ρ_L and ρ_g, respectively; and the acceleration due to gravity, g. In addition it should depend on the liquid and gas viscosities, μ_L and μ_g. Recall that viscosity, with units lb_m/ft sec, is a measure of the resistance of a fluid to flow. A gas, then, generally has a much lower viscosity than a liquid, and water is much less viscous than glycerine. The variables and dimensions are

$$u\ [=]\ L\theta^{-1}$$

$$D\ [=]\ L$$

$$\rho_L\ [=]\ ML^{-3}$$

$$\rho_g\ [=]\ ML^{-3}$$

$$g\ [=]\ L\theta^{-2}$$

$$\mu_L\ [=]\ ML^{-1}\theta^{-1}$$

$$\mu_g\ [=]\ ML^{-1}\theta^{-1}$$

We have seven variables and three dimensions, hence four dimensionless groups. By inspection it is clear that two groups will be the density and viscosity ratios, ρ_g/ρ_L and μ_g/μ_L. Neither of these ratios will ever be very different from zero, and we might make an engineering judgment at this point that the properties of the gas are probably much less important than the liquid. Thus we will neglect the contribution of ρ_g and μ_g and consider only the remaining five variables. With $\mathcal{V} = 5$, $\mathcal{D} = 3$, we have $G = 2$, so there will be two dimensionless groups.

A term in the dimensionless equation will have the form

$$u^\alpha D^\beta \rho_L{}^\gamma g^\delta \mu_L{}^\varepsilon$$

with dimensions

$$[L\theta^{-1}]^\alpha [L]^\beta [ML^{-3}]^\gamma [L\theta^{-2}]^\delta [ML^{-1}\theta^{-1}]^\varepsilon = L^{\alpha+\beta-3\gamma+\delta-\varepsilon} M^{\gamma+\varepsilon} \theta^{-\alpha-2\delta-\varepsilon}$$

The condition for no dimensional dependence is the vanishing of exponents of each dimension

$$\alpha + \beta - 3\gamma + \delta - \varepsilon = 0$$

$$\gamma + \varepsilon = 0$$

$$-\alpha - 2\delta - \varepsilon = 0$$

Solving in terms of γ and δ we obtain

$$\alpha = -2\delta + \gamma$$

$$\beta = \gamma + \delta$$

$$\varepsilon = -\gamma$$

and the typical term has the form

$$u^{-2\delta+\gamma}D^{\gamma+\delta}\rho_L{}^{\gamma}g^{\delta}\mu_L^{-\gamma} = \left[\frac{u^2}{gD}\right]^{-\delta}\left[\frac{Du\rho_L}{\mu_L}\right]^{\gamma}$$

These two dimensionless groups arise frequently and have special names:

$$\text{Reynolds Number:} \quad N_{\text{Re}} = \frac{Du\rho_L}{\mu_L}$$

$$\text{Froude Number:} \quad N_{\text{Fr}} = \frac{u^2}{gD}$$

In principle the single dimensionless equation can be solved for the Froude Number in terms of the Reynolds Number

$$N_{\text{Fr}} = f(N_{\text{Re}})$$

or

$$\frac{u^2}{gD} = f\left(\frac{Du\rho_L}{\mu_L}\right)$$

We must find the function $f(N_{\text{Re}})$ by performing experiments at various values of the Reynolds Number.

Now, when the liquid is quite inviscid and its resistance to motion small we would expect μ_L to play a small role. That is, for large N_{Re} (small μ_L) we expect f to approach a constant value and the rise velocity to satisfy the relation

$$u = C_2\sqrt{gD}$$

Figure 3.10 shows data for air in water plotted on logarithmic coordinates as Froude Number versus Reynolds Number. For $N_{\text{Re}} > 3000$ we see that the Froude Number does indeed approach a constant. Furthermore, at smaller N_{Re} the relation has a slope of approximately unity on the log coordinates, or

$$N_{\text{Fr}} = C_1 N_{\text{Re}}, \quad N_{\text{Re}} < 500$$

$$u = C_1\frac{gD^2\rho_L}{\mu_L}$$

The experimental values of C_1 and C_2 are 0.019 and 0.5, respectively.

Although the data shown are for water, the functional dependence of the Froude Number on the Reynolds Number should apply to any gas-liquid system where density and viscosity ratios are small. The dimensional analysis did not, after all, specify water in any way. Data are much more limited on other materials, but the same dependence does seem to apply, as expected.

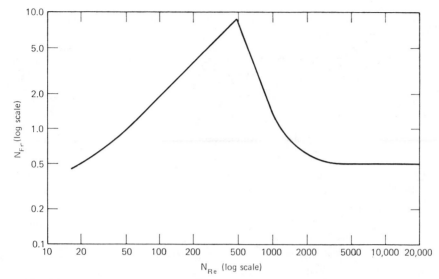

FIGURE 3.10 Froude Number, u^2/gD, versus Reynolds Number, $Du\rho_L/\mu_L$, for air bubbles rising in water. Limited data for other liquids lie on the same curve.

3.5.4 Final Observations

Some final observations on the use of dimensional analysis are helpful. First, we reiterate that the only variables which can appear in the final relationship are those which are assumed to belong. Assuming too few will lead to an incorrect result, too many to unnecessary complexity. Second, we can observe that the formal manipulations which we have carried out are really unnecessary in many situations as long as we keep the proper number of groups (variables less dimensions) in mind.

To illustrate this latter point, consider the orifice problem again, with variables

$$q \;[=]\; L^3\theta^{-1}$$
$$\Delta p \;[=]\; FL^{-2}$$
$$\rho \;[=]\; ML^{-3}$$
$$A_0 \;[=]\; L^2$$
$$g_c \;[=]\; MF^{-1}L\theta^{-2}$$

It is obvious that g_c and Δp must go together, for they are the only variables involving F. Furthermore, only g_c and ρ contain M, so they must be in a ratio to one another. Thus we can see immediately that a necessary combination is

$$\frac{g_c\,\Delta p}{\rho} \;[=]\; \frac{M}{F}\frac{L}{\theta^2}\cdot\frac{F}{L^2}\cdot\frac{L^3}{M} = \frac{L^2}{\theta^2}$$

The only remaining time is in q, which must enter as q^{-2} to cancel the θ^2 already there:

$$\frac{g_c \Delta p}{q^2 \rho} \; [=] \; \frac{L^2}{\theta^2} \cdot \frac{\theta^2}{L^6} = \frac{1}{L^4}$$

Finally, then, we need A_0:

$$\frac{A_0{}^2 g_c \Delta p}{q^2 \rho} \; [=] \; \frac{L^4}{L^4} = \text{dimensionless}$$

There can be only one group (five variables, four dimensions), so we must have

$$\frac{A_0{}^2 g_c \Delta p}{q^2 \rho} = \text{constant}$$

which is, of course, the result obtained earlier. Some care must be taken in this approach, as is generally true of shortcut methods, but if done properly it will save time over the more formal approach.

Finally, we note that the observations concerning dimensions which we have exploited to carry out dimensional analysis are sometimes grouped under the name *Buckingham Pi Theorem*.

3.6 CONCLUDING REMARKS

Developing the ability to write mathematical descriptions depends on your thoroughly understanding the logical structure of the analysis process. This logic is summarized in the sequence of figures leading up to Figure 3.8 and forms the base for everything that follows in the book. Be sure that you understand the idea of a control volume, its use in the application of a conservation principle, and the influence that control volume selection has on the number of independent variables. Recognize that there are only four independent variables and three fundamental dependent variables of possible concern in chemical engineering problems, and that the most frequently encountered situations are those in which the only independent variable is time or one spatial direction. Carefully review the important role of the constitutive equation.

Dimensional analysis is introduced in this chapter because of the central role played by dimensions at each step of the analysis process. A check for dimensional consistency is a remarkably useful tool for finding errors, and the value of dimensional analysis as a guide to experiment and the formulation of constitutive relations cannot be overemphasized. Notice how the treatment of the tank-draining problem in Section 3.4 leads immediately to $n = 1/2$ in the equation $q = kh^n$, instead of searching over all values of n

between zero and one, and how the constant k obtained from the one experiment can be expressed in terms of variables that characterize the orifice. You may find it helpful to refer to the discussion of dimensional analysis in Section 2 of the *Chemical Engineers' Handbook* and the additional references contained there.

3.1 J. H. Perry, *Chemical Engineers' Handbook*, 4th ed., McGraw-Hill, New York, 1963.
See also
3.2 J. W. Mullin, "SI Units in Chemical Engineering," *AIChE Journal*, **18**, 222 (1972).

3.7 PROBLEMS

3.1 The volumetric flow rate, q, is a commonly employed quantity and various units are in common usage. Compute the numerical conversion factors which, when multiplied by a value of q in ft^3/min, will produce the correct value for q with the following units.

(a) $\dfrac{cm^3}{hour}$

(b) $\dfrac{m^3}{sec}$

(c) $\dfrac{ft^3}{hour}$

(d) $\dfrac{gal}{min}$

(e) $\dfrac{liters}{sec}$

(f) $\dfrac{qt}{hr}$

3.2 For a liquid with density 1.00 g/cm^3 and molecular weight of 18.0, compute the numerical conversion factor which, when multiplied by a value of q in ft^3/min, will produce the correct value for the mass flow rate in the following units.

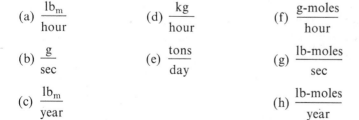

(a) $\dfrac{lb_m}{hour}$

(b) $\dfrac{g}{sec}$

(c) $\dfrac{lb_m}{year}$

(d) $\dfrac{kg}{hour}$

(e) $\dfrac{tons}{day}$

(f) $\dfrac{g\text{-moles}}{hour}$

(g) $\dfrac{lb\text{-moles}}{sec}$

(h) $\dfrac{lb\text{-moles}}{year}$

3.3 The concentration of a species in a mixture is expressed with the dimensions ML^{-3}. Denoting the molecular weight by M_w, compute the conversion factor which, when multiplied by a concentration in

lb_m/ft^3, will produce the correct value of the concentration in the following units.

(a) $\dfrac{\text{lb-moles}}{ft^3}$ (d) $\dfrac{lb_m}{\text{gal}}$ (g) $\dfrac{g}{\text{liter}}$

(b) $\dfrac{\text{g-moles}}{ft^3}$ (e) $\dfrac{g}{cm^3}$ (h) $\dfrac{\text{g-moles}}{cm^3}$

(c) $\dfrac{\text{g-moles}}{\text{liter}}$ (f) $\dfrac{kg}{m^3}$ (i) $\dfrac{\text{lb-moles}}{\text{gal}}$

3.4 A spherical tank of radius 5 ft is used as a feed tank for a batch processing unit. The correct amount of raw material is fed from this tank to a reactor by an operator who has a graph showing the mass of liquid as a function of height. He measures the liquid level before he withdraws material and then stops when he has fed the correct amount. Construct the graphs he will need if the tank may be used for liquid methanol, ethanol, and butanol at 20°C.

3.5 The specific reaction rate constant, k, for the chemical reaction between sulfuric acid and diethyl sulfate in an aqueous solution is calculated in Chapter 5 to be equal to 6.05×10^{-4} liters/g-mole min. Find its value in the following units.

(a) $\dfrac{ft^3}{\text{lb-mole min}}$ (d) $\dfrac{\text{gal}}{\text{lb-mole min}}$

(b) $\dfrac{cm^3}{\text{g-mole hr}}$ (e) $\dfrac{m^3}{\text{g-mole sec}}$

(c) $\dfrac{cm^3}{\text{g-mole min}}$ (f) $\dfrac{qt}{\text{lb-mole sec}}$

3.6 What temperature has the same numerical value in °F and °C?

3.7 The holding time in a vessel is defined as the volume divided by the volumetric throughput. Find the holding time, θ, in minutes for the following situations.

Volume	Throughput of water
10,000 liters	3 gal/min
1,000 gal	10^3 cm³/sec
500 ft³	2 liters/min
10,000 ft³	5000 gal/hr

3.8 Pressure, p, is a force per unit area. At sea level the air exerts a pressure of 14.7 $lb_f/in.^2$ Calculate this quantity in the following units.

(a) n/m^2

(d) $lbl/in.^2$

(b) d/cm^2

(e) n/cm^2

(c) lb_f/ft^2

(f) d/m^2

3.9 Work or energy has the dimensions FL or $ML^2\theta^{-2}$. Find the conversion factor which, when multiplied by a quantity measured in lb_f ft, will produce the correct value in the following units.

(a) g cm^2 sec^{-2} (ergs)

(d) calories

(b) kg m^2 sec^{-2} (joules)

(e) BTU

(c) lb_m ft^2 sec^{-2}

3.10 The gas constant, R, has a value 1545 ft lb_f/lb-mole°R. Compute R in the following units.

(a) atm ft^3/lb-mole°R (note: 1 atm = 14.7 lb_f/in^2)

(b) joules/g-mole°K

(c) BTU/lb-mole°R (note: 1 BTU = 778 ft lb_f)

3.11 The following properties of liquid water are given in cgs units. Convert to English engineering units (lb_m, ft, BTU).

(a) viscosity $\mu = 0.01$ g/cm sec

(b) heat capacity $c = 1$ cal/g °C (1 cal = 4.186 Joule)

(c) thermal conductivity $k = 0.0014$ cal/cm sec °C

3.12 The Prandtl Number is defined

$$N_{Pr} = c\mu/k$$

Show that it is dimensionless. By obtaining values for c, μ, and k from a handbook, calculate its value for liquid water, liquid glycerol, methanol, and the following gases at 1 atm and room temperature: oxygen, ethylene, ammonia, carbon dioxide, hydrogen. Can you anticipate any general conclusions?

3.13 In the tank-emptying problem assume (mistakenly) that the variables are q (ft^3/min), h (ft), g_c ($lb_m ft/lb_f sec^2$), Δp (lb_f/ft^2), and ρ (lb_m/ft^3). How does q depend on the other variables?

3.14 A rock of mass M (lb_m) falls in a vacuum under the acceleration of gravity g (ft/sec^2). Express the distance S (ft) fallen in time t (sec) in terms of M, g, and t. Compare with the result obtained in elementary physics.

3.15 The pertinent variables in describing behavior of a gas are taken as p (lb$_f$/ft^2), V (ft^3), n (moles), and RT (ft lb$_f$/mole). What is the relation between the variables?

3.16 In Problem 3.15 suppose that the pressure also depends on the volume per mole occupied by the gas molecules, V_0 (ft^3/mole). What is the resulting relation, expressed as a correction to the ideal gas? Compare with the special case

$$p(V - nV_0) = nRT$$

where the hypothesis is made that the volume that is important is the unoccupied volume.

3.17 When a liquid flows in a pipe the volumetric flow rate Q (ft^3/min) depends on density ρ (lb$_m$/ft^3), pipe radius R (ft), viscosity μ (lb$_m$/ft sec), and pressure gradient, or pressure change per length $\Delta p/L$ (lb$_f$/ft^2/ft). How does Q depend on the problem variables? (Note that lb$_f$ and lb$_m$ appear!)

At low flow an experiment is carried out holding ρ, μ, and R constant. It is found that Q varies linearly with $\Delta p/L$. How will Q depend on R?

A (very) crude approximation to high flow rate behavior is that Q is independent of μ. How will Q depend on R and $\Delta p/L$?

3.18 Repeat Example 3.6 including viscosity, $\mu(LM^{-1}\theta^{-1})$, as a variable.

PART II

Isothermal Systems

CHAPTER 4

Nonreacting Liquid Systems

4.1 INTRODUCTION

We have outlined by a simple example the basic concepts of analysis and have illustrated the essence of the model development step by discussing the source of the model equations. The remainder of this book is devoted to illustrating applications of the principles of analysis in a variety of situations. We begin by stressing the model development step with simple applications of the law of conservation of mass. By gradually increasing the problem complexity we illustrate in some detail how the logical processes illustrated on the various diagrams introduced in Chapter 3 are utilized. Although our major emphasis is on the model development step in these initial chapters, we will also keep the total analysis process in mind by indicating the engineering use of our model developments.

4.2 A SIMPLE MASS BALANCE

Let us begin by returning to the simplest problem, the filling and emptying of a cylindrical tank. The tank is shown in Figure 4.1. It fills at a constant rate q_f ft³/min, empties through a pump at constant rate q ft³/min, has cross-sectional area A ft², and liquid level h ft. The liquid density is ρ lb$_m$/ft³. (Before reading further it is instructive to see if a relationship showing h as a function of t can be derived for this simple situation on intuitive grounds. Note that the elementary equation for h as a function of t obtained in this way depends on the fact that q_f and q are constants. Try to reconstruct the logic you used to derive the relationship and think about extending the procedure to more complicated problems.)

The fundamental variable of interest is clearly mass, and the logical control volume is the tank. With the density fixed the mass is completely characterized by the height of liquid in the tank. Thus we observe the following

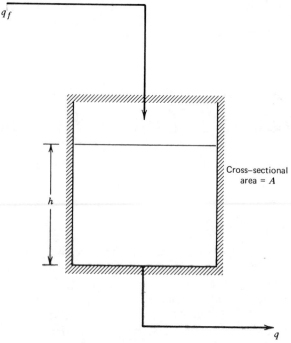

FIGURE 4.1 Tank filling at volumetric flow rate q_f and emptying at rate q.

identities between the fundamental and characterizing variables:

mass in tank at any time $t = \rho A h(t)$

$$lb_m = \left[\frac{lb_m}{ft^3}\right][ft^2][ft]$$

mass which entered between t
and some short time later, $t + \Delta t = \rho q_f \Delta t$

$$lb_m = \left[\frac{lb_m}{ft^3}\right]\left[\frac{ft^3}{min}\right][min]$$

mass which left between t and $t + \Delta t = \rho q\, \Delta t$

$$lb_m = \left[\frac{lb_m}{ft^3}\right]\left[\frac{ft^3}{min}\right][min]$$

The principle of conservation of mass, applied to the control volume of the tank, is then

$$\begin{Bmatrix} \text{mass in tank} \\ \text{at time } t + \Delta t \\ \rho Ah(t + \Delta t) \end{Bmatrix} = \begin{Bmatrix} \text{mass in tank} \\ \text{at time } t \\ \rho Ah(t) \end{Bmatrix}$$

$$+ \begin{Bmatrix} \text{mass which entered} \\ \text{between } t \text{ and } t + \Delta t \\ \rho q_f \, \Delta t \end{Bmatrix} - \begin{Bmatrix} \text{mass which left between} \\ t \text{ and } t + \Delta t \\ \rho q \, \Delta t \end{Bmatrix}$$

Dividing by ρA

$$h(t + \Delta t) = h(t) + \frac{1}{A} [q_f - q] \Delta t \qquad (4.1)$$

Since q_f and q are constants in this case, Equation 4.1 is valid for any value of Δt, even if it is not small. Thus, if t_0 refers to the time at which we begin to fill the tank and t refers to any later time we obtain

$$h(t) = h(t_0) + \frac{1}{A} [q_f - q][t - t_0] \qquad (4.2)$$

Equation 4.2 is an algebraic relationship that allows us to obtain the height of liquid in the tank at any time, $h(t)$, provided we know the area of the tank, A; the flow rates in and out, q_f and q; the time interval $t - t_0$; and the height of liquid at t_0, $h(t_0)$. *The restriction on Equation 4.2 is that q_f and q must be constant over the interval $t - t_0$.* We can illustrate the more important case in which q_f and q are variable in time by the following development:

rate at which mass leaves $= \rho q$

$$\frac{\text{lb}_m}{\text{min}} = \left[\frac{\text{lb}_m}{\text{ft}^3} \right] \left[\frac{\text{ft}^3}{\text{min}} \right]$$

Over a very small time interval Δt we can treat q as essentially constant. Notice the parallel to developments in the calculus in Chapter 15. Thus

amount which leaves in $= \rho q \, \Delta t$

a small time interval Δt

$$\text{lb}_m = \left[\frac{\text{lb}_m}{\text{ft}^3} \right] \left[\frac{\text{ft}^3}{\text{min}} \right] [\text{min}]$$

$$\begin{array}{ll} \text{total mass leaving} & = \text{sum over} = \int_{t_1}^{t_2} \rho q(t) \, dt \\ \text{between any times} & \text{all time} \\ t_1 \text{ and } t_2 & \text{intervals} \end{array}$$

Similarly,

$$\begin{array}{l} \text{total mass entering} = \int_{t_1}^{t_2} \rho q_f(t) \, dt \\ \text{between } t_1 \text{ and } t_2 \end{array}$$

Thus, application of the law of conservation of mass for this problem yields

$$\rho A h(t_2) = \rho A h(t_1) + \int_{t_1}^{t_2} \rho q_f(t)\, dt - \int_{t_1}^{t_2} \rho q(t)\, dt$$

or, since ρ is a constant,

$$h(t_2) = h(t_1) + \frac{1}{A} \int_{t_1}^{t_2} [q_f(t) - q(t)]\, dt \qquad (4.3)$$

The "name" of the dummy integration variable is unimportant, so to avoid future confusion we call it τ rather than t:

$$h(t_2) = h(t_1) + \frac{1}{A} \int_{t_1}^{t_2} [q_f(\tau) - q(\tau)]\, d\tau \qquad (4.4)$$

Equation 4.4, written for variable flow rates into and out of the tank, can be used to calculate h if q_f and q are known functions of the independent variable, time. As we saw in Chapter 2, however, we may have q available only as a function of h and *for this situation Equation 4.4 is not in a useful form.* We can illustrate an important manipulation in model construction and review some basic calculus by rewriting Equation 4.4 for the case in which we are interested in the times t and $t + \Delta t$:

$$h(t + \Delta t) - h(t) = \frac{1}{A} \int_{t}^{t+\Delta t} [q_f(\tau) - q(\tau)]\, d\tau$$

Applying the mean value theorem to the right-hand side we obtain

$$h(t + \Delta t) - h(t) = \frac{1}{A} \overline{[q_f - q]} \Delta t$$

where $\overline{q_f - q}$ represents some average value somewhere between t and $t + \Delta t$. This is simply Equation 4.1, except that we have now been careful to note that the flow rates are average values over the short interval Δt. Then

$$\frac{h(t + \Delta t) - h(t)}{\Delta t} = \frac{1}{A} \overline{[q_f - q]}$$

In the limit at $\Delta t \to 0$ the left-hand side is the derivative, dh/dt, while on the right $\overline{q_f - q}$ must simply become $q_f - q$ evaluated at t, since all points between t and $t + \Delta t$ collapse to t. Thus

$$\frac{dh}{dt} = \frac{1}{A} [q_f - q] \qquad (4.5)$$

and, since t is an arbitrary time, Equation 4.5 holds for all times.

If we remultiply each side of Equation 4.5 by ρA, the relationship with the conservative principle is more clearly illustrated.

$$\frac{d}{dt}[\rho A h] = \rho q_f - \rho q \qquad (4.6)$$

$$\left[\frac{1}{\min}\right]\left[\frac{lb_m}{ft^3}\right][ft^2][ft] = \left[\frac{lb_m}{ft^3}\right]\left[\frac{ft^3}{\min}\right] - \left[\frac{lb_m}{ft^3}\right]\left[\frac{ft^3}{\min}\right]$$

$$\frac{lb_m}{\min} = \frac{lb_m}{\min}$$

The left-hand side is the rate of change of mass in the tank, while the right is the net mass flow rate.

Let us summarize the manipulative steps that led to this equation.

1. The word statement of the law of conservation of mass was expressed in symbols using the characterizing variables of the problem (Equations 4.1 and 4.3). The concept of an integral relationship to represent the mass entering and leaving the system was illustrated.
2. The equation representing the conservation law was rearranged so that the right-hand side contained the change in characterizing variables of interest (h in this case) divided by the change in the independent variable.
3. The mean value theorem was applied.
4. The limit as Δt approached zero was taken and the final model equation expressed in derivative form.

If q_f and q are constant, this very simple physical situation serves to illustrate clearly the detailed steps in model development as shown by Figure 3.1. Only one conservation law was needed, time was the only independent variable of concern, and a constitutive relation was not required. By changing the geometry of the tank we can show how a simple constitutive relation, available to us from geometric consideration, is used.

A wedge-shaped tank of overall height H, width B, and length L is now considered (Figure 4.2). The flow rates in and out are again represented by q_f and q and the liquid density is ρ. Application of the word statement produces the following equation, equivalent to Equation 4.4:

$$\rho V(t_2) = \rho V(t_1) + \rho \int_{t_1}^{t_2} [q_f(\tau) - q(\tau)]\, d\tau$$

where V is the volume of liquid in the tank at time t, ft^3. Dividing by ρ, applying the mean value theorem, and taking the limit as Δt approaches zero yields

$$\frac{dV}{dt} = q_f - q \qquad (4.7)$$

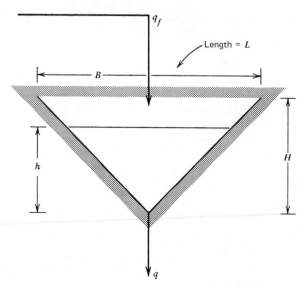

FIGURE 4.2   Wedge-shaped tank filling at volumetric flow rate q_f and emptying at rate q.

Equation 4.7 is the basic model equation. If we refer to Figure 3.3 and check to determine if we have sufficient equations we find that we must supply an additional relationship if we seek to find h as a function of time. (It is, of course, obvious that this relationship cannot be generated from the conservation laws.) The needed constitutive relationship can be obtained by using our knowledge of the system geometry.

$\quad V(t) =$ area of wedge filled with liquid multiplied by the length of wedge-shaped tank

$$= \left[\frac{hw}{2} \right] L$$

where

$\quad w =$ width of the wedge at the liquid surface

$$= \frac{hB}{H}$$

$$V(t) = \frac{BLh^2}{2H}$$

This is the desired relationship between V and h. In the cylindrical tank this geometrical relation was too simple to warrant mention ($V = Ah$).

Substitution into Equation 4.7 produces, after some minor manipulation,

$$\frac{dh^2}{dt} = \frac{2H}{BL}[q_f - q] \tag{4.8}$$

If q_f and q are constant, Equation 4.8 can be integrated and the relationship between h and t established.

$$\int_{h_0}^{h} dy^2 = \frac{2H}{BL}[q_f - q]\int_{t_0}^{t} d\tau$$

$$h^2 - h_0^2 = \frac{2H}{BL}[q_f - q][t - t_0]$$

$$h = \sqrt{h_0^2 + \frac{2H}{BL}[q_f - q][t - t_0]}$$

Here, h_0 is the liquid height at time t_0.

4.3 APPLICATION OF THE MODEL EQUATIONS

Before we return to model equation development by means of more complex examples it is helpful for our understanding of the total analysis process if we see how the model equations for a draining tank can be used in an elementary but interesting engineering situation. A secondary purpose of this section will be to gain some practice with the basic calculus so that we sharpen our skills at correctly interpreting mathematical descriptions. We will develop in detail some basic ideas in level control using our tested mathematical description of the filling and emptying tank.

If we are interested in controlling the liquid level in a tank that is part of some process, we first need to consider how the system behaves if a change occurs in the inlet flow rate. If the process is operating under constant conditions for some period of time, q_f, the inlet flow rate, will be equal to q, the outlet flow rate. Equation 4.5 shows us what we know intuitively: the rate of change of the liquid level is zero and the level remains constant. We shall assume that the outlet flow from the tank in this process under consideration is maintained at a fixed rate, q, by a pump (a very common situation).

Let us assume that because of some upstream upset at time T_1 the inlet flow rate increases by an amount Q^*. At some later time T_2 the process returns to normal and Q^* becomes 0. This type of temporary upset is shown graphically in Figure 4.3(a).

The equation governing the level, Equation 4.5, is

$$\frac{dh}{dt} = \frac{1}{A}[q_f(t) - q(t)]$$

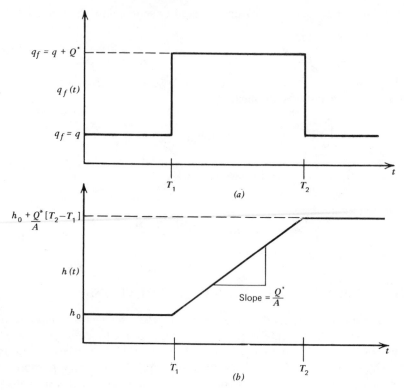

FIGURE 4.3 (a) Change in the inlet flow rate between T_1 and T_2. (b) Change in the liquid level.

From time $t = 0$ until $t = T_1$ we have

$$q_f = q, \qquad 0 \le t < T_1$$

Thus

$$\frac{dh}{dt} = 0, \qquad 0 \le t < T_1$$

$$h = \text{constant}, \qquad 0 \le t < T_1$$

We shall call this constant h_0.

At $t = T_1$ and until $t = T_2$

$$q_f = q + Q^*, \qquad T_1 \le t < T_2$$

$$\frac{dh}{dt} = \frac{Q^*}{A}, \qquad T_1 \le t < T_2$$

The derivative of h is a constant, so h is a linear function of t passing through h_0 at $t = T_1$:

$$h(t) = h_0 + \frac{Q^*}{A}[t - T_1], \qquad T_1 \le t \le T_2$$

In particular, when $t = T_2$

$$h(T_2) = h_0 + \frac{Q^*}{A}[T_2 - T_1]$$

Finally, at $t = T_2$, q_f returns permanently to its original value,

$$q_f = q, \qquad t \ge T_2$$

$$\frac{dh}{dt} = 0, \qquad t \ge T_2$$

$$h = \text{constant}, \qquad t \ge T_2$$

$$h = h_0 + \frac{Q^*}{A}[T_2 - T_1], \qquad t \ge T_2$$

The situation is shown graphically in Figure 4.3(b).

Our goal is to maintain a reasonably constant level in the tank despite inlet fluctuations. It is obvious that we will want to design a system which will adjust the outlet flow to compensate for level variations. The simplest device that we might conceive is one which monitors the level continuously (e.g., a float) and sends electrical or pneumatic instructions to a valve which opens wider when the level is above that desired, closes when lower. Figure 4.4 is a schematic diagram illustrating the arrangement. For example, if h^* is the desired level and q^* the design flow rate we might adjust the exit flow according to the relation

$$q = q^* + K[h - h^*]$$

In this way, known as "proportional feedback control," the control effort depends directly on the "error," $h - h^*$. When h exceeds h^* the exit flow is increased and when h is lower than h^* the exit flow is decreased. A large error dictates a large control effort, a small error requires little control.

The flow system is still governed by the equation

$$\frac{dh}{dt} = \frac{1}{A}[q_f - q]$$

The system with a level controller obeys this equation with $q^* + K[h - h^*]$ substituted for q

$$\frac{dh}{dt} = \frac{1}{A}\{q_f - q^* - K[h - h^*]\} \qquad (4.9)$$

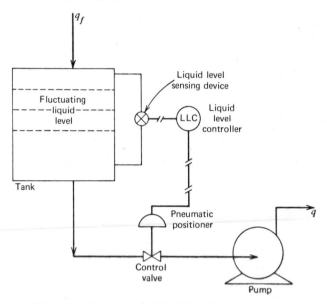

FIGURE 4.4   Schematic diagram of a liquid level control system.

If we normally expect $q_f = q^*$ and denote the disturbance by $Q(t)$, we can write

$$q_f = q^* + Q(t)$$

Equation 4.9 for the tank system with control can be written

$$\frac{dh}{dt} + \frac{K}{A}[h - h^*] = \frac{1}{A}Q(t)$$

Finally, since we are interested in the difference $h - h^*$ and since h^* is a constant, we can write

$$\frac{d[h - h^*]}{dt} = \frac{dh}{dt} - \frac{dh^*}{dt} = \frac{dh}{dt}$$

so

$$\frac{d}{dt}[h - h^*] + \frac{K}{A}[h - h^*] = \frac{1}{A}Q(t) \tag{4.10}$$

We seek a solution to this equation starting at time $t = 0$, where the system is presumed to be at desired level $h = h^*$.

Equation 4.10 is not separable, but it can be integrated by means of the following trick, which is discussed more fully in Section 15.10:

$$\frac{d}{dt}[h - h^*] + \frac{K}{A}[h - h^*] = e^{-Kt/A}\frac{d}{dt}\{e^{Kt/A}[h - h^*]\}$$

Thus Equation 4.10 can be written

$$\frac{d}{dt}\{e^{Kt/A}[h - h^*]\} = \frac{e^{Kt/A}}{A} Q(t)$$

Integrating both sides gives

$$e^{K\tau/A}[h - h^*]\Big|_0^t = \frac{1}{A} \int_0^t e^{K\tau/A} Q(\tau)\, d\tau$$

and making use of the fact that when $t = 0$, $h - h^* = 0$, we finally obtain the height as a function of time:

$$h(t) = h^* + \frac{e^{-Kt/A}}{A} \int_0^t e^{K\tau/A} Q(\tau)\, d\tau \tag{4.11}$$

To see the effect of putting a control system on our tank let us return to the step input disturbances we previously discussed. The feed to the system varies as follows

$$Q(t) = \begin{cases} 0, & 0 \le t < T_1 \\ Q^* = \text{constant}, & T_1 \le t < T_2 \\ 0, & T_2 \le t \end{cases}$$

Then

$$\int_0^t e^{K\tau/A} Q(\tau)\, d\tau = 0, \qquad t < T_1$$

$$\int_0^t e^{K\tau/A} Q(\tau)\, d\tau = \int_{T_1}^t e^{K\tau/A} Q^*\, d\tau = \frac{AQ^*}{K} [e^{Kt/A} - e^{KT_1/A}],$$

$$T_1 \le t < T_2$$

$$\int_0^t e^{K\tau/A} Q(\tau)\, d\tau = \int_{T_1}^{T_2} e^{K\tau/A} Q^*\, d\tau = \frac{AQ^*}{K} [e^{KT_2/A} - e^{KT_1/A}],$$

$$T_2 \le t$$

The level in the tank as a function of time is then

$$h = h^* + 0 = h^*, \qquad 0 \le t < T_1$$

$$h = h^* + \frac{Q^*}{K} [1 - e^{-K[t-T_1]/A}], \qquad T_1 \le t < T_2$$

$$h = h^* + \frac{Q^*}{K} [e^{-K[t-T_2]/A} - e^{-K[t-T_1]/A}], \qquad T_2 \le t$$

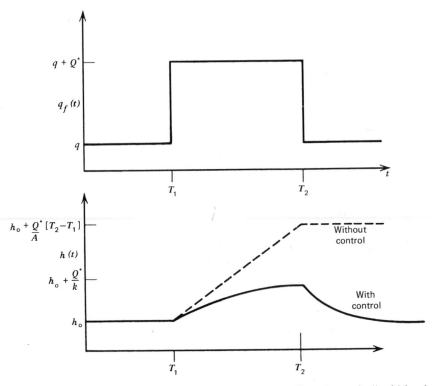

FIGURE 4.5 Change in inlet flow rate and corresponding change in liquid level with and without proportional feedback control.

The system behavior is shown in Figure 4.5 where it is compared with the behavior of the system without control ($K = 0$). The difference is striking, and it is clear that by proper choice of K the fluctuations in tank level can be kept within desired limits.

As was discussed in Chapter 2 and indicated graphically in Figure 2.1, the ultimate reason that we engage in engineering analysis is to use our mathematical model for useful prediction and design. For the simple level control system discussed above we can indicate here how the mathematical model enables us to specify the "proportional gain" parameter K in order to ensure that whatever the disturbance, the level will remain within required bounds. We suppose that fluctuations in the inlet feed will never exceed a fraction f of the total design flow rate q^*, that is,

$$\text{maximum of } |Q(t)| \leq fq^*$$

We wish to choose K so that fluctuations in the liquid level never exceed a fraction ϕ of the total design level, that is,

$$\text{maximum of } |h(t) - h^*| \leq \phi h^*$$

We can find the design value for K by manipulation of Equation 4.11. First, we rewrite the equation as

$$\frac{h - h^*}{h^*} = \frac{e^{-Kt/A}}{Ah^*} \int_0^t e^{K\tau/A} Q(\tau) \, d\tau$$

The design condition $|h - h^*| \leq \phi h^*$ then leads to

$$\frac{e^{-Kt/A}}{Ah^*} \left| \int_0^t e^{K\tau/A} Q(\tau) \, d\tau \right| \leq \phi$$

The integral takes on its largest value when $Q(t)$ is as large as possible for all time. Thus, we take $Q(t) = fq^*$ and obtain the design inequality

$$\frac{fq^*}{Ah^*} e^{-Kt/A} \int_0^t e^{K\tau/A} \, d\tau \leq \phi$$

or, carrying out the integration,

$$\frac{fq^*}{Kh^*} [1 - e^{-Kt/A}] \leq \phi$$

The left-hand side takes on its largest value when t gets very large, so we have

$$\frac{fq^*}{Kh^*} \leq \phi$$

or, solving for K,

$$K \geq \frac{fq^*}{\phi h^*}$$

The smallest allowable value of K will require the least activity on the part of the control system. Thus, for design purposes,

$$K = \frac{fq^*}{\phi h^*} \tag{4.12}$$

and the exit flow rate is required to depend on the liquid level by the equation

$$q = q^* \left[1 + \frac{f}{\phi} \frac{h - h^*}{h^*} \right]$$

If feed rate fluctuations can go as high as 10 percent of the design feed flow rate, and it is required that the liquid level stay constant to within 3 percent, then $f = 0.10$, $\phi = 0.03$, and

$$q = q^* \left[1 + \frac{10}{3} \frac{h - h^*}{h^*} \right]$$

That is, for each percent increase or decrease in liquid level the exit flow rate is increased or decreased by 3-1/3 percent in order to ensure that the design tolerance is never exceeded. Notice that this is a conservative design and that generally the fluctuations in h will be significantly less than 3 percent.

4.4 COMPONENT MASS BALANCES

Most engineering situations in which we will be interested involve several distinct mass species, and for the remainder of this chapter we shall focus on the model development part of analysis for these systems. The simplest such application of the principle of conservation of mass is to the flow system show in Figure 4.6, in which we have a cylindrical tank with two streams entering and one leaving. The entering streams are numbered 1 and 2, with mass concentrations c_1 and c_2 of a dissolved substance (a salt) in water, while the concentration is c_3 in the exit stream 3. Densities are ρ_1, ρ_2, and ρ_3, respectively, and the volumetric flow rates are maintained by pumps at q_1, q_2, and q_3. Tank cross-sectional area is A and liquid level h. We wish to

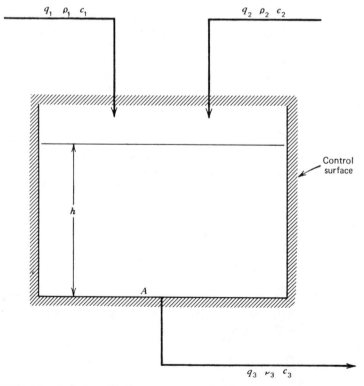

FIGURE 4.6 Tank in which two streams of different concentration are mixed continuously.

relate the liquid level and exit concentration to the flows and inlet concentrations.

Referring to the logic flow chart in Figure 3.4 we see that the first several steps are obvious. The fundamental variable is mass, and this is characterized by the variables noted above. Let us examine in more detail the question of the choice of a control volume. We have drawn the shaded line in Figure 4.6 as though we intended to use the piece of process equipment, the mixing tank, as the control volume. Since the two streams entering the tank are different in composition and density we would expect in the absence of agitation that there would be variations in the salt concentration and liquid density from point to point within the tank. Hence the control volume designated could not be characterized by unique values of the variables with respect to spatial position. To circumvent this difficulty we make a physical assumption that the tank is sufficiently well agitated to be of completely uniform composition. Thus a sample drawn from one location will be identical to a sample drawn at the same time from any other location. That is, we assume that a lumped representation will adequately describe the problem, *an assumption which will require ultimate experimental verification.* An important consequence of this *well-stirred,* or *perfectly mixed* assumption is that the composition at any point in the tank is identical to the exit composition. Thus, the concentration and density of material in the tank at any time are, respectively, c_3 and ρ_3.

Application of the principle of conservation of mass to the total mass is as follows:

$$\begin{pmatrix} \text{mass in tank} \\ \text{at time } t_2 \\ \rho_3(t_2)Ah(t_2) \end{pmatrix} = \begin{pmatrix} \text{mass in tank} \\ \text{at time } t_1 \\ \rho_3(t_1)Ah(t_1) \end{pmatrix} + \begin{Bmatrix} \text{mass in by flow} \\ \text{in stream 1} \\ \int_{t_1}^{t_2} \rho_1(\tau)q_1(\tau)\, d\tau \end{Bmatrix}$$

$$+ \begin{Bmatrix} \text{mass in by flow} \\ \text{in stream 2} \\ \int_{t_1}^{t_2} \rho_2(\tau)q_2(\tau)\, d\tau \end{Bmatrix} - \begin{Bmatrix} \text{mass out by flow} \\ \text{in stream 3} \\ \int_{t_1}^{t_2} \rho_3(\tau)q_3(\tau)\, d\tau \end{Bmatrix}$$

or

$$\rho_3(t_2)Ah(t_2) = \rho_3(t_1)Ah(t_1)$$
$$+ \int_{t_1}^{t_2} [\rho_1(\tau)q_1(\tau) + \rho_2(\tau)q_2(\tau) - \rho_3(\tau)q_3(\tau)]\, d\tau \quad (4.13)$$

There is little information here about the amount of salt in the system. To obtain that we need to observe that conservation of mass applies not only to the total mass of the system, but also to the *component species,* salt and water.

It must be true, for example, that

$$\begin{Bmatrix} \text{mass of } salt \text{ in} \\ \text{tank at time } t_2 \end{Bmatrix} = \begin{Bmatrix} \text{mass of } salt \text{ in} \\ \text{tank at time } t_1 \end{Bmatrix} + \begin{Bmatrix} \text{mass of } salt \\ \text{in by flow} \\ \text{in stream 1} \end{Bmatrix}$$

$$+ \begin{Bmatrix} \text{mass of } salt \\ \text{in by flow} \\ \text{in stream 2} \end{Bmatrix} - \begin{Bmatrix} \text{mass of } salt \\ \text{out by flow} \\ \text{in stream 3} \end{Bmatrix}$$

Since the concentration of salt, c_1, c_2, and c_3, is in units lb_m salt/ft^3, we note

mass of salt in tank $= c_3 A h$

$$[lb_m] = \left[\frac{lb_m}{ft^3} \right] [ft^2][ft]$$

mass flow rate of salt in stream $1 = c_1 q_1$

$$\left[\frac{lb_m}{min} \right] = \left[\frac{lb_m}{ft^3} \right] \left[\frac{ft^3}{min} \right]$$

$$\begin{Bmatrix} \text{mass flow of salt in stream 1} \\ = \text{integral of mass flow rate} \end{Bmatrix} = \int_{t_1}^{t_2} c_1(\tau) q_1(\tau) \, d\tau$$

Thus

$$c_3(t_2) A h(t_2) = c_3(t_1) A h(t_1) + \int_{t_1}^{t_2} c_1(\tau) q_1(\tau) \, d\tau$$

$$+ \int_{t_1}^{t_2} c_2(\tau) q_2(\tau) \, d\tau - \int_{t_1}^{t_2} c_3(\tau) q_3(\tau) \, d\tau \quad (4.14)$$

It is similarly true that

$$\begin{Bmatrix} \text{mass of } water \text{ in} \\ \text{tank at time } t_2 \end{Bmatrix}$$

$$= \begin{Bmatrix} \text{mass of } water \text{ in} \\ \text{tank at time } t_1 \end{Bmatrix} + \begin{Bmatrix} \text{mass of } water \text{ in} \\ \text{by flow in stream 1} \end{Bmatrix}$$

$$+ \begin{Bmatrix} \text{mass of } water \text{ in} \\ \text{by flow in stream 2} \end{Bmatrix} - \begin{Bmatrix} \text{mass of } water \text{ out} \\ \text{by flow in stream 3} \end{Bmatrix}$$

However, the concentration of water in stream 1 is simply $\rho_1 - c_1$, total pounds per cubic foot less pounds of salt per cubic foot, and similarly for streams 2 and 3. Thus the equation for conservation of mass applied to water is

$$[\rho_3(t_2) - c_3(t_2)] A h(t_2) = [\rho_3(t_1) - c_3(t_1)] A h(t_1)$$

$$+ \int_{t_1}^{t_2} \{ [\rho_1(\tau) - c_1(\tau)] q_1(\tau) + [\rho_2(\tau) - c_2(\tau)] q_2(\tau)$$

$$- [\rho_3(\tau) - c_3(\tau)] q_3(\tau) \} \, d\tau$$

and this is simply the difference between those for the total mass and salt. Hence no new information will be obtained. Of the three equations, total mass, salt, and water, only two can be independent and provide useful information.

We have, then, two equations governing the behavior of the flow system, which we can rewrite in terms of times t and $t + \Delta t$:

salt + water:

$$\rho_3(t + \Delta t)Ah(t + \Delta t) = \rho_3(t)Ah(t)$$
$$+ \int_t^{t+\Delta t} [\rho_1(\tau)q_1(\tau) + \rho_2(\tau)q_2(\tau) - \rho_3(\tau)q_3(\tau)]\, d\tau$$

salt:

$$c_3(t + \Delta t)Ah(t + \Delta t) = c_3(t)Ah(t)$$
$$+ \int_t^{t+\Delta t} [c_1(\tau)q_1(\tau) + c_2(\tau)q_2(\tau) - c_3(\tau)q_3(\tau)]\, d\tau$$

In terms of mean values, by application of the mean value theorem,

salt + water:

$$\rho_3(t + \Delta t)Ah(t + \Delta t) = \rho_3(t)Ah(t) + [\overline{\rho_1 q_1} + \overline{\rho_2 q_2} - \overline{\rho_3 q_3}]\Delta t$$

salt:

$$c_3(t + \Delta t)Ah(t + \Delta t) = c_3(t)Ah(t) + [\overline{c_1 q_1} + \overline{c_2 q_2} - \overline{c_3 q_3}]\Delta t$$

where the overbar indicates that the function is evaluated somewhere between t and $t + \Delta t$. Thus

salt + water: $\dfrac{\rho_3(t + \Delta t)h(t + \Delta t) - \rho_3(t)h(t)}{\Delta t}$

$$= \frac{1}{A}[\overline{\rho_1 q_1} + \overline{\rho_2 q_2} - \overline{\rho_3 q_3}]$$

salt: $\dfrac{c_3(t + \Delta t)h(t + \Delta t) - c_3(t)h(t)}{\Delta t} = \dfrac{1}{A}[\overline{c_1 q_1} + \overline{c_2 q_2} - \overline{c_3 q_3}]$

In the limiting process $\Delta t \to 0$ the term on the left becomes the derivative evaluated at t, while those on the right are evaluated at t. (All times between t and $t + \Delta t$ become t as $\Delta t \to 0$.) Since t is any time the following two equations are then always true:

salt + water: $\quad \dfrac{d}{dt}\rho_3 h = \dfrac{1}{A}[\rho_1 q_1 + \rho_2 q_2 - \rho_3 q_3]$ (4.15)

salt: $\quad \dfrac{d}{dt}c_3 h = \dfrac{1}{A}[c_1 q_1 + c_2 q_2 - c_3 q_3]$ (4.16)

Notice that it is essential to have the salt equation in this differential equation form, for the original integral formulation developed directly from the principle of conservation of mass, Equation 4.14, involved the (probably unknown) function $c_3(t)$ both on the left and under an integral on the right. The integration to find $c_3(t)$ could not be carried out without knowing how c_3 depended on time, and the dependence of c_3 on t, of course, is part of the problem.

4.5 MODEL BEHAVIOR

The flow equations for the salt-water system derived above do not contain adequate information for problem solution. We have a total of eleven variables; three concentrations, c_1, c_2, and c_3; three densities, ρ_1, ρ_2, and ρ_3; three flow rates, q_1, q_2, and q_3; the height of liquid, h; and the area, A. Typically we would be given A, q_1, q_2, q_3, and perhaps c_1 and c_2. The two equations will be adequate to find h and c_3 only when we have further information about the densities. That is, we require a *constitutive relation* to establish the relation between density and concentration and thus to provide us with three further equations.

At this point we shall make a simplifying constitutive assumption that is not unreasonable and simplifies the analysis, though we shall ultimately find that we need not make it to obtain the results given. We shall assume that the density of the salt water solutions in this problem is essentially independent of concentration and hence a constant in all streams for all time. We denote this constant value as ρ. Then

$$\frac{d}{dt}\rho_3 h = \rho \frac{dh}{dt} = \frac{\rho}{A}[q_1 + q_2 - q_3]$$

or

$$\frac{dh}{dt} = \frac{1}{A}[q_1 + q_2 - q_3] \tag{4.17}$$

When the net volumetric flow rate is zero ($q_1 + q_2 = q_3$), the level does not change.

4.5.1 Steady State Behavior

A case of interest is that in which the inflow and outflow are equal ($q_1 + q_2 = q_3$) and constant. Then $dh/dt = 0$ and $h = $ constant. The salt equation becomes

$$\frac{d}{dt}c_3 h = h\frac{dc_3}{dt} = \frac{1}{A}[c_1 q_1 + c_2 q_2 - c_3 q_3]$$

or

$$\frac{dc_3}{dt} = \frac{1}{\theta}[\lambda_1 c_1 + \lambda_2 c_2 - c_3] \tag{4.18}$$

where

$$\theta = Ah/q_3 \; [=] \text{ time}$$

$$\lambda_1 = q_1/q_3 \qquad \lambda_2 = q_2/q_3 = 1 - \lambda_1$$

If we are to carry out the integration to find $c_3(t)$ we will need to evaluate one constant of integration, and thus we need one piece of information about the system at some time. Suppose that it is the value of concentration, c_{30}, at time $t = 0$. If c_1 and c_2 are constant, then Equation 4.18 is rearranged to yield

$$\frac{dc_3}{c_3 - \lambda_1 c_1 - \lambda_2 c_2} = -\frac{dt}{\theta}$$

Integrating from $t = 0$ to some later time

$$\ln \frac{c_3 - \lambda_1 c_1 - \lambda_2 c_2}{c_{30} - \lambda_1 c_1 - \lambda_2 c_2} = -\frac{t}{\theta}$$

$$c_3(t) = \lambda_1 c_1 + \lambda_2 c_2 + [c_{30} - \lambda_1 c_1 - \lambda_2 c_2]e^{-t/\theta} \tag{4.19}$$

As t gets large the exponential goes to zero, in which case

$$t \text{ large:} \quad c_3 \rightarrow \lambda_1 c_1 + \lambda_2 c_2$$

That is, c_3 approaches a *steady state*. This steady state is characteristic of most process equipment. In the steady state, of course, nothing is changing in time and time derivatives must be zero. We could find the steady state value directly from the equation

$$Ah \frac{dc_3}{dt} = c_1 q_1 + c_2 q_2 - c_3 q_3$$

by setting dc_3/dt to zero. At steady state the rate at which salt flows in, $c_1 q_1 + c_2 q_2$, is equal to the rate at which it leaves, $c_3 q_3$.

We note further that "clock time" t is not the pertinent quantity determining the steady state value, but the dimensionless ratio t/θ, for only when t/θ is sufficiently large (say, $t/\theta \geq 3$) is c_3 essentially constant. The quantity $\theta = Ah/q_3$, the ratio of volume to volumetric flow rate, is called the *residence time*. Any general analysis of a system must take into account both the transient interval over several residence times, when the system is changing substantially, and the subsequent steady state, during which there is no variation in time.

Finally, let us observe here that the analysis process is not yet complete. We have made two assumptions, perfect mixing and constant density. We shall show subsequently that the second of these is not required for the results obtained here and ask the reader to accept that fact for the present. The first, however, needs to be verified if the results of this and similar analyses are to be used for predictive purposes. The simplest experiment that can be performed to study the validity of the perfect mixing assumption is a washing experiment, in which a tank with an initial known salt concentration is washed by an entering pure stream. In that case $q_1 = q_3$, $q_2 = 0$, $c_1 = 0$, and the time dependence of $c_3(t)$ from Equation 4.19 is

$$\frac{c_3(t)}{c_{30}} = e^{-t/\theta}$$

or

$$\ln \frac{c_3}{c} = -t/\theta$$

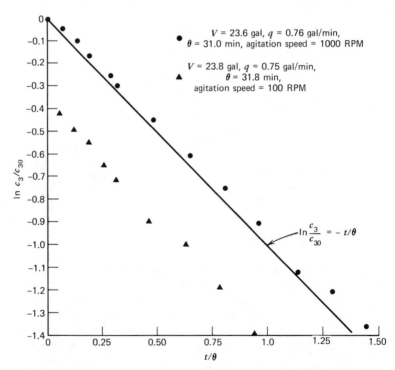

FIGURE 4.7   Natural logarithm of c_3/c_0 versus t/θ for salt being washed from a tank. Agitation was by a 2-1/2 inch diameter marine propellor in a liquid height of 10 inches. This was (intentionally) extremely inefficient agitation. Data of D. T. T. Chiang.

The data shown in Figure 4.7 indicate that with sufficient agitation the theoretical predictions based on perfect mixing agree well with the experimental observations.

4.5.2 Unsteady State Behavior

A second situation of interest which really includes the one just studied as a special case is that in which q_1, q_2, and q_3 are constants but $q_1 + q_2 \neq q_3$. The governing equations, Equations 4.16 and 4.17, are then

$$\text{salt + water:} \quad \frac{dh}{dt} = \frac{1}{A}[q_1 + q_2 - q_3] = \text{constant} \qquad (4.20)$$

$$\text{salt:} \quad \frac{d}{dt}c_3 h = \frac{1}{A}[c_1 q_1 + c_2 q_2 - c_3 q_3]$$

$$= h\frac{dc_3}{dt} + c_3\frac{dh}{dt}$$

$$= h\frac{dc_3}{dt} + \frac{1}{A}c_3[q_1 + q_2 - q_3] \qquad (4.21)$$

The total mass balance equation can be directly integrated to give

$$h = h_0 + \frac{1}{A}[q_1 + q_2 - q_3]t \qquad (4.22)$$

where h_0 is the level at $t = 0$, in which case the salt equation simplifies to

$$\{Ah_0 + [q_1 + q_2 - q_3]t\}\frac{dc_3}{dt} = q_1 c_1 + q_2 c_2 - [q_1 + q_2]c_3$$

or

$$\left[\frac{Ah_0}{q_1 + q_2 - q_3} + t\right]\frac{dc_3}{dt} = \frac{q_1 c_1 + q_2 c_2}{q_1 + q_2 - q_3} - \frac{q_1 + q_2}{q_1 + q_2 - q_3}c_3$$

This equation is integrated by noting that

$$\frac{dt}{t + \dfrac{Ah_0}{q_1 + q_2 - q_3}} = d[\ln \tau]$$

where

$$\tau = t + \frac{Ah_0}{q_1 + q_2 - q_3}$$

Thus

$$
\frac{dc_3}{c_3 - \dfrac{q_1 c_1 + q_2 c_2}{q_1 + q_2}} = - \frac{q_1 + q_2}{q_1 + q_2 - q_3} \, d \ln \tau
$$

or

$$
\ln \left[\frac{c_3 - \dfrac{q_1 c_1 + q_2 c_2}{q_1 + q_2}}{c_{30} - \dfrac{q_1 c_1 + q_2 c_2}{q_1 + q_2}} \right]
= - \frac{q_1 + q_2}{q_1 + q_2 - q_3} \ln \left[\frac{t + \dfrac{A h_0}{q_1 + q_2 - q_3}}{0 + \dfrac{A h_0}{q_1 + q_2 - q_3}} \right]
$$

$$
c_3(t) = \frac{q_1 c_1 + q_2 c_2}{q_1 + q_2} + \left[c_{30} - \frac{q_1 c_1 + q_2 c_2}{q_1 + q_2} \right]
$$

$$
\times \left[\frac{1}{1 + \dfrac{[q_1 + q_2 - q_3]t}{A h_0}} \right]^{[q_1+q_2]/[q_1+q_2-q_3]} \tag{4.23}
$$

Notice that for small h_0 the approach to a steady state for concentration (height varies steadily!) is rapid

$$
c_3 \to \frac{q_1 c_1 + q_2 c_2}{q_1 + q_2}
$$

In fact, for an initially empty tank ($h_0 = 0$) the square bracket on the right is identically zero and the concentration is always constant, a result which might at first seem surprising.

The case in which inflow equals outflow ($q_1 + q_2 - q_3 = 0$) is obtained by a limiting process, as follows:

$$
\left[\frac{1}{1 + \dfrac{[q_1 + q_2 - q_3]t}{A h_0}} \right]^{[q_1+q_2]/[q_1+q_2-q_3]}
= \left[\frac{1}{1 + \varepsilon [t/\theta]} \right]^{1/\varepsilon}
$$

where $\theta = \dfrac{A h_0}{q_1 + q_2}$ (when $q_1 + q_2 = q_3$ and $h = $ constant this is as defined above) and

$$
\varepsilon = \frac{q_1 + q_2 - q_3}{q_1 + q_2}
$$

In the limit as $\varepsilon \to 0$ we then obtain the exponential by first taking the logarithm and then applying l'Hôpital's rule,

$$\lim_{\varepsilon \to 0} \left\{ \frac{1}{1 + \varepsilon[t/\theta]} \right\}^{1/\varepsilon} = e^{-t/\theta}$$

in which case the concentration becomes that given in Equation 4.19

$$\lim q_1 + q_2 \to q_3:$$

$$c_3(t) = \frac{q_1 c_1 + q_2 c_2}{q_1 + q_2} + \left[c_{30} - \frac{q_1 c_1 + q_2 c_2}{q_1 + q_2} \right] e^{-t/\theta}$$

4.5.3 Density Assumption

The preceding results were based on the assumption that in the two-component system the density was always a constant independent of concentration, leading to the equations

$$\text{total mass:} \quad \frac{dh}{dt} = \frac{1}{A} [q_1 + q_2 - q_3] \tag{4.17}$$

$$\text{salt:} \quad \frac{d}{dt} c_3 h = \frac{1}{A} [q_1 c_1 + q_2 c_2 - q_3 c_3] \tag{4.16}$$

Figure 4.8 shows data for the density of aqueous sodium chloride solutions. Over the concentration range shown there is a 20 percent change in the density, so clearly a constant density assumption can be a poor one at times.

The important consequence of the constant density assumption is Equation 4.17, which states that the rate of change of the volume is equal to the net volumetric flow rate. We know from experience that when we mix materials of different density the total volume is not generally equal to the sum of the volumes mixed together, so that volume conservation of the type expressed in Equation 4.17 would appear to be a rather limited special case. Equation 4.17 is a substantial simplification over the more general Equation 4.15 and is the form of the overall mass balance usually employed for liquid systems despite density variations. We show here now that Equation 4.17 is derivable under less restrictive assumptions than those used previously.

The density data in Figure 4.8 indicate that over a wide range the density of salt-water solutions can be closely approximated by

$$\rho = \rho_0 + bc \tag{4.24}$$

In general, ρ_0 is simply an intercept without any particular physical significance, although if we are fitting only data at low concentration ρ_0 would be the density of pure water. This linear relation is a *constitutive equation*, for

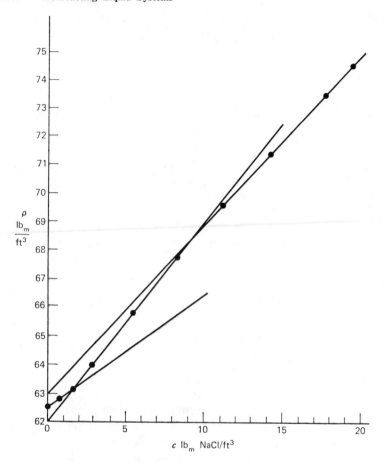

FIGURE 4.8   Density of aqueous NaCl solutions at 20°C.

although it is not a universal principle, it is always true for the salt-water system within a given concentration range.

The overall mass balance, Equation 4.15, is

$$\frac{d}{dt} \rho_3 h = \frac{1}{A} [\rho_1 q_1 + \rho_2 q_2 - \rho_3 q_3] \tag{4.15}$$

From Equation 4.24 we can write for each density

$$\rho_1 = \rho_0 + bc_1$$

$$\rho_2 = \rho_0 + bc_2$$

$$\rho_3 = \rho_0 + bc_3$$

Thus, Equation 4.15 becomes

$$\frac{d}{dt}[\rho_0 + bc_3]h$$

$$= \frac{1}{A}\{[\rho_0 + bc_1]q_1 + [\rho_0 + bc_2]q_2 - [\rho_0 + bc_3]q_3\}$$

Since

$$\frac{d}{dt}[\rho_0 + bc_3]h = \rho_0\frac{dh}{dt} + b\frac{d}{dt}c_3h$$

we can break the equation up as follows:

$$\left\{ \begin{array}{c} \rho_0\dfrac{dh}{dt} \\[2ex] + b\dfrac{d}{dt}c_3h \end{array} \right\} = \left\{ \begin{array}{c} \rho_0\dfrac{1}{A}[q_1 + q_2 - q_3] \\[2ex] + b\dfrac{1}{A}[c_1q_1 + c_2q_2 - c_3q_3] \end{array} \right\}$$

The second line is simply Equation 4.16, the salt balance, multiplied by b, so the term on the left equals the term on the right, leaving

$$\rho_0\frac{dh}{dt} = \rho_0\frac{1}{A}[q_1 + q_2 - q_3]$$

or

$$\frac{dh}{dt} = \frac{1}{A}[q_1 + q_2 - q_3]$$

This is Equation 4.17, which was derived previously by assuming a constant density. We have thus obtained the important result that *the volume rate of change equals the net volumetric flow rate when the density-concentration relation is linear.* That volume is conserved in a case of varying density comes as something of a surprise, and even in this simple problem a mathematical model provides information which is not readily apparent. Some reflection on the physical meaning of a linear density function will ultimately lead to the same conclusion, but for most people that is an a posteriori process that would not have been motivated without the mathematical analysis.

Let us note, finally, that the gross simplification in the analysis is possible only for a linear density function. In general we may write

$$\rho = \rho_0 + bc + \phi(c)$$

where $\phi(c)$ contains the nonlinear part of the relation. The resulting equation for h, which depends now explicitly on concentration, is

$$\frac{dh}{dt} = \frac{1}{A}(\psi_1 q_1 + \psi_2 q_2 - q_3)$$

where

$$\psi_1 = \frac{\rho_0 + \phi(c_1) - c_1\phi'(c_3)}{\rho_0 + \phi(c_3) - c_3\phi'(c_3)}, \qquad \psi_2 = \frac{\rho_0 + \phi(c_2) - c_2\phi'(c_3)}{\rho_0 + \phi(c_3) - c_3\phi'(c_3)}$$

$\phi'(c_3)$ is the function $d\phi/dc$ evaluated at $c = c_3$. In only the most exceptional cases will ψ_1 and ψ_2 differ from unity by enough to have any noticeable effect on the results.

4.6   CONCLUDING REMARKS

The physical problem studied in this chapter is an elementary one which you understand well. This should have allowed you to concentrate your attention on the model development part of the analysis process. Go back through Sections 4.2, 4.4, and 4.5 and compare them with the logic outlined in Figure 3.8. Notice particularly the importance of establishing the density-concentration constitutive relation in Section 4.5.

The mathematical statement of the principle of conservation of mass, which first arises in the chapter as Equation 4.3, is an expression in terms of an integral. Students are sometimes confused over the need to manipulate the equation into the form involving derivatives, Equation 4.5, which is clearly equivalent. Recall that unless $q(t)$ is known explicitly as a function of time the integration in Equation 4.3 cannot be carried out. We could not do the integration, for example, for the case $q = kh^{1/2}$, since $h(t)$ is known only in terms of an integral of $h(t)$! On the other hand, Equation 4.5 can be manipulated further using the calculus to obtain a solution. The control problem in Section 4.3 is a case in point, where the derivative form, Equation 4.10, could be solved as Equation 4.11. Had we put the control equation

$$q = q^* + K[h - h^*]$$

into Equation 4.3, the integral form, we would have been unable to proceed further.

Section 4.3, where the model equations are used for a control system design, is included between the sections in model development to provide some practice in reading mathematical statements. The mathematics is simplified in form by the new nomenclature introduced in that section, but it is quite easy to lose sight of the physical meaning of the variables. It is usually helpful in material of this type to keep a list of symbols and their meanings as you reread the section.

The assumption of the perfectly mixed tank, first introduced in Section 4.4, is a concept which is always used in the engineering analysis of tank-type devices. Even when it is not an accurate description of physical reality the perfectly mixed case provides as important limiting case of possible performance. Notice the experimental justification of the perfect mixing assumption for the easily stirred salt-water system. The approach to the steady state of the salt-water system following several residence times is quite important. Most industrial processes operate at the steady state and the steady state is assumed for all equipment design procedures.

There are several textbooks on simulation and applied mathematics written especially for chemical engineers. The topics are generally at a level beyond that needed for most of this text, but you might want to become acquainted with them at this time.

4.1 C. M. Crowe, A. E. Hamielec, T. W. Hoffman, A. I. Johnson, P. T. Shannon, and D. R. Woods, *Chemical Plant Simulation*, Prentice-Hall, Englewood Cliffs, 1971.

4.2 R. G. E. Franks, *Mathematical Modeling in Chemical Engineering*, Wiley, New York, 1966.

4.3 V. G. Jenson and G. V. Jefferies, *Mathematical Methods in Chemical Engineering*, Academic, New York, 1963.

4.4 W. R. Marshall and R. L. Pigford, *The Application of Differential Equations to Chemical Engineering Problems*, Univ. of Delaware Press, Newark, Del., 1947.

4.5 H. S. Mickley, T. K. Sherwood, and C. E. Reed, *Applied Mathematics in Chemical Engineering*, McGraw-Hill, New York, 1957.

It is a necessary part of the analysis process to have access to the experiments of others. Such data are available in the following standard references, with which you should familiarize yourself.

4.6 J. H. Perry, *Chemical Engineers' Handbook*, 4th ed., McGraw-Hill, New York, 1963.

4.7 *International Critical Tables*, McGraw-Hill, New York, 1926–1930.

4.8 *Handbook of Chemistry and Physics*, 52nd ed., Chemical Rubber Publishing Co., Cleveland, 1971.

You will also want to be aware of a text that deals with the estimation of physical properties based on molecular structure:

4.9 R. C. Reid and T. K. Sherwood, *The Properties of Liquids and Gases*, 2nd ed., McGraw-Hill, New York, 1966.

4.7 PROBLEMS

4.1 A cylindrical tank of radius R is cut in half to make a trough of length L, which when mounted horizontally is part of a food processing unit. If the tank is fed with a stream of constant volumetric flow rate q_f, develop an algebraic expression relating h (height of liquid in the trough) to q_f, R, L, t (time in seconds), and the initial height h_0.

4.2 A spherical stainless steel tank is used to supply raw material for a semibatch processing unit. Liquid is removed from the tank through a valve at the bottom by pressurizing the system with nitrogen. The volumetric flow rate from the tank, q, is to be kept constant by a system that will vary the pressure of the nitrogen above the liquid. In order to design this system a relationship between h, the height of fluid at any time, and q must be derived. Find this relationship if the tank radius is 5 ft.

4.3 A vessel which is shaped so that when water drains through an orifice the level changes at a constant rate can be used for a water clock. Assuming that flow rate through the orifice is proportional to the square root of liquid height, determine the shape of the vessel.

4.4 A holding tank designed to accept the effluent from a small chemical plant operates such that the flow from the tank, q, is proportional to $h(q = bh)$. The feed to the tank is intermittent, but the flow rate is constant at 80 ft³/sec when liquid does enter. The cylindrical tank is 30 ft in diameter and 10 ft deep.

(a) Derive the mathematical description for this situation and express h as a function of the inlet flow q_f, b, t, the tank area A, and the initial height of liquid h_0.

(b) b is found experimentally to be equal to 8 ft²/sec when the tank drain valve is fully open. If the tank is initially empty and the drain valve open, how long can the feed stream flow into the tank before it overflows?

(c) If the flow rate of the feed stream is doubled, how long will it take for an initially empty tank to overflow if the drain valve is fully open?

(d) If the tank contains 8 ft of liquid when the drain valve is opened, how long will it take for the level to reach 4 ft with no liquid entering?

(e) If the tank contains 8 ft of liquid when the drain valve is opened, how long can the feed stream be allowed to enter the tank without overflow?

4.5 The system shown in the figure might be used to load ore onto barges for shipment. Let W_1 = flow rate of ore to barge, tons/hour, and M = mass of ore at any time, tons. After the pile has reached a mass M_0,

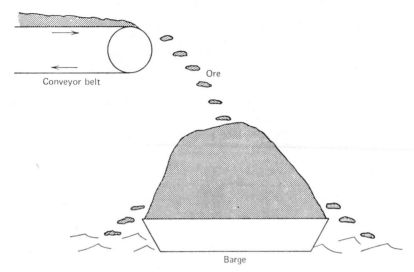

Ore

Conveyor belt

Barge

ore is lost over the side of the barge at a rate W_2 tons/hour. The loss rate is about proportional to the amount of ore in excess of M_0 at any time.

(a) Derive the mathematical description for this situation.

(b) Express the time required for the ore on the barge to reach a mass M_0 in terms of the constant flow rate W_1.

(c) Derive an algebraic relationship showing M as a function of t for $M > M_0$. What happens as $t \to \infty$?

4.6 A circular cone of radius $R = 5$ ft and height $H = 20$ ft is filled to the top so that it may feed a conveyor line carrying gallon cans. The

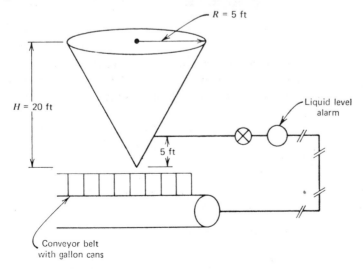

$R = 5$ ft

$H = 20$ ft

Liquid level
alarm

5 ft

Conveyor belt
with gallon cans

system has been designed with an automatic shutoff that stops the operation when the height of liquid is 5 ft above the tip of the cone. This was done to assure a constant flow from the tank. The flow rate from the tank is 30 gal/min and the conveyor speed is adjusted such that each can receives 1 gal.

(a) How long will this system run before the operation is stopped by the level controller?

(b) Derive an expression for the liquid height h as a function of time.

4.7 A cylindrical tank of cross-sectional area A_1 contains liquid to a height h_{10}. At time $t = 0$ a line connecting this tank to an empty cylindrical tank of cross-sectional area A_2 is opened and the first tank is allowed to drain into the second. The configuration is shown in the figure.

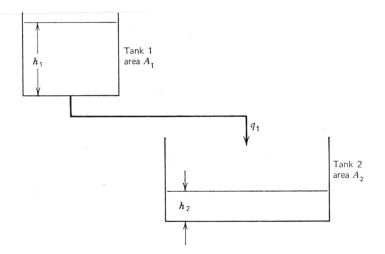

Designate the height in the first and second tank by h_1 and h_2, respectively, and develop a mathematical model that relates h_2 to A_1, A_2, h_{10}, and t. Compute $h_2(t)$ when q_1 depends on h_1 in each of the following ways:

(a) $q_1 = bh_1$

(b) $q_1 = kh_1^{1/2}$

4.8 Two tanks are connected as shown in the figure. The flow rates q_1 and q_2 may be assumed to be

$$q_1 = b_1[h_1 - h_2]$$
$$q_2 = b_2h_2$$

(a) Develop the mathematical description for this situation as two first-order differential equations in h_1 and h_2.

(b) Why can you not solve this set of equations in the same way you did for Problem 4.7?

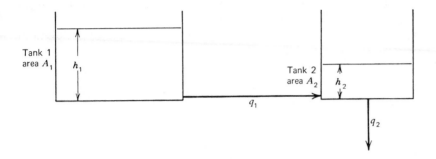

4.9 Repeat the calculations and figures in Section 4.3 when the feed stream disturbance has the form

(a) $q_f = \begin{cases} 0, & 0 \le t < T_1 \\ \beta[t - T_1], & T_1 \le t < T_2 \\ 0, & T_2 \le t \end{cases}$

(b) $q_f = \begin{cases} 0, & 0 \le t < T_1 \\ \alpha \sin \omega[t - T_1] & T_1 \le t \end{cases}$

4.10 The following data are available for the population of Iceland and the Faroe Islands.

Year	Births	Deaths	Mean Population
1921	3215	1708	116,000
1922	3214	1491	118,000
1923	3264	1542	119,000
1924	3150	1730	120,000
1925	3153	1457	122,000
1926	3267	1320	124,000
1927	3221	1449	126,000
1928	3162	1318	128,000
1929	3219	1490	130,000
1930	3441	1522	132,000

Neglecting immigration and emigration, construct a mathematical model for the population as a function of time and compare with the actual population.

(a) Assume that birth (inflow) rate and death (outflow) rate are constants, using an average value.

(b) Assume that birth and death rates are constant fractions of the total population.

4.11 The following data show the population of the United States during the period 1930–1960.

Year	Births/Thousand Population	Deaths/Thousand Population	Population in Millions
1930	21.3	11.3	123
1940	19.4	10.8	132
1950	24.1	9.6	151
1960	23.7	9.5	179

Immigration and emigration have been neglected. Construct a model for population growth and check your prediction against the actual population in 1970 of 202 million.

4.12 Banks commonly advertise continuous compounding of interest, by which they mean that interest is added to an account at a rate equal to the product of the account balance and the interest rate divided by 100. Develop a model for account balance as a function of time with continuous compounding at 5 percent per year, and compare with annual, semiannual, and quarterly compounding.

4.13 A well-stirred process vessel contains 450 gal of a solution of water and 75 lb of sodium chloride at the time a pure water stream is introduced. Find the salt concentration at the end of 60 min of operation if the water flow to the tank is maintained at 7 gal/min and brine is removed at the same rate.

4.14 During start-up of the process containing the 450-gal vessel described in the preceding problem, the following steps are taken:

(a) 75 lb of salt are mixed with water in the tank until all the salt is dissolved and a uniform mixture of 100 gal of brine is obtained.

(b) When this is achieved a pure water stream is introduced to the vessel at a rate of 7 gal/min. When the liquid capacity of the tank is reached (450 gal) an overflow line is employed that keeps the outflow from the tank equal to the inflow.

Develop a mathematical description of the process assuming that brine density is a linear function of salt concentration, and obtain the salt concentration as a function of time both before and after the tank is filled with liquid.

4.15 A process vessel operating at steady state to mix brine streams has feed flow rates of 10 and 15 ft³/min, with salt concentrations of 10 and 17 lb_m/ft^3, respectively. What is the salt concentration in the exit stream?

4.16 The Mediterranean Sea exchanges water with the Atlantic Ocean. Fresh water inflow to the Mediterranean is approximately 30,000 m³/sec and evaporation occurs at a rate of approximately 80,000 m³/sec. The salt content of the Mediterranean is 37 g salt/1000 g solution and it is 36 g/1000 g in the Atlantic. Estimate the flow from sea to ocean and from ocean to sea.

4.17 A tank contains 100 gal of brine with 50 lb_m of dissolved salt. Pure water runs into the tank at a rate of 2 gal/min, while the effluent flows into a second tank which is initially empty at a rate of 3 gal/min. The second tank is emptied at a constant rate of 2 gal/min. Develop the mathematical description that would enable you to compute the concentration of salt in the second tank as a function of time. Solve the equation if you are familiar with linear first-order differential equations, Section 15.10.

4.18 A well-stirred tank containing dissolved salt is fed with a pure water stream of flow rate $q_1(ft^3/min)$ and a salt water stream of concentration $c_2(lb_m/ft^3)$ and flow rate $q_2(ft^3/min)$. If the flow rate out of the tank is $q_3(ft^3/min)$, develop expressions for c_3, the concentration of salt in the tank, and h, the height of liquid in the tank, as functions of time. The initial liquid height is h_0 and the initial salt concentration in the tank c_{30}. You may assume that the relationship between density and concentration is of the form $\rho = \rho_0 + bc$.

4.19 A well-dispersed mixture of solids and liquid is to be concentrated by being pumped through a tank with a porous bottom through which pure liquid may pass, as shown schematically in the figure. Because the dispersion is good you may consider the mixture to be a homogeneous liquid with solids concentration c. (See figure on page 110.)

    (a) Show that the mixture density is a linear function of c. The total volume V is the sum of solids volume V_S and liquid volume V_L.

    (b) Assume that the volumetric flow rate q_3 depends on the liquid density ρ_L and viscosity μ_L, plate area A, and pressure change

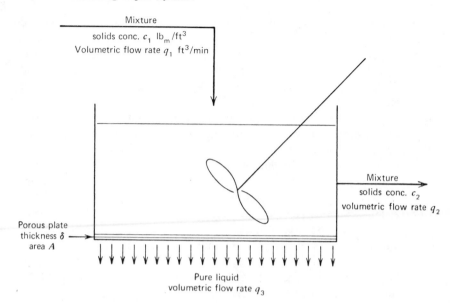

Mixture
solids conc. c_1 lb_m/ft^3
Volumetric flow rate q_1 ft^3/min

Mixture
solids conc. c_2
volumetric flow rate q_2

Porous plate
thickness δ
area A

Pure liquid
volumetric flow rate q_3

per unit thickness across the plate, $\Delta p/\delta$. Use dimensional analysis and the experimental observation that for pure water a plot of log height versus time is linear for flow through a porous plate to obtain a constitutive equation for q_3.

(c) Obtain a steady-state design equation relating tank area and volume to liquid throughput and desired degree of concentration.

CHAPTER 5

Reacting Liquid Systems

5.1 INTRODUCTION

In Chapter 4 we developed overall and component mass balances for tank-type systems in which the contents were well mixed. This is an important class of problems, for tank-type systems are used both in the laboratory to obtain data and as continuous processing units. The complete mixing of vessel contents is generally easy to achieve with low viscosity materials, so our model equations have considerable practical significance. In this chapter we again consider tank-type systems and extend our development to those cases in which there is a chemical reaction occurring.

The rational design and operating control of vessels in which reaction is occurring can only be carried out if we carefully follow the complete analysis process. We initiate our discussion here by developing the model equations. In Chapter 7 we consider some simple reactor design problems showing how the equations may be used.

5.2 BASIC MODEL EQUATIONS FOR A TANK-TYPE REACTOR

We shall consider a tank with liquid volume V into which two streams flow, each containing a different chemically active species. The two chemical species are designated as A and B, and they will react when brought into contact with one another. The system is sketched in Figure 5.1. The volumetric flow rate of the feed stream containing A is q_{Af}, with density ρ_{Af} and a concentration c_{Af} of A. The volumetric flow rate of the feed stream containing B is q_{Bf}, with density ρ_{Bf}, and concentration c_{Bf} of B. The exit stream has a volumetric flow rate q, density ρ, and concentrations c_A and c_B of A and B, respectively. The tank is assumed to be well mixed, so that samples drawn at some time from any location in the tank, including the exit, will be indistinguishable from samples drawn at the same time from any other location. Thus time is the only independent variable and we can use the tank itself as a control volume.

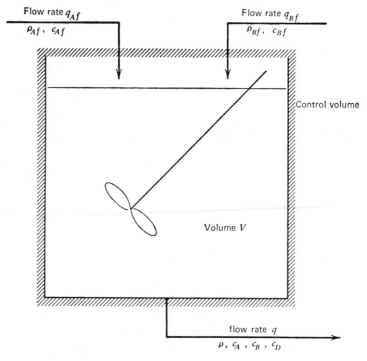

FIGURE 5.1   Well-stirred reactor with feed streams containing pure A and pure B.

The overall mass balance may be written as before:

$$\rho(t_2)V(t_2) = \rho(t_1)V(t_1) + \int_{t_1}^{t_2} \rho_{Af}(\tau)q_{Af}(\tau)\,d\tau$$

$$+ \int_{t_1}^{t_2} \rho_{Bf}(\tau)q_{Bf}(\tau)\,d\tau - \int_{t_1}^{t_2} \rho(\tau)q(\tau)\,d\tau$$

When the times are taken as t and $t + \Delta t$ and the mean value theorem is applied to each integral we obtain

$$\frac{\rho(t + \Delta t)V(t + \Delta t) - \rho(t)V(t)}{\Delta t} = \overline{\rho_{Af}q_{Af} + \rho_{Bf}q_{Bf} - \rho q}$$

The overbar denotes a quantity evaluated at some time $\bar{t}$ between t and $t + \Delta t$.

In the limit at $\Delta t \to 0$ the left-hand side becomes the derivative of ρV, while the quantity on the right is evaluated at t, since $\bar{t} \to t$ as $\Delta t \to 0$. Thus

$$\frac{d\rho V}{dt} = \rho_{Af}q_{Af} + \rho_{Bf}q_{Bf} - \rho q \tag{5.1}$$

We now turn to the application of the law of conservation of mass to the components of the system. The natural unit of measurement in a reacting system is the *mole*, since chemical compounds form and react on a molar basis. Thus, we shall designate concentrations in lb-moles/ft^3 or g-moles/cm^3. We can then write for substance A, with concentration c_A,

$$c_A(t_2)V(t_2) = c_A(t_1)V(t_1) + \int_{t_1}^{t_2} q_{Af}(\tau)c_{Af}(\tau)\,d\tau - \int_{t_1}^{t_2} q(\tau)c_A(\tau)\,d\tau$$

+ [moles of A which appear in the control
   volume by any other process during the
   interval from t_1 to t_2]

− [moles of A which leave the control volume
   by any other process during the interval
   from t_1 to t_2]   (5.2)

The two integrals represent, as usual, the convective flow of A in and out of the system. The expression in the brackets is needed because we must now allow for the possibility that A appears or disappears within the control volume as a result of formation or loss by chemical reaction. This word statement, of course, needs to be expressed quantitatively before a working equation can be developed.

To be specific, let us assume that A and B react to form products according to the following reaction:

$$A + B \rightarrow nD$$

That is, *one mole* of A and *one mole* of B react to n moles of the product, D. In this situation n will generally be 1 or 2. Two examples of this form that we shall deal with subsequently in detail are $A = $ sodium ethoxide ($NaOC_2H_5$), $B = $ ethyl dimethyl sulfonium iodide ($C_2H_5S(CH_3)_2I$), and $A = $ sulfuric acid (H_2SO_4), $B = $ diethyl sulfate (($C_2H_5)_2SO_4$).

We have found it convenient previously to denote the inflow and outflow terms by integrals of rates at which these processes occur. It is convenient to do the same for the reaction terms. We shall designate the rate at which a substance Y appears in the control volume because of reaction as r_{Y+}, with units of moles per volume per time. The rate is written on a volumetric basis because it is evident from the outset that, given identical concentrations in two reactors of different volumes, the number of moles formed per time in each will be proportional to the reactor volume. If compound Y disappears because of reaction the rate of disappearance is denoted as r_{Y-}. Both r_{Y+} and r_{Y-} are positive numbers. In a more complex situation than the one which we are considering here a compound might be created and depleted in the same reactor. Notice that the r_{Y+} or r_{Y-} is simply a quantitative symbol for a word statement and at this point of development has no obvious physical meaning.

We can now return to the word-equation 5.2. The first bracketed group, "Moles of A which appear ... ," is zero, since A is not formed in the reaction. The second, "moles of A which disappear ... ," is formed as follows:

$$\text{rate at which } A \text{ disappears by reaction} = r_{A-}V$$

$$\text{amount of } A \text{ that leaves because of reaction} = \int_{t_1}^{t_2} r_{A-}V \, d\tau$$

Thus, Equation 5.2 becomes

$$c_A(t_2)V(t_2) = c_A(t_1)V(t_1) + \int_{t_1}^{t_2} q_{Af}c_{Af} \, d\tau$$

$$- \int_{t_1}^{t_2} qc_A \, d\tau - \int_{t_1}^{t_2} r_{A-}V \, d\tau$$

After application of the mean value theorem and the usual limiting argument we obtain the final equation for conservation of mass of A

$$\frac{dc_A V}{dt} = q_{Af}c_{Af} - qc_A - r_{A-}V \tag{5.3}$$

In a similar manner we obtain equations for B and D

$$\frac{dc_B V}{dt} = q_{Bf}c_{Bf} - qc_B - r_{B-}V \tag{5.4}$$

$$\frac{dc_D V}{dt} = -qc_D + r_{D+}V \tag{5.5}$$

We emphasize again that concentrations are expressed in molar units.

Equations 5.3 to 5.5 involve three different reaction rates, r_{A-}, r_{B-}, and r_{D+}. Without any further information about the nature of this symbol we can simplify the equations considerably by some elementary use of the reaction chemistry. The chemical equation tells us that every time a mole of A vanishes a mole of B must also vanish; thus, it is evident that

$$r_{A-} = r_{B-}$$

Furthermore, when a mole of A or B vanishes, n moles of D appear. Thus

$$nr_{A-} = nr_{B-} = r_{D+}$$

Clearly, then, only a *single* rate is needed to describe the reaction. This single rate is sometimes called the *intrinsic rate of reaction* and denoted simply as r,

$$r_{A-} = r_{B-} = \frac{r_{D+}}{n} \equiv r \tag{5.6}$$

The component equations then become

$$\frac{dc_A V}{dt} = q_{Af} c_{Af} - q c_A - rV \tag{5.7}$$

$$\frac{dc_B V}{dt} = q_{Bf} c_{Bf} - q c_B - rV \tag{5.8}$$

$$\frac{dc_D V}{dt} = -q c_D + nrV \tag{5.9}$$

Equations 5.1 and 5.7 to 5.9 are the basic model equations for a well-mixed tank in which the liquid phase reaction between A and B to form D is taking place. For the discussion that follows we can make some simplifying assumptions. First, we will assume that the densities of all streams are equal

$$\rho_{Af} = \rho_{Bf} = \rho$$

ρ will then be a constant. (As in the preceding chapter this assumption is not really necessary to obtain the simplifications that follow from it, and we shall return to this point subsequently.) As a consequence of the equal and constant density assumption the overall mass balance, Equation 5.1, reduces to

$$\frac{dV}{dt} = q_{Af} + q_{Bf} - q \tag{5.10}$$

If, for purposes of illustration, we confine our attention to those cases in which the volumetric flow into the reactor is equal to the volumetric flow out

$$q_{Af} + q_{Bf} = q$$

then dV/dt is equal to zero and the liquid volume is a constant. For this constant volume operation the remaining three conservation equations simplify to

$$V \frac{dc_A}{dt} = q_{Af} c_{Af} - q c_A - rV \tag{5.11}$$

$$V \frac{dc_B}{dt} = q_{Bf} c_{Bf} - q c_B - rV \tag{5.12}$$

$$V \frac{dc_D}{dt} = -q c_D + nrV \tag{5.13}$$

In order to consider the important features of the analysis of reacting systems without introducing too much complexity we shall confine ourselves to isothermal systems. The principle of conservation of energy does not generally need to be employed for reacting isothermal liquid systems, a consideration which we take up again in Chapter 11, so Equations 5.11 to 5.13 represent all of the information that we can conveniently obtain from the conservation laws. Since the rate of reaction, r, is contained in the basic model equations it is clear that a relationship between r and the other dependent variables of the problem must be obtained if we expect to have equations sufficient to solve for the unknown quantities.

5.3 THE REACTION RATE

The *reaction rate* terms r_{A-}, r_{B-}, and r_{D+}, or equivalently through Equation 5.6 for this example, the single term r, arise as a mathematical device enabling us to write equations describing the changes in concentrations of materials A, B, and D. We need, however, a further relation between r and the concentrations c_A, c_B, and c_D. This constitutive equation may be deduced in part by physical considerations, but it ultimately requires experimentation for a complete determination.

It is clear that the reaction cannot proceed at all if either A or B is missing, so that r must be a function of c_A and c_B which goes to zero if either c_A or c_B vanishes. Furthermore, we know from elementary physical chemistry that the chemical reaction is the result of an interaction between a molecule of A and a molecule of B. Clearly, the more molecules of a substance in a given volume the more likely a collision. Thus the rate at which the reaction proceeds should increase with increased c_A or c_B. The simplest functional form which embodies both of these features is

$$r = kc_A c_B \tag{5.14}$$

We can, however, postulate many other forms that meet our simple requirements of vanishing when c_A or c_B does and of increasing when c_A or c_B increases. Some examples are

$$r = kc_A^n c_B^m$$

$$r = \frac{kc_A^2 c_B^3}{1 + \beta c_B}$$

There are an infinite number of expressions that meet the simple requirements, so we obviously need additional information to determine the correct constitutive relation between r and c_A and c_B. The rate of a reaction can only be determined by allowing the reaction to occur. Hence it is necessary to plan an experiment or to analyze experimental results that someone else has obtained. The basic problem is not unlike that encountered when we searched for a constitutive relation between q and h in the tank draining experiment. We postulated a relationship, solved the basic model equation, then compared the model prediction with data. We follow the same procedure for the reacting system, using as our basic model Equations 5.11 to 5.13. We can simplify the analysis and the experimental procedure if we use as our laboratory reactor a vessel into which the reactants are placed at time zero and allowed to react with no inflow or outflow, since in that case the q_{Af}, q_{Bf}, and q terms in the model equations vanish.

5.4 THE BATCH REACTOR

A reaction vessel in which there is no inflow or outflow is called, for obvious reasons, a *batch* reactor. Such reactors have extensive commercial use, although our immediate purpose here is to consider the use of the batch reactor as an experimental device. At time zero we will fill the reactor with known amounts of materials A, B, and D. We will then remove small samples at frequent intervals and determine the concentrations of A, B, and D as time progresses. Recall that because of our lumped representation there is no spatial dependence.

Since there is no flow in or out $(q_{Af} = q_{Bf} = 0)$ Equations 5.11 to 5.13 reduce to

$$\frac{dc_A}{dt} = -r \tag{5.15}$$

$$\frac{dc_B}{dt} = -r \tag{5.16}$$

$$\frac{dc_D}{dt} = +nr \tag{5.17}$$

Equations 5.15 to 5.17 are our basic model equations for the batch reactor system. We need to examine the behavior of this set of equations in order to plan a rational experimental program.

The first important relation follows from Equations 5.15 and 5.16, which state

$$\frac{dc_A}{dt} = \frac{dc_B}{dt}$$

This equation is integrated to yield

$$c_A(t) - c_{A0} = c_B(t) - c_{B0} \tag{5.18}$$

Here, c_{A0} and c_{B0} are the initial concentrations of A and B, respectively. Equation 5.18 states that the number of moles of A that have reacted equals the number of moles of B that have reacted. This conclusion also follows from the chemical equation, since nothing enters or leaves a batch reactor.

Similarly, Equations 5.15 and 5.17 taken together yield

$$\frac{dc_D}{dt} = -n\frac{dc_A}{dt}$$

which yields, on integration,

$$c_D(t) - c_{D0} = n[c_{A0} - c_A(t)] \tag{5.19}$$

c_{D0} is the initial concentration of D, and is probably equal to zero. Thus, the concentrations of B and D can be expressed in terms of A, so we need solve only Equation 5.15 to determine the full concentration-time behavior of the batch reactor.

In order to integrate Equation 5.15 it is now necessary to postulate the functional form of the rate, r. Then we can solve Equation 5.15 for $c_A(t)$ and check the model predictions with experiment. As suggested in the preceding section, the *simplest* expression which we can postulate is

$$r = kc_A c_B \tag{5.20}$$

Equation 5.15 then becomes

$$\frac{dc_A}{dt} = -kc_A c_B$$

or, using Equation 5.18 to relate c_A to c_B,

$$\frac{dc_A}{dt} = -kc_A[c_A + M] \tag{5.21}$$

Here, $M = c_{B0} - c_{A0}$.

A solution to Equation 5.21 can be obtained directly, although it is necessary to consider separately the two cases $M = 0$ and $M \neq 0$. In the former case, corresponding to equal starting concentrations of reactants A and B, the solution is obtained by the sequence of steps shown in the first column of Table 5.1, giving

$$M = 0: \quad \frac{1}{c_A(t)} = \frac{1}{c_{A0}} + kt \tag{5.22}$$

For $M \neq 0$ the solution steps are shown in the second column of Table 5.1, giving

$$M \neq 0: \quad \ln\left[\frac{c_{A0}}{c_A(t)} \cdot \frac{M + c_A(t)}{M + c_{A0}}\right] = Mkt \tag{5.23}$$

or, equivalently,

$$M \neq 0: \quad \ln\frac{c_B(t)}{c_A(t)} = \ln\frac{c_{B0}}{c_{A0}} + Mkt \tag{5.24}$$

TABLE 5.1 Integration of the equation for c_A in a batch reactor with second-order reaction $A + B \rightarrow nD$ for equal $(M = 0)$ and unequal $(M \neq 0)$ starting concentrations

$M = 0$	$M \neq 0$
$\dfrac{dc_A}{dt} = -kc_A^2$	$\dfrac{dc_A}{dt} = -kc_A[M + c_A]$
$\dfrac{dc_A}{c_A^2} = -k\,dt$	$\dfrac{dc_A}{c_A[M + c_A]} = -k\,dt$
$\displaystyle\int_{c_{A0}}^{c_A(t)} \dfrac{dc}{c^2} = -k \int_0^t d\tau$	$\displaystyle\int_{c_{A0}}^{c_A(t)} \dfrac{dc}{c[M + c]} = -k \int_0^t d\tau$
$-\dfrac{1}{c_A(t)} + \dfrac{1}{c_{A0}} = -kt$	$\dfrac{1}{M} \ln \dfrac{c_A(t)}{c_{A0}} - \dfrac{1}{M} \ln \left[\dfrac{c_A(t) + M}{c_{A0} + M} \right] = -kt$

Thus, if the rate is truly represented by Equation 5.20 then, according to Equation 5.22, for equal starting concentrations a plot of $1/c_A$ versus t will yield a straight line with slope k. Similarly, for unequal initial concentrations, Equation 5.23 indicates that a plot of a more complicated function of c_A versus t will give a straight line with slope Mk.

Example 5.1

The data in Table 5.2 were obtained for the reaction of sulfuric acid with diethyl sulfate in aqueous solution at 22.9°C. The initial concentrations of

TABLE 5.2 Concentration of H_2SO_4 versus time for the reaction of sulfuric acid with diethyl sulfate in aqueous solution at 22.9°C

Time (minutes) t	Concentration of H_2SO_4 (gram-moles/liter) c_A
0	5.50
41	4.91
48	4.81
55	4.69
75	4.38
96	4.12
127	3.84
162	3.59
180	3.44
194	3.34

Source. Data of Hellin and Jungers, *Bull. Soc. Chim. France*, p. 386 (1957).

H_2SO_4 and $(C_2H_5)_2SO_4$ were each 5.5 g-moles/liter. The reaction can be represented as follows

$$H_2SO_4 + (C_2H_5)_2SO_4 \rightarrow 2C_2H_5SO_4H$$

In our symbolic nomenclature H_2SO_4 is designated as A, $(C_2H_5)_2SO_4$ is designated as B, n is 2, and $C_2H_5SO_4H$ is designated as D.

The data are in a form that makes it convenient before going any further to use the procedure outlined in Section 2.4 to examine the reaction order (exponent of c_A in the rate constitutive equation). In an analogous manner to that used for the tank-draining problem, if we start with

$$\frac{dc_A}{dt} = -kc_A^n$$

then we obtain the relation

$$\ln\left(-\frac{\Delta c_A}{\Delta t}\right) \simeq \ln k + n \ln c_A$$

TABLE 5.3   Sulfuric acid data from Table 5.2 used to compute $\ln(-\Delta c_A/\Delta t)$ versus $\ln c_A$ for estimation of reaction order

t	c_A	$-\Delta c_A$	Δt	$\ln\left(-\dfrac{\Delta c_A}{\Delta t}\right)$	$\ln c_A$
0	5.50				1.70
		0.59	41	−4.25	
41	4.91				1.59
		0.10	7	−4.25	
48	4.81				1.57
		0.12	7	−4.06	
55	4.69				1.54
		0.31	20	−4.17	
75	4.38				1.48
		0.26	21	−4.38	
96	4.12				1.41
		0.28	31	−4.70	
127	3.84				1.35
		0.25	35	−4.84	
162	3.59				1.28
		0.15	18	−4.79	
180	3.44				1.24
		0.10	14	−4.94	
194	3.34				1.21

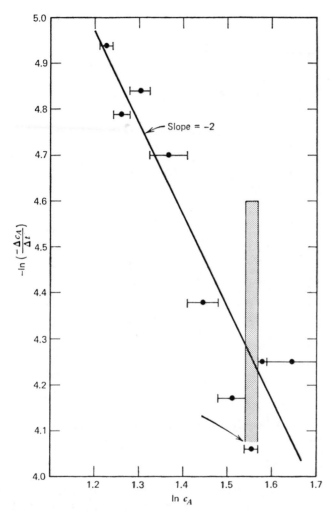

FIGURE 5.2   Estimation of the order of the reaction between sulfuric acid and diethyl sulfate in aqueous solution. The shaded area shows the effect of an error of -1 percent in measuring c_A.

The data are tabulated in this way in Table 5.3 and plotted in Figure 5.2. There is a great deal of scatter in the data, reflecting the extreme sensitivity of this type of plot to small errors. For example, if the error in determining c_A were only -1 percent, the variation in the data marked with an arrow in Figure 5.2 would be over the shaded region shown in the figure. If it were $+1$ percent the region of uncertainty would extend nearly as far in the opposite direction. It is nearly impossible, then, to make any definitive statement about n. The line shown has slope -2 and is clearly as good a fit to the data

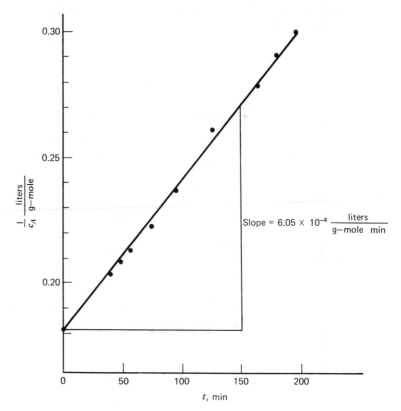

FIGURE 5.3   Computation of the second-order rate constant for the reaction between sulfuric acid and diethyl sulfate in aqueous solution.

as any that might be drawn, and on this basis we proceed, since Equation 5.21 and its consequences are consistent with this first check. (By taking the data point to be the midpoint of the band, as shown, and by using a relation derived in the next chapter, the "best" line through the data can be shown to have slope -2.06.)

As required by Equation 5.22, the data are plotted in Figure 5.3 as $1/c_A$ versus t. There is still some experimental scatter, but a straight line can easily be drawn through the data, confirming the adequacy of the rate expression, Equation 5.20, for the data shown. The parameter k is determined from the slope in Figure 5.3 as

$$k = 6.05 \times 10^{-4} \frac{\text{liters}}{\text{g-mole min}}$$

In engineering units

$$k = \left[6.05 \times 10^{-4} \frac{\text{liter}}{\text{g-mole min}}\right]\left[\frac{1 \text{ g-mole}}{454 \text{ lb-mole}}\right]$$

$$\times \left[\frac{1000 \text{ cm}^3}{\text{liter}}\right]\left[\frac{1 \text{ ft}^3}{\{12 \times 2.54\}^3 \text{ cm}^3}\right]$$

$$k = 9.7 \times 10^{-3} \frac{\text{ft}^3}{\text{lb-mole min}}$$

Example 5.2

The data in Table 5.4 were taken in a batch reactor for the reaction between sodium ethoxide and ethyl dimethyl sulfonium iodide in solution in absolute ethanol:

$$\text{NaOC}_2\text{H}_5 + \text{C}_2\text{H}_5\!\!-\!\!\overset{\displaystyle \text{CH}_3}{\underset{\displaystyle \text{CH}_3}{\text{S}}}\!\!-\!\!\text{I}$$

$$\longrightarrow \text{products} \quad \left[\begin{array}{c} \text{NaI} + (\text{C}_2\text{H}_5)_2\text{O} + \text{S}(\text{CH}_3)_2 \\[4pt] \text{or} \\[4pt] \text{NaI} + \text{C}_2\text{H}_5\text{OH} + \text{C}_2\text{H}_4 + \text{S}(\text{CH}_3)_2 \end{array}\right]$$

TABLE 5.4   Concentrations of sodium ethoxide and ethyl dimethyl sulfonium iodide in ethanol solution versus time

Time (minutes)	Concentration of NaOC_2H_5, c_A (gram-moles/liter)	Concentration of $\text{C}_2\text{H}_5\text{S}(\text{CH}_3)_2\text{I}$, c_B (gram-moles/liter)	$c_B - c_A$ (gram-moles/liter)
0	0.0961	0.0472	−0.0489
12	0.0857	0.0387	−0.0470
20	0.0805	0.0334	−0.0471
30	0.0749	0.0278	−0.0471
42	0.0698	0.0228	−0.0470
51	0.0671	0.0200	−0.0471
63	0.0638	0.0168	−0.0470
∞	0.0470	0	−0.0470

Source. Data of Hughes et al., *J. Chem. Soc.*, p. 2072 (1948).

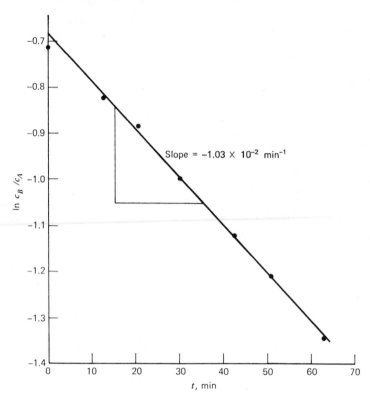

FIGURE 5.4 Computation of the second-order rate constant for the reaction between sodium ethoxide and ethyl dimethyl sulfonium iodide in absolute ethanol.

We denote sodium ethoxide by A and ethyl dimethyl sulfonium iodide by B. It is believed that the reaction rate is adequately represented by $r = kc_Ac_B$, in which case, according to Equation 5.24, a plot of $\ln c_B(t)/c_A(t)$ versus t should give a straight line. Notice from Table 5.4 that the value of $M = -0.0470$ g-mole/liter and that there seems to be some discrepancy in the measurement of the initial concentration of c_A. The data are plotted in Figure 5.4 and it is evident that they do follow a straight line with the exception of the intercept point. The rate expression is thus verified and we can calculate the parameter k from the slope

$$k = \frac{1}{M} \times \text{slope} = \frac{-1.03 \times 10^{-2}}{-4.7 \times 10^{-2}} = 0.21 \, \frac{\text{liters}}{\text{g-mole min}}$$

It is interesting to note how different the rate constant is in this example compared to the previous one.

5.5 PSEUDO FIRST-ORDER REACTIONS

Chemical reactions of the form

$$A + B \rightarrow nD$$

in solution will often be carried out in practice with c_B much larger than c_A. This would occur, for example, when B is the solvent in which A is dissolved. In that case the role of B is effectively masked in the mathematical description of the reaction and the resulting equations are somewhat simpler.

Consider Equation 5.21

$$\frac{dc_A}{dt} = -kc_A[c_A + M] \tag{5.21}$$

$$M = c_{B0} - c_{A0}$$

If $c_{B0} \gg c_{A0}$ then $M + c_A = c_{B0} - [c_{A0} - c_A]$ is only slightly different from c_{B0} throughout the entire course of the reaction. To a very good approximation, then, we can write Equation 5.21 for this special case in the form

$$\frac{dc_A}{dt} = -kc_{B0}c_A$$

or

$$\frac{dc_A}{dt} = -k'c_A \tag{5.25}$$

where $k' = kc_{B0}$ has units of inverse time, say min^{-1}. The form $r = -k'c_A$ is first order ($n = 1$), and k' is often called a pseudo first-order rate constant and the reaction a pseudo first-order reaction. Solution of Equation 5.25 leads to

$$\ln \frac{c_A(t)}{c_{A0}} = -k't \tag{5.26}$$

so a plot of the logarithm of c_A versus t will be linear with slope $-k'$. The pseudo first-order relation can also be obtained directly from Equation 5.23, the solution of Equation 5.21 for arbitrary nonzero M.

$$\ln \left[\frac{c_{A0}}{c_A(t)} \cdot \frac{M + c_A(t)}{M + c_{A0}} \right] = Mkt \tag{5.23}$$

Since $c_{B0} \gg c_A$, $M + c_A(t) \simeq M + c_{A0}$ and $M \simeq c_{B0}$, in which case Equation 5.23 reduces directly to Equation 5.26.

There is another manner in which a reaction in solution might appear to be first order. A second species is often required for a reaction to occur, although the second species, or *catalyst*, is not used up in the reaction. That is, the chemical equation will have the form

$$A + B \rightarrow nD + B$$

and hence will appear to be of the form

$$A \rightarrow nD$$

The concentration of B is then unchanged throughout the reaction and Equation 5.25 is an exact representation.

A good example of a catalyzed reaction is the cleavage of diacetone alcohol at 25°C:

$$CH_3-\underset{\underset{OH}{|}}{\overset{\overset{CH_3}{|}}{C}}-CH_2-\overset{\overset{O}{||}}{C}-CH_3 \longrightarrow 2CH_3-\overset{\overset{O}{||}}{C}-CH_3$$

The reaction is carried out in the presence of hydroxide ion (e.g., aqueous NaOH) and really occurs as

$$CH_3-\underset{\underset{OH}{|}}{\overset{\overset{CH_3}{|}}{C}}-CH_2-\overset{\overset{O}{||}}{C}-CH_3 + OH^- \longrightarrow 2CH_3-\overset{\overset{O}{||}}{C}-CH_3 + OH^-$$

The experimentally determined rate of reaction of diacetone alcohol (A) in the presence of NaOH (B) is

$$r = -0.47 c_A c_B$$

where the concentrations are in gram-moles/liter and time in minutes. However, since c_B is a constant in any experiment the apparent rate is

$$r = -k' c_A, \qquad k' = 0.47 c_B$$

where units of k' are min^{-1}. Data for this reaction at various concentrations of hydroxide are shown in Table 5.5.

TABLE 5.5   Pseudo first-order and actual second-order rate constant versus NaOH concentration for the cleavage of diacetone alcohol at 25°C

NaOH (c_B) (gram-moles/liter)	k', (min^{-1})	$k = k'/c_B$ (liter/gram-mole min)
5×10^{-3}	2.32×10^{-3}	0.465
10×10^{-3}	4.67×10^{-3}	0.467
20×10^{-3}	9.40×10^{-3}	0.470
40×10^{-3}	19.2×10^{-3}	0.479
100×10^{-3}	47.9×10^{-3}	0.479

Source. Data of French, J. Am. Chem. Soc., **51**, 3215 (1929).

Example 5.3

Data for the decomposition of dibromo succinic acid (2,3-dibromo butane-dioic acid), $C_2H_2Br_2(COOH)_2$, in a batch reactor are shown in Table 5.6

TABLE 5.6   Grams of dibromo succinic acid versus time for the decomposition of the acid

Time (minutes)	Grams of Acid (m_A)	$\ln m_A/m_{A0}$
0	5.11	0
10	3.77	-0.302
20	2.74	-0.624
30	2.02	-0.930
40	1.48	-1.24
50	1.08	-1.55

as grams of acid remaining versus time. The reaction is believed to follow an apparent first-order relation, which in the batch reactor will have the form

$$\frac{dc_A}{dt} = -k'c_A$$

Since $c_A = m_A/V$, where V is the volume and m_A the mass of acid, it follows that Equation 5.26 can be written

$$\ln \frac{m_A(t)}{m_{A0}} = -k't$$

The data are plotted in Figure 5.5 as $\ln m_A(t)/m_{A0}$ versus time, and it is evident that a linear relation for a first-order reaction is obtained. The rate constant obtained from the slope is

$$k' = 0.031 \text{ min}^{-1}$$

5.6   REVERSIBLE REACTIONS

Thus far we have considered only the case of an irreversible reaction. In many cases the reverse reaction must also be considered. That is, we have two reactions taking place simultaneously

$$A + B \rightarrow nD$$

$$nD \rightarrow A + B$$

or, equivalently,

$$A + B \rightleftharpoons nD$$

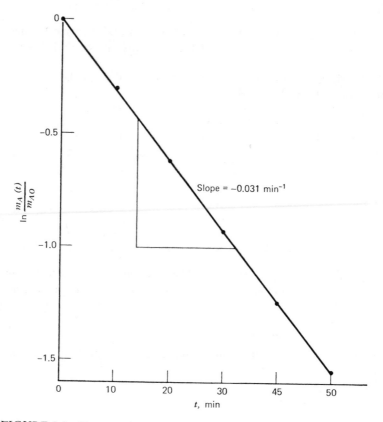

FIGURE 5.5 Computation of the first-order rate constant for the decomposition of dibromo succinic acid.

It then follows that in the expression for conservation of mass applied to the chemical components there will be *two* terms, one representing appearance and the other disappearance.

Consider component A. We denote the rate of disappearance of A per unit volume as a result of the forward reaction ($A + B \rightarrow nD$) by r_{A-} and the rate of formation by the reverse reaction as r_{A+}. Then Equation 5.2 for the general flow reactor can be written

$$c_A(t_2)V(t_2) = c_A(t_1)V(t_1) + \int_{t_1}^{t_2} q_{Af}(\tau)c_{Af}(\tau)\,d\tau - \int_{t_1}^{t_2} q(\tau)c_A(\tau)\,d\tau$$

$$+ \int_{t_1}^{t_2} r_{A+}V\,d\tau - \int_{t_1}^{t_2} r_{A-}V\,d\tau$$

or, after applying the mean value theorem, dividing by Δt, and taking the limit as $\Delta t \to 0$,

$$\frac{d}{dt} c_A V = q_{Af} c_{Af} - q c_A + V r_{A+} - V r_{A-}$$

For the special case of a batch reactor and constant density

$$\frac{dc_A}{dt} = r_{A+} - r_{A-} \tag{5.27}$$

Similarly, for B and D,

$$\frac{dc_B}{dt} = r_{B+} - r_{B-} \tag{5.28}$$

$$\frac{dc_D}{dt} = r_{D+} - r_{D-} \tag{5.29}$$

Many of the arguments used previously for the irreversible reaction remain valid. Clearly, a mole of B still vanishes whenever a mole of A vanishes, and n moles of D appear. Thus

$$n r_{A-} = n r_{B-} = r_{D+} \tag{5.30}$$

By identical reasoning

$$n r_{A+} = n r_{B+} = r_{D-} \tag{5.31}$$

For algebraic convenience we can then define a single rate, denoted by r, as

$$r \equiv r_{A-} - r_{A+} = r_{B-} - r_{B+} = \frac{r_{D+} - r_{D-}}{n} \tag{5.32}$$

Equations 5.27 to 5.29 can then be written

$$\frac{dc_A}{dt} = -r$$

$$\frac{dc_B}{dt} = -r$$

$$\frac{dc_D}{dt} = nr$$

These are identical to Equations 5.15 to 5.17, which were derived for the irreversible case. Clearly, then, these basic equations and their consequences do not depend on the reversibility or irreversibility of the reaction. Two important consequences are Equations 5.18 and 5.19 which relate the concentrations to one another,

$$c_A(t) - c_{A0} = c_B(t) - c_{B0} \tag{5.18}$$

$$c_D(t) - c_{D0} = n[c_{A0} - c_A(t)] \tag{5.19}$$

We can gain some useful insight into the reversible reaction by considering the simplest relations for the rates. As before, we take the rate of the forward reaction as $r_{A-} = k_1 c_A c_B$, where we now denote the rate coefficient by k_1. We will take $n = 2$ and assume for the reverse reaction $(D + D \rightarrow A + B)$ an analogous rate, $r_{A+} = r_{B+} = r_{D-}/2 = k_2 c_D{}^2$. Then from Equations 5.27 to 5.31

$$\frac{dc_A}{dt} = \frac{dc_B}{dt} = -\frac{1}{n}\frac{dc_D}{dt} = -k_1 c_A c_B + k_2 c_D{}^2 \tag{5.33}$$

If the concentrations c_A, c_B, and c_D are such that

$$k_1 c_A c_B = k_2 c_D{}^2 \tag{5.34}$$

the rates of change of concentrations are zero and the system is at *equilibrium*. Thus, if we wait a sufficiently long period of time the concentrations will always satisfy Equation 5.34. If we denote concentrations at equilibrium by c_{Ae}, c_{Be}, and c_{De}, then we note that the ratio

$$K_e = \frac{k_1}{k_2} = \frac{c_{De}^2}{c_{Ae}c_{Be}} \tag{5.35}$$

is a constant that can be determined by a single measurement at the end of the experiment. K_e is, of course, the equilibrium constant that is familiar from courses in chemistry. The meaning of chemical equilibrium, according to Equation 5.34, is that rates of the forward and reverse reactions are equal, so there is no net change in the concentration of any species.

We can draw a further conclusion from Equation 5.33. If we begin an experiment with no D present, then the term $k_2 c_D{}^2$ will be negligible until sufficient product has formed for the reverse reaction to be important. Thus, if we are careful, it will be possible to treat the system as irreversible at least for small values of c_D, which in a batch experiment corresponds to small time.

Example 5.4

The reaction between sulfuric acid and diethyl sulfate studied in Example 5.1 for time up to 194 minutes is, in fact, reversible. The complete data are shown in Table 5.7, where it can be seen that there is an equilibrium concentration of acid of 2.60 g-moles/liter. Based on the previous analysis the forward rate appears to be adequately described by $r_{A-} = k_1 c_A c_B$. We shall assume that

TABLE 5.7 Concentration of H_2SO_4 versus time for the reaction of sulfuric acid with diethyl sulfate in aqueous solution at 22.9°C

Time (minutes) t	Concentration of H_2SO_4 (gram-moles/liter) c_A
0	5.50
41	4.91
48	4.81
55	4.69
75	4.38
96	4.12
127	3.84
146	3.62
162	3.59
180	3.44
194	3.34
212	3.27
267	3.07
318	2.92
379	2.84
410	2.79
∞	2.60

Source. Data of Hellin and Jungers, *Bull. Soc. Chim. France*, p. 386 (1957).

the reverse reaction is described by the rate $r_{A+} = k_2 c_D^2$, in which case Equation 5.33 applies. Since $c_{A0} = c_{B0}$ and $c_{D0} = 0$, Equations 5.18 and 5.19 give $c_A(t) = c_B(t)$ and $c_D(t) = 2[c_{A0} - c_A(t)]$, so that Equation 5.33 can be written

$$\frac{dc_A}{dt} = -k_1 c_A^2 + k_2\{2[c_{A0} - c_A]\}^2 \tag{5.36}$$

The equilibrium relationship for c_{Ae}, when $dc_A/dt = 0$, is

$$\frac{k_1}{k_2} = 4\left[\frac{c_{A0}}{c_{Ae}} - 1\right]^2 \equiv 4\beta \tag{5.37}$$

For the data given in Table 5.7,

$$\beta = \left[\frac{5.50}{2.60} - 1\right]^2 \simeq \frac{5}{4}$$

Using Equation 5.37, Equation 5.36 can be written

$$\frac{dc_A}{dt} = k_1 \left\{ \frac{1}{\beta} [c_{A0} - c_A]^2 - c_A^2 \right\}$$

This can be formally separated as

$$\frac{\beta \, dc_A}{[c_{A0} - c_A]^2 - \beta c_A^2} = k_1 \, dt$$

or, integrating from time zero to some later time,

$$\int_{c_{A0}}^{c_A} \frac{\beta \, dc}{[c_{A0} - c]^2 - \beta c^2} = \int_0^t k_1 \, d\tau = k_1 t$$

When the integration on the left is carried out we obtain

$$\ln \left\{ \frac{c_A[1 - \sqrt{\beta}] - c_{A0}}{c_A[-1 - \sqrt{\beta}] + c_{A0}} \right\} = \frac{2c_{A0}k_1}{\sqrt{\beta}} t \qquad (5.38)$$

The data in Table 5.7 are plotted in Figure 5.6 as indicated in Equation 5.38, using the values $\beta = 1.25$, $c_{A0} = 5.50$. The data do follow a straight

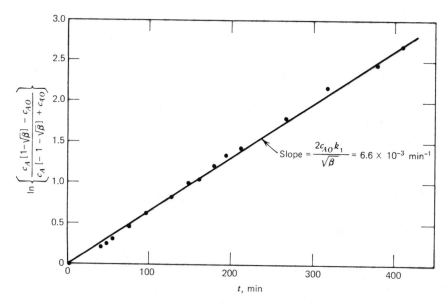

FIGURE 5.6   Computation of the second-order rate constant for the reversible reaction between sulfuric acid and diethyl sulfate in aqueous solution.

line and are consistent with the model. Calculation of k_1 from the slope yields

$$k_1 = 6.7 \times 10^{-4} \, \frac{\text{liter}}{\text{g-mole min}}$$

This differs only by 10 percent from the value obtained by assuming irreversibility over the first 194 min.

5.7   MULTIPLE REACTIONS

In most practical applications the reaction chemistry is more complex than the simple irreversible and reversible reactions considered thus far. Reaction rate determination is an essential feature of the analysis of a reacting system, and the need to find rate expressions has led to a number of clever experimental techniques. Studies of this type, designated as chemical kinetics, are an active area for research in both chemical engineering and chemistry. We cannot, in this elementary text, discuss chemical kinetics in any depth, but the following sections are intended to suggest the behavior of somewhat more complex situations.

5.7.1   Consecutive Reactions

Consecutive reactions occur when the product of one reaction undergoes further reaction to form another compound. For example, we might have a situation like the following:

$$A + B \rightarrow D$$

$$D + S \rightarrow T$$

In a batch reactor the principle of conservation of mass will then lead to the following equations:

$$\frac{dc_A}{dt} = -r_{A-} \tag{5.39a}$$

$$\frac{dc_B}{dt} = -r_{B-} \tag{5.40a}$$

$$\frac{dc_D}{dt} = r_{D+} - r_{D-} \tag{5.41a}$$

$$\frac{dc_S}{dt} = -r_{S-} \tag{5.42a}$$

$$\frac{dc_T}{dt} = r_{T+} \tag{5.43a}$$

The chemical equation implies

$$r_{A-} = r_{B-} = r_{D+} \equiv r_1 \tag{5.44}$$

$$r_{D-} = r_{S-} = r_{T+} \equiv r_2 \tag{5.45}$$

where we have used r_1 and r_2 to denote intrinsic rates of the first $(A + B \rightarrow D)$ and second $(D + S \rightarrow T)$ reactions. Equations 5.39a to 5.43a then become

$$\frac{dc_A}{dt} = -r_1 \tag{5.39b}$$

$$\frac{dc_B}{dt} = -r_1 \tag{5.40b}$$

$$\frac{dc_D}{dt} = r_1 - r_2 \tag{5.41b}$$

$$\frac{dc_S}{dt} = -r_2 \tag{5.42b}$$

$$\frac{dc_T}{dt} = r_2 \tag{5.43b}$$

Since $dc_A/dt = dc_B/dt$, $dc_S/dt = dc_T/dt$, and $dc_D/dt = -dc_A/dt + dc_S/dt$, it further follows that

$$c_A - c_{A0} = c_B - c_{B0} \tag{5.46}$$

$$c_S - c_{S0} = c_T - c_{T0} \tag{5.47}$$

$$c_D - c_{D0} = c_{A0} - c_A + c_S - c_{S0} \tag{5.48}$$

If we take the simplest possible form for the reaction rates r_1 and r_2

$$r_1 = k_1 c_A c_B \tag{5.49}$$

$$r_2 = k_2 c_D c_S \tag{5.50}$$

then Equations 5.39 to 5.43 become

$$\frac{dc_A}{dt} = -k_1 c_A c_B \tag{5.39c}$$

$$\frac{dc_B}{dt} = -k_1 c_A c_B \tag{5.40c}$$

$$\frac{dc_D}{dt} = k_1 c_A c_B - k_2 c_D c_S \tag{5.41c}$$

$$\frac{dc_S}{dt} = -k_2 c_D c_S \tag{5.42c}$$

$$\frac{dc_T}{dt} = k_2 c_D c_S \tag{5.43c}$$

Equations 5.39c and 5.40c have already been solved in Section 5.4 for cases of equal and unequal starting compositions, and we have discussed the experimental determination of k_1.[ ] The *qualitative* behavior of the remaining equations can be obtained by noting that if $c_{D0} = 0$, then when $t = 0$, $dc_D/dt = k_1 c_A c_B > 0$, while after a long time, when c_A and c_B are nearly zero, $dc_D/dt \simeq 0 - k_2 c_D c_S < 0$. A derivative that is first positive and then negative passes through zero at some time, indicating a maximum concentration of D at that time.

The existence of a maximum concentration of D at some time is of key importance if we are considering the design of an industrial reactor to produce D. If we are certain that the expression of r_2 is correct, then we can also use the maximum to estimate k_2. We denote the values of all concentrations by a subscript m at the time when c_D takes on its maximum. At that point $dc_D/dt = 0$, so Equation 5.41c gives

$$0 = k_1 c_{Am} c_{Bm} - k_2 c_{Dm} c_{Sm} \tag{5.51}$$

Using Equations 5.46 to 5.48 and assuming for simplicity that $c_{A0} = c_{B0} = c_{S0}$, $c_{D0} = 0$, Equation 5.51 can be rearranged to give

$$k_2 = \frac{k_1 c_{Am}^2}{c_{Dm}[c_{Dm} + c_{Am}]} \tag{5.52}$$

Notice that this estimate from one measurement of c_D is useful only if Equation 5.50 is known to be correct.

By some manipulation we can obtain a relation to verify the constitutive equation for r_2. For simplicity we assume $c_{A0} = c_{B0} = c_{S0}$, $c_{D0} = 0$. Then Equation 5.42c, together with Equations 5.46 and 5.48, becomes

$$\frac{dc_S}{dt} = k_2 c_A c_S - k_2 c_S^2$$

and using Equation 5.22 for c_A we have

$$\frac{dc_S}{dt} = \left[\frac{k_2 c_{A0}}{1 + k_1 c_{A0} t}\right] c_S - k_2 c_S^2 \tag{5.53}$$

By the trick of setting $y = 1/c_S$ we obtain the equation

$$\frac{dy}{dt} = -\left[\frac{k_2 c_{A0}}{1 + k_1 c_{A0} t}\right] y + k_2, \qquad y(0) = y_0 = 1/c_{S0} \tag{5.54}$$

This is of the form of Equation 15.25 whose solution is obtained in Section 15.10 as

$$y(t) = \frac{y_0}{[1 + k_1 c_{A0} t]^{k_2/k_1}}$$
$$+ \frac{k_2[1 + k_1 c_{A0} t]}{c_{A0}[k_1 + k_2]} \left\{1 - [1 + k_1 c_{A0} t]^{-[k_1 + k_2]/k_1}\right\}$$

or

$$\frac{1}{c_S(t)} = \frac{1}{c_{S0}}\left[\frac{c_A}{c_{A0}}\right]^{k_2/k_1} + \frac{k_2}{c_{A0}[k_1 + k_2]}\left\{\frac{c_{A0}}{c_A} - \left[\frac{c_A}{c_{A0}}\right]^{k_2/k_1}\right\} \qquad (5.55)$$

At long time $(c_A/c_{A0} \to 0)$. A plot of c_S versus c_A will have slope $1 + k_1/k_2$, providing an estimate of k_2. The rate expression can then be checked over the entire time interval by plotting according to Equation 5.55.

5.7.2 Parallel Reactions

Parallel reactions occur when A reacts with two different species to produce two distinct products. For purposes of discussion we will assume that the following reactions occur:

$$A + B \to D \qquad \text{(reaction 1)}$$

$$A + R \to S \qquad \text{(reaction 2)}$$

We denote the rate at which A disappears in reaction 1 by r_{A1-} and in reaction 2 by r_{A2-}. Then in a batch reactor the component mass balances are

$$\frac{dc_A}{dt} = -r_{A1-} - r_{A2-} = -r_1 - r_2 \qquad (5.56a)$$

$$\frac{dc_B}{dt} = -r_{B-} = -r_{A1-} = -r_1 \qquad (5.57a)$$

$$\frac{dc_D}{dt} = r_{D+} = r_{A1-} = +r_1 \qquad (5.58a)$$

$$\frac{dc_R}{dt} = -r_{R-} = -r_{A2-} = -r_2 \qquad (5.59a)$$

$$\frac{dc_S}{dt} = r_{S+} = r_{A2-} = +r_2 \qquad (5.60a)$$

Here we have included the information provided by the chemical equations and defined r_1 and r_2 as r_{A1-} and r_{A2-}, respectively. Taking the simplest forms for r_1 and r_2,

$$r_1 = k_1 c_A c_B \qquad (5.61)$$

$$r_2 = k_2 c_A c_R \qquad (5.62)$$

Equations 5.56 to 5.60 become

$$\frac{dc_A}{dt} = -k_1 c_A c_B - k_2 c_A c_R \tag{5.56b}$$

$$\frac{dc_B}{dt} = -k_1 c_A c_B \tag{5.57b}$$

$$\frac{dc_D}{dt} = k_1 c_A c_B \tag{5.58b}$$

$$\frac{dc_R}{dt} = -k_2 c_A c_R \tag{5.59b}$$

$$\frac{dc_S}{dt} = k_2 c_A c_R \tag{5.60b}$$

Equations 5.56, 5.57, and 5.59 can be combined to give one relation between the concentrations,

$$c_A - c_{A0} = c_B - c_{B0} + c_R - c_{R0} \tag{5.63}$$

A second relation is obtained by writing Equations 5.57b and 5.59b as

$$\frac{1}{k_1 c_B} \frac{dc_B}{dt} = \frac{1}{k_2 c_R} \frac{dc_R}{dt}$$

or, integrating,

$$\frac{k_2}{k_1} \ln \frac{c_B}{c_{B0}} = \ln \frac{c_R}{c_{R0}} \tag{5.64}$$

Thus, a plot of $\ln c_R/c_{R0}$ versus $\ln c_B/c_{B0}$ will yield a straight line with slope k_2/k_1. Equations 5.63 and 5.64 can then be substituted into Equation 5.57b to obtain an equation in terms of c_B only:

$$\frac{dc_B}{dt} = -k_1 [c_B + M] c_B - k_1 c_{R0} \left[\frac{c_B}{c_{B0}} \right]^{k_2/k_1} c_B \tag{5.65}$$

Here $M = c_{A0} - c_{B0} - c_{R0}$. Equation 5.65 is separable (Section 15.9) and can be written as

$$\int_{c_{B0}}^{c_B} \frac{dc}{c[c + M] + c_{R0} \left[\dfrac{c}{c_{B0}} \right]^{k_2/k_1} c} = -k_1 t \tag{5.66}$$

A plot of the left-hand side versus t will give a straight line with slope $-k_1$ if the assumed rate expression is correct. The integration can be performed analytically only for special values of k_2/k_1, and numerical integration as outlined in Section 15.6 will generally be required.

5.7.3 Complex Reactions

In many situations of interest a number of parallel and consecutive reactions occur simultaneously. The following are typical:

$$A + B \rightarrow R \qquad\qquad\qquad A + B \rightarrow R$$

$$R + B \rightarrow S \qquad\qquad\qquad R + S \rightarrow T$$

$$S + B \rightarrow T \qquad\qquad\qquad T \rightarrow W + Z$$

$$Z + B \rightarrow M$$

$$\text{etc.}$$

We shall return to the first of these complex reactions in Chapter 7. At this point we shall not develop any model equations, and in general we shall deal henceforth with only simple chemical equations, since the simplest reaction chemistry is often adequate to demonstrate important points of analysis.

5.8 CONSTANT DENSITY ASSUMPTION

We have carried out all of the analysis of reacting systems with the hypothesis that density is a constant. In fact we know that density will be a function of concentration. We shall show here that the result obtained in Section 4.5.3 for simple mixtures applies to reacting systems as well, namely, that the governing equations are the same as for constant density when the density is a linear function of concentration.

Consider, for simplicity, the specific reaction

$$A + B \rightleftarrows nD$$

and let us suppose that the reactants are mixed together in a solution of some solvent with density ρ_0 just prior to entering the reactor. The mass balance equations for a well-stirred flowing system are

$$\text{overall:} \quad \frac{d\rho V}{dt} = \rho_f q_f - \rho q \tag{5.67}$$

$$A: \quad \frac{dc_A V}{dt} = c_{Af} q_f - c_A q - Vr \tag{5.68}$$

$$B: \quad \frac{dc_B V}{dt} = c_{Bf} q_f - c_B q - Vr \tag{5.69}$$

$$R: \quad \frac{dc_D V}{dt} = c_{Df} q_f - c_D q + nVr \tag{5.70}$$

where r is the rate of disappearance of component A. If ρ is a linear function of the concentrations, we may write

$$\rho = \rho_0 + \phi_A c_A + \phi_B c_B + \phi_D c_D \tag{5.71}$$

Substituting Equation 5.71 into 5.67 gives

$$\rho_0 \frac{dV}{dt} + \phi_A \frac{dc_A V}{dt} + \phi_B \frac{dc_B V}{dt} + \phi_D \frac{dc_D V}{dt}$$
$$= \rho_0[q_f - q] + \phi_A[c_{Af}q_f - c_A q] + \phi_B[c_{Bf}q_f - c_B q]$$
$$+ \phi_D[c_{Df}q_f - c_D q] \tag{5.72}$$

Multiplying Equations 5.68, 5.69, and 5.70, respectively, by ϕ_A, ϕ_B, and ϕ_D and adding gives

$$\phi_A \frac{dc_A V}{dt} + \phi_B \frac{dc_B V}{dt} + \phi_D \frac{dc_D V}{dt}$$
$$= \phi_A[c_{Af}q_f - c_A q] + \phi_B[c_{Bf}q_f - c_B q] + \phi_D[c_{Df}q_f - c_D q]$$
$$+ Vr[n\phi_D - \phi_A - \phi_B] \tag{5.73}$$

Subtracting Equation 5.73 from 5.72 we obtain

$$\frac{dV}{dt} = q_f - q + \frac{Vr}{\rho_0}[\phi_A + \phi_B - n\phi_D] \tag{5.74}$$

The last term in Equation 5.74 represents the contribution to volume change from the chemical reaction. The rate of change of density with composition will usually be nearly proportional to the molecular weight of the dissolved species. Thus we can write

$$\phi_A = \phi M_{wA} \qquad \phi_B = \phi M_{wB} \qquad \phi_D = \phi M_{wD}$$

and Equation 5.74 becomes

$$\frac{dV}{dt} = q_f - q + \frac{Vr\phi}{\rho_0}[M_{wA} + M_{wB} - nM_{wD}]$$

But from the reaction stoichiometry $M_{wA} + M_{wB} = nM_{wD}$. Thus

$$\frac{dV}{dt} = q_f - q \tag{5.75}$$

which is the same result as for constant density independent of concentration. Notice that this result, which will usually be a reasonable approximation for liquid systems, depends on *two* assumptions:

1. Density is linear in concentration:

$$\rho = \rho_0 + \phi_A c_A + \phi_B c_B + \phi_D c_D$$

2. Rate of change of density with concentration is proportional to molecular weight.

Subject to these restrictions the derivation given here can be generalized to any number of flowing streams, chemical species, and reactions.

5.9 ORDER AND STOICHIOMETRY

In all of the reaction examples that we have considered thus far the stoichiometry of the reaction, or the chemical equation, has in fact determined the rate, as implied by the very simple reasoning at the start of the chapter. That is, for the reaction

$$\alpha A + \beta B \rightarrow \text{products}$$

we have always found the rate to be of the form

$$r_{A-} = k c_A{}^\alpha c_B{}^\beta$$

Although this will often be the case, it would be a serious error to assume that overall stoichiometry is really related to reaction rate in a consistent, meaningful fashion.

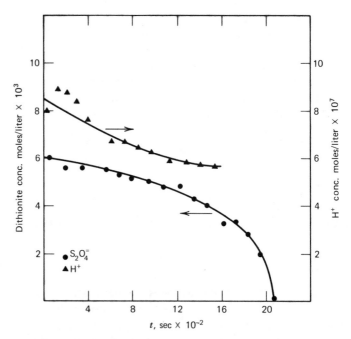

FIGURE 5.7   Concentration of H^+ and $S_2O_4^=$ as functions of time in an unbuffered system at 70°C. Data of Rinker et al., *Ind. Eng. Chem. Fundamentals*, **4**, 282 (1964). Reproduced by permission.

Consider, for example, the data shown in Figure 5.7 for the reaction

$$2S_2O_4^= + H_2O \rightarrow S_2O_3^= + 2H_2SO_3^-$$

We would expect from a naive assumption based on stoichiometry that the rate would be

$$r = k[c_{S_2O_4^=}]^2[c_{H_2O}]$$

In fact, the data are represented by

$$r = k[c^*_{S_2O_4^=}]^{3/2}[c_{H^+}]^{1/2}$$

where $c^*_{S_2O_4^=}$ refers to the concentration of disulfide ion and anything else present that can reduce methylene blue in a basic medium.

Attempts at explanation of this apparent contradiction between the simple approach based on counting collisions of molecules and the actual data are beyond the scope of our treatment here, and are dealt with in some detail in courses in physical chemistry and kinetics and reactor design. For our purposes it suffices to note that, although we might sometimes assume simple rate expressions for illustrative purposes, the rate for any real system must be obtained from careful experiments over the entire range of interest.

5.10 CONCLUDING REMARKS

Review again the development of the basic reactor model equations in terms of the logical structure outlined in Figure 3.8. Compare the model equations with those in Chapter 4 and note particularly how the chemical equation is used to express all of the reaction information in terms of the single rate expression, r. Review Section 5.6 with regard to what you have learned in chemistry concerning reactions at equilibrium.

You should be sure that you understand exactly what the reaction rate means physically. The chapter is mostly concerned with the determination of the constitutive relationship between rate and composition using experimental observations from a batch reactor. If you really understand the physical significance of the rate, then you will recognize that the constitutive relationship between rate and composition is applicable in all reactors. The published literature in many areas such as sewage treatment is often confused on this basic and rather obvious point. Make certain that the mathematical manipulations do not obscure your understanding of the physical significance of what you are doing. The integrations performed in this chapter are more complicated than those in Chapter 4, and you should review integration methods in your calculus text if they are causing you any trouble.

The comments in Section 5.9 need to be emphasized. The data in this chapter have been selected for demonstration purposes because the rates follow simple relations with concentration. In many important situations the rate-composition relation is more complex. This is brought out only to a limited extent in the problems. The same basic principles obviously apply, but integration of the batch reactor equations will require more effort.

A very readable discussion of reaction rates and batch experimentation is contained in a small paperback.

5.1 E. L. King, *How Chemical Reactions Occur*, Benjamin, New York, 1964.

The following chemistry texts contain more extensive treatments:

5.2 S. W. Benson, *Foundations of Chemical Kinetics*, McGraw-Hill, New York, 1960.

5.3 A. A. Frost and R. G. Pearson, *Kinetics and Mechanisms*, 2nd ed., Wiley, New York, 1961.

5.4 K. J. Laidler, *Chemical Kinetics*, McGraw-Hill, 1950.

Texts on physical chemistry contain at least one chapter on the rates of chemical reactions. See, for example,

5.5 G. M. Barrow, *Physical Chemistry*, 2nd ed., McGraw-Hill, New York, 1966.

5.6 F. Daniels and R. A. Alberty, *Physical Chemistry*, Wiley, New York, 1961.

5.7 E. A. Moelwyn Hughes, *Physical Chemistry*, 2nd ed., Pergamon, New York, 1961.

5.8 W. J. Moore, *Physical Chemistry*, 2nd ed., Prentice-Hall, Englewood Cliffs, 1962.

An early text in which the subject of rate determination is examined from a chemical engineering viewpoint is

5.9 O. A. Hougen and K. M. Watson, *Chemical Process Principles, Part III: Kinetics and Catalysis*, Wiley, New York, 1947.

Most chemical engineering texts on reactor design contain at least one chapter on batch experimentation. The compilation at the end of Chapter 7 should be consulted. Extensive tabulations of reaction rate constants are contained in a series of National Bureau of Standards publications,

5.10 "Tables of Chemical Kinetics: Homogeneous Reactions," NBS Circular 510, 1951, plus supplements 1956, 1960, and NBS Monograph 34, 1961.

In most cases it will be necessary to consult the periodical literature for particular data. *Chemical Abstracts* can be used as a guide for particular reactions.

5.11 PROBLEMS

5.1 Suppose that the reaction $A + B \rightarrow nD$ taking place in a batch reactor has a rate constitutive equation

$$r_{A-} = kc_A^2 c_B$$

Denote the initial concentrations of A and B as c_{A0} and c_{B0}, and suppose that there is initially no D in the reactor. Determine the model behavior for the case when $c_{A0} = c_{B0}$ and when $c_{A0} \neq c_{B0}$. How would you use experimental data to determine k?

5.2 The hydrolysis of acetic anhydride in excess water to form acetic acid

$$(CH_3CO)_2O + H_2O \rightarrow 2CH_3COOH$$

has been studied by Eldridge and Piret [*Chem. Eng. Progress*, **46**, 290 (1950)] who find that the rate at 15°C can be expressed as

$$r = kc_A, \qquad k = 0.0567 \text{ min}^{-1}$$

If a batch reactor initially contains 100 lb of anhydride, how long will it take for 90 lb to be converted to acid? 99 lb? Does the fraction of anhydride converted depend on the initial amount? Why?

5.3 The data in Table 5.2 for the reaction between sulfuric acid and diethyl sulfate were shown in Example 5.1 to be represented by the rate expression

$$r = kc_A c_B$$

Because of the scatter in experimental data it may often be difficult or impossible to distinguish between models. Using the data in Table 5.2 check the use of the constitutive equation

$$r = k'c_A$$

What conclusions do you draw?

5.4 The reaction

$$ClO_3^- + 3H_2SO_3 \rightarrow Cl^- + 3SO_4^= + 6H^+$$

has been studied in 0.2N H_2SO_4 by Nixon and Krauskopf [*J. Am. Chem. Soc.*, **54**, 4606 (1932)]. The following data are reported:

t (minutes)	ClO_3^- (moles/liter)	H_2SO_3 (moles/liter)
0	0.0160	0.0131
2	0.0146	0.0090
3.5	0.0141	0.0074
5	0.0136	0.0060
7.5	0.0131	0.0044
10	0.0127	0.0034
12.5	0.0125	0.0026
15	0.0123	0.0020
17.5	0.0121	0.0015

Find a rate expression that is consistent with the data.

5.5 The reaction

$$\text{Acetic acid } (A) + \text{Butanol } (B) \rightarrow \text{ester} + \text{water}$$

has been studied by Leyes and Othmer [*Ind. Eng. Chem.*, **37**, 968 (1945)], who report the following data at 100°C in the presence of 0.03 percent sulfuric acid. The initial concentrations are 0.2327 moles acetic acid/100 g of solution and 1.1583 moles butanol/100 g of solution.

Time (hours)	Moles of Acid Converted/100 g Solution
0	0
1	0.1552
2	0.1876
3	0.2012
4	0.2067
5	0.2089
6	0.2099
7	0.2109

Leyes and Othmer claim that the reaction is second order in acetic acid and zero order in butanol up to 75 percent conversion, that is,

$$r = kc_A^2$$

Do the data support this contention? Can you distinguish this relation from $r = kc_A c_B$ with the data given here? (You may assume that the density of the solution is constant over the course of the reaction, with a value of approximately 0.75 g/ml. Why?)

5.6 The formation of glucose from cellulose has been studied by Harris and Kline [*J. Phys. and Colloid Chem.*, **53**, 344 (1949)]. Using Douglas Fir cellulose they carried out the reaction at a number of temperatures in the presence of HCl. The rate was found to be first order in cellulose concentration ($r = k_1 c_A$), but dependent on HCl concentration as follows.

HCl concentration (moles/liter)	k_1, min^{-1}	
	160°C	190°C
0.055	0.00203	0.0627
0.11	0.00486	0.149
0.22	0.01075	0.357
0.44	0.0261	
0.88	0.0672	

Find a rate expression including the dependence on HCl.

(Hint: try $k_1 = k_1' c_{HCl}^n$)

Glucose decomposes in the presence of HCl. Harris and Kline also reported this reaction to be first order in glucose ($r = k_2 c_B$), with the rate constant dependent on HCl as follows at 190°C.

HCl concentration (moles/liter)	k_2, min^{-1}
0.055	0.0488
0.11	0.107
0.22	0.218
0.44	0.406
0.88	0.715

Can you find a rate expression in a simple form?

5.7 The isomerization of the ester triethylphosphite to diethyl ethyl phosphorate is catalyzed by ethyl iodide:

$$P(OC_2H_5)_3 + C_2H_5I \rightarrow C_2H_5PO(OC_2H_5)_2 + C_2H_5I$$

Isbell, Watson, and Zerweth [*J. Phys. Chem.*, **61**, 89 (1957)] report the following data at 90°C.

Mole %C$_2$H$_5$I at $t = 0$	Time to Isomerize a Fraction of Ester (minutes)						
	0.10	0.20	0.30	0.40	0.50	0.60	0.70
56.0	14	33	52	72	92	112	136
50.0	21	44	67	89	112	135	158
37.5	30	64	96	129	161	192	
24.9	51	105	158	212	255	318	

How does the rate depend on triethyl phosphite concentration?

5.8 The saponification of the ester propargyl acetate (2 propyn-*i*-ol acetate)

$$HC \equiv C-CH_2-\overset{\overset{\displaystyle O}{\|}}{C}-O-CH_3 + OH^- \longrightarrow$$

$$CH_3OH + HC \equiv C-CH_2-\overset{\overset{\displaystyle O}{\|}}{C}-OH$$

was studied by Myers, Collett, and Lazzell [*J. Phys. Chem.*, **56**, 461 (1952)] using a conductivity technique to measure the course of the reaction. Conductivity $C(t)$ is related to conversion of the base through the relation

$$\frac{\text{fraction base}}{\text{converted}} = \frac{c_B}{c_{B0}} = \frac{C(t) - C(\infty)}{C(0) - C(\infty)}$$

Starting with a solution of 0.00873N ester, 0.00673N base, they report the following data.

t (minute)	C
0	1.087
0.200	1.042
0.417	1.000
0.737	0.952
1.047	0.909
1.397	0.870
1.787	0.833
2.187	0.800
2.637	0.769
3.147	0.741
3.667	0.714
∞	0.490

Find a rate expression for the reaction consistent with the above data.

5.9 For the reversible reaction $A + B \rightleftharpoons nD$ taking place in a batch reactor, determine model behavior and methods of evaluating the constants for the following rate expressions.

(a) $r_{A-} = k_1 c_A c_B$ $r_{A+} = k_2 c_D$

(b) $r_{A-} = k_1 c_A$ $r_{A+} = k_2 c_D$

(c) $r_{A-} = k_1 c_A$ $r_{A+} = k_2 c_D^2$

5.10 Using the data in Table 5.7 for the reaction between sulfuric acid and diethyl sulfate check the use of the constitutive equations in Problem 5.9. What is the significance of the result?

5.11 The reaction

$$CH_3COCH_3 + HCN \rightleftharpoons (CH_3)_2C\overset{\displaystyle CN}{\underset{\displaystyle OH}{\big|}}$$

was studied in $0.0441N$ acetic acid, $0.049N$ sodium acetate by Svirebly and Roth [*J. Am. Chem. Soc.*, **75**, 3106 (1953)]. They report the following data from a batch experiment.

t (minutes)	HCN (moles/liter)	CH$_3$COCH$_3$ (moles/liter)
0	0.0758	0.1164
4.37	0.0748	
73	0.0710	
173	0.0655	
265	0.0610	
347	0.0584	
434	0.0557	
∞	0.0366	0.0772

Find a constitutive equation for the rate.

5.12 Reconsider the acetic acid butanol reaction in Problem 5.5 taking the equilibrium and reverse reaction into account. What can you say about the rate expression from the data given?

5.13 For the reaction sequence

$$A \rightarrow R \qquad r_{A-} = k_1 c_A$$
$$R \rightarrow S \qquad r_{R-} = k_2 c_R$$

carried out in a batch reactor with no R or S initially present, the concentration of R will go through a maximum as a function of time. Derive the model equations for this situation and solve them for $c_R(t)$. Compute the time at which the maximum occurs, t_M, and the value $c_R(t_M)$. (It will be necessary to solve an equation of the form $dx/dt + kx = b(t)$. The solution is given in the Mathematical Review Notes, Example 15.9).

5.14 The conversion of cellulose to glucose follows a reaction scheme like that in Problem 5.13, where A is cellulose and R glucose. Thus, there is an optimum time to operate a batch reactor for maximum glucose production. Using the data in Problem 5.6, plot the maximum conversion and optimum operating time at 190°C as a function of HCl concentration. Can you draw any conclusions about the way in which the process should be operated?

5.15 Show that when c_A is close to c_{A0} Equation 5.38 reduces to Equation 5.22 for the irreversible reaction.

CHAPTER 6

Treatment of Experimental Data

6.1 INTRODUCTION

We have, on several occasions, been faced with the necessity of passing a line through experimental data. In Chapter 2 we compared a model equation with the tank-draining data, and from the slope of the best straight line through the data we were able to calculate a parameter required in the analysis. Similarly, in Chapter 5 we compared model predictions with batch reactor data and computed a rate constant from the slope of the line. This step of comparison of model prediction with experiment and determination of model parameters is an essential feature of the analysis process.

In Chapter 5 we needed a functional relationship between salt concentration and density, while we found in the literature only a series of experimental points. Thus, we chose to pass a straight line through the density-concentration data and to use this linear relationship in the conservation equations. The important distinction to be noted is that when comparing model equations with experiment we have been guided by theory and the form of the equation is fixed. When fitting a line to data to determine a functional relationship we have simply been proceeding empirically, and the form of the function is arbitrary and guided by convenience. In both circumstances, because of error in experimental data, approximations in theory, or because the empirical form that we have chosen is only approximate in its representation of the data, there will rarely be a unique line to be drawn through the data. Hence, there will be some uncertainty about the proper slope and intercept and, therefore, uncertainty about any parameters that we might calculate. We therefore need to develop a rational procedure for evaluating the proper, or *best* line through a set of experimental data.

6.2 CRITERIA FOR BEST FIT

Let us suppose that we have a set of data points taken at several values of an independent variable, say time, and we have reason to believe that the data should be represented by a straight line passing through zero, that is, by the equation

$$y = bt \qquad (6.1)$$

Our data are measured values y_1 at time t_1, y_2 at time t_2, and so on. In general, we shall denote the value measured at time t_k as y_k and suppose that we have a total of N data points. We wish to find the value of the slope, b, in Equation 6.1 which best represents the data.

Since the data are expected to scatter, no straight line will pass exactly through every point and there will be an error at each point associated with the use of Equation 6.1. The difference between bt_k, the value of y predicted from Equation 6.1 at time t_k, and the value y_k actually measured at time t_k is

$$(\text{difference})_k = y_k - bt_k \equiv e_k \qquad (6.2)$$

where the notation e_k is chosen to denote the error at point k. We are concerned, not with individual points, but with the *overall* error taking all points into consideration. This overall error must then be defined.

It is helpful to think in terms of the *magnitude* of the point error, $|e_k|$, since in that way positive and negative differences, equally objectionable, show up in the same way. A reasonable means of defining the "goodness of fit" of a line to the data is then to add up all of the absolute values of point errors. We denote this quantity by e_{AV}, for total absolute value error

$$e_{AV} = \sum_{k=1}^{N} |e_k| = \sum_{k=1}^{N} |y_k - bt_k| \qquad (6.3)$$

An alternative that we might choose to examine is based on the *square* of the point error, which is also the same for positive and negative differences. We denote this by e_{SS}, for total sum of squares error

$$e_{SS} = \left[\sum_{k=1}^{N} e_k^2 \right]^{1/2} = \left\{ \sum_{k=1}^{N} [y_k - bt_k]^2 \right\}^{1/2} \qquad (6.4)$$

Notice that e_{SS} has the form of a distance in Cartesian coordinates.

The important point to note is that all of the y_k and t_k are known numbers, so that only the slope, b, is required in order to calculate either e_{AV} or e_{SS}. We can, therefore, evaluate the overall error as a function of b and find the "best" value of b as the one that minimizes the error. To illustrate this point simply, let us extract some representative data from the tank-draining experiment in Table 2.1.

h	t	Average t
10	10.9, 11.5, 11.6	11.3
5	43.8, 42.8, 43.8	43.5
1	85.0, 84.0, 85.2	84.7

For illustrative purposes we will use only these data and, in fact, only the average value of t. The theoretical relation which we are examining is Equation 2.12 for the special case $n = 1/2$, which can be written as

$$h_0^{1/2} - h^{1/2} = \frac{k}{2A} t$$

where $h_0 = 12.0$. We denote $h_0^{1/2} - h^{1/2}$ by y and $k/2A$ by b. The data are plotted in Figure 6.1, where the shaded area shows the possible range of slopes. Notice that no single straight line can pass through all of the data.

The data are tabulated in Table 6.1 in a form suitable for calculating the overall error defined by Equations 6.3 and 6.4. The computation then proceeds in a straightforward manner. For example, for $b = 0.027$,

$$e_{AV}(0.027) = |e_1| + |e_2| + |e_3|$$
$$= |0.30 - 11.3 \times 0.027| + |1.23 - 43.5 \times 0.027|$$
$$+ |2.46 - 84.7 \times 0.027| = 0.23$$

$$e_{SS}(0.027) = [e_1^2 + e_2^2 + e_3^2]^{1/2}$$
$$= \{[0.30 - 11.3 \times 0.027]^2 + [1.23 - 43.5 \times 0.027]^2$$
$$+ [2.46 - 84.7 \times 0.027]^2\}^{1/2} = 0.18$$

Figure 6.2 shows a plot of e_{AV} and e_{SS} versus the slope, b, and the minimum is quite clearly defined in each case at a value close to 0.029.

It should be noted that there is a certain degree of ambiguity here. We have defined *two* criteria for goodness of fit and, although they give approximately the same value for the slope of the best line, the slopes do differ somewhat even on these data that have little scatter. We shall not try to resolve the ambiguity, for it is not clear that one criterion is any better than the other or, for that matter, any better than several others that we might propose. The sum of squares has some significance in the statistical interpretation of the data and, probably of greater importance, the square has convenient analytical properties (e.g., differentiability everywhere) that the absolute

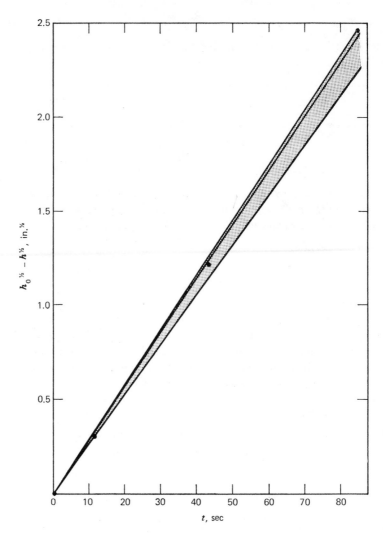

FIGURE 6.1 Selected data from the emptying tank experiment. The shaded region shows the range of possible straight lines through the data.

TABLE 6.1 Error at selected points for the tank-emptying problem as a function of slope, b

	$y_k = h_0^{1/2} - h_k^{1/2}$	t_k	$e_k = y_k - bt_k$
$k = 1$:	0.30	11.3	$e_1 = 0.30 - 11.3b$
$k = 2$:	1.23	43.5	$e_2 = 1.23 - 43.5b$
$k = 3$:	2.46	84.7	$e_3 = 2.46 - 84.7b$

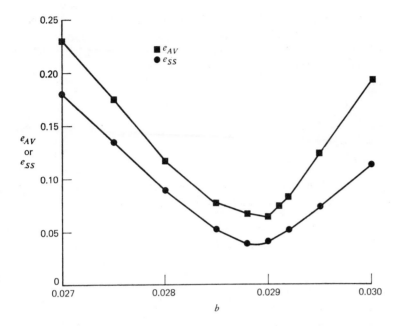

FIGURE 6.2   Total absolute value error (e_{AV}) and total sum of squares error (e_{SS}) versus slope for selected data from the emptying tank experiment.

value lacks. Thus, the minimum sum of squares is the criterion in common use and it is the criterion of best fit which we shall use henceforth, though emphasizing that it is not a unique choice.

6.3   BEST SLOPE—I

Using the sum of squares criterion for overall error we can obtain an explicit analytical expression for the best slope. For convenience we shall drop the subscript "SS." Then

$$e^2 = \sum_{k=1}^{N} [y_k - bt_k]^2 = \sum_{k=1}^{N} y_k^2 - 2b \sum_{k=1}^{N} y_k t_k + b^2 \sum_{k=1}^{N} t_k^2 \tag{6.5}$$

By completing the square, Equation 6.5 can be rewritten

$$e^2 = \left\{ \sum_{k=1}^{N} y_k^2 - \frac{\left[\sum_{k=1}^{N} y_k t_k \right]^2}{\sum_{k=1}^{N} t_k^2} \right\} + \left\{ b \left[\sum_{k=1}^{N} t_k^2 \right]^{1/2} - \frac{\sum_{k=1}^{N} y_k t_k}{\left[\sum_{k=1}^{N} t_k^2 \right]^{1/2}} \right\}^2$$

The first bracketed term is independent of the slope, b, while the second depends on it. Thus we make e^2 (and hence e) as small as possible by making the second term as small as possible. Since the second term is a square its minimum value is zero, which we obtain when the slope is equal to

$$b = \frac{\sum\limits_{k=1}^{N} y_k t_k}{\sum\limits_{k=1}^{N} t_k^2} \qquad (6.6)$$

Example 6.1

The selected data plotted in Figure 6.1 and recorded in Table 6.1 for the tank-draining example are repeated in Table 6.2 with the relevant calculations

TABLE 6.2   Computation of the best slope through selected points for the tank-emptying problem using Equation 6.6

y_k	t_k	t_k^2	$y_k t_k$
0.30	11.3	128	3.4
1.23	43.5	1892	53.5
2.46	84.7	7174	208.4
		$\sum\limits_{k} t_k^2 = 9194$	$\sum\limits_{k} y_k t_k = 265.3$

$$b = \frac{\sum\limits_{k} y_k t_k}{\sum\limits_{k} t_k^2} = \frac{265.3}{9194} = 0.0289$$

to compute the best slope according to Equation 6.6. The result is seen to be in perfect agreement with the minimum in the plot of e_{SS} versus b in Figure 6.2. It is this ability to calculate the minimum point directly from the data without trial and error that makes the sum of squares criterion the more useful.

The complete set of data for the tank-draining problem, including all three experimental runs, is contained in Table 6.3. It is clear that inclusion of all 33 data points adds no complexity whatever to the computation of the slope, which is found to be $b = k/2A = 0.0287$, the result used in Chapter 2 and shown in Figure 2.6.

In Chapter 2 it was emphasized that over a time span short relative to the characteristic time, the behavior of all models is the same, and estimates were made of slopes over the first few data points only. Computing the slope only for the first twelve points, to a height of 8 in., $\sum t_k = 2990$, $\sum y_k t_k = 81.0$, and $b = 0.0271$, close to the value shown as the upper line on Figure 2.6.

TABLE 6.3 Computation of the best slope for the tank-emptying problem using Equation 6.6

h_k	$y_k = h_0^{1/2} - h_k^{1/2}$	t_k	t_k^2	$y_k t_k$
11.0	0.15	5.8	33.6	0.87
	0.15	6.1	37.2	0.92
	0.15	5.9	34.8	0.89
10.0	0.30	10.9	118.8	3.27
	0.30	11.5	132.3	3.45
	0.30	11.6	134.6	3.48
9.0	0.46	16.6	275.6	7.64
	0.46	17.8	316.8	8.19
	0.46	17.2	295.8	7.91
8.0	0.64	23.0	529.0	14.7
	0.64	23.5	552.3	15.0
	0.64	23.0	529.0	14.7
7.0	0.82	30.0	900.0	24.6
	0.82	29.2	852.6	23.9
	0.82	29.8	888.0	24.4
6.0	1.01	36.4	1325	36.8
	1.01	35.8	1282	36.2
	1.01	36.4	1325	36.8
5.0	1.23	43.8	1918	53.9
	1.23	42.8	1832	52.6
	1.23	43.8	1918	53.9
4.0	1.46	51.0	2601	74.5
	1.46	50.5	2550	73.7
	1.46	51.6	2663	75.3
3.0	1.73	60.2	3624	104.1
	1.73	59.2	3505	102.4
	1.73	60.6	3672	104.8
2.0	2.05	71.0	5041	145.6
	2.05	69.8	4872	143.1
	2.05	71.4	5098	146.4
1.0	2.46	85.0	7225	209.1
	2.46	84.0	7056	206.6
	2.46	85.2	7259	209.6
			$\sum_{k=1}^{33} t_k^2 = 70{,}396$	$\sum_{k=1}^{33} y_k t_k = 2019$

$$b = \frac{\sum_k y_k t_k}{\sum_k t_k^2} = \frac{2019}{70{,}396} = 0.0287$$

Example 6.2

A second example of the determination of the best value of a parameter from the slope of an experimental straight line is to be found in the data for the reaction of sulfuric acid and diethyl sulfate. The concentration of sulfuric acid is denoted by c_A. According to Equation 5.38 a plot of

$$y = \ln \left\{ \frac{c_A[1 - \sqrt{\beta}] - c_{A0}}{c_A[-1 - \sqrt{\beta}] + c_{A0}} \right\}$$

versus time should be linear with slope $2c_{A0}k_1/\sqrt{\beta}$. Here $c_{A0} = 5.50$ and $\sqrt{\beta} = 1.115$. The concentration-time data from Table 5.7 are shown in Table 6.4,

TABLE 6.4 Computation of the best value of the forward rate coefficient, k_1, for the reaction between sulfuric acid and diethyl sulfate using Equation 6.6 Concentration versus time data are from Table 5.7

t_k	c_A	$y_k = \ln \left\{ \dfrac{c_A[1 - \sqrt{\beta}] - c_{A0}}{c_A[-1 - \sqrt{\beta}] + c_{A0}} \right\}$	$y_k t_k$	t_k^2
0	5.50	0	0	0
41	4.91	0.216	8.9	1681
48	4.81	0.259	12.4	2304
55	4.69	0.312	17.2	3025
75	4.38	0.467	35.0	5625
96	4.12	0.620	59.5	9216
127	3.84	0.818	103.9	16,129
146	3.62	1.009	147.3	21,316
162	3.59	1.039	168.3	26,244
180	3.44	1.200	216.0	32,400
194	3.34	1.325	257.1	37,636
212	3.27	1.423	301.7	44,944
267	3.07	1.774	473.7	71,289
318	2.92	2.156	685.6	101,124
379	2.84	2.442	925.5	143,641
410	2.79	2.676	1097.2	168,100
(∞)	2.60		$\sum t_k y_k = 4509$	$\sum t_k^2 = 684{,}674$

$$b = \frac{4.509 \times 10^3}{6.847 \times 10^5} = 6.59 \times 10^{-3} = \frac{2c_{A0}k_1}{\sqrt{\beta}}$$

$$c_{A0} = 5.50, \qquad \sqrt{\beta} = \frac{c_{A0}}{c_{Ae}} - 1 = \frac{5.50}{2.60} - 1 = 1.115$$

$$k_1 = 6.7 \times 10^{-4}$$

together with the relevant computations for the best slope and value of the forward rate coefficient, k_1. The straight line shown with the data in Figure 5.6 is the one computed here.

6.4 BEST SLOPE—II

In the preceding section we computed the best slope by using algebra, which is all that is really necessary for this problem. In preparation for slightly more general functions it is helpful to resolve the best slope problem with the use of calculus. Calculus is never really needed for what we are doing, but the algebra can get quite tedious and the use of calculus is generally a shortcut to the solution.

We make use of the fact that the minimum of e^2 with respect to b occurs at a point where the following two relations are satisfied:

$$\frac{d}{db} e^2 = 0$$

$$\frac{d^2 e^2}{db^2} > 0$$

From Equation 6.5

$$\frac{de^2}{db} = -2 \sum_{k=1}^{N} y_k t_k + 2b \sum_{k=1}^{N} t_k^2$$

$$\frac{d^2 e^2}{db^2} = 2 \sum_{k=1}^{N} t_k^2 > 0$$

Setting the first derivative to zero we obtain, as before,

$$b = \frac{\displaystyle\sum_{k=1}^{N} y_k t_k}{\displaystyle\sum_{k=1}^{N} t_k^2}$$

The second derivative is always positive.

6.5 BEST STRAIGHT LINE

We now turn to the problem of finding the best straight line through a set of data without the restriction that the line must pass through the origin. That is, given data $y_1, y_2, \ldots, y_N$ at points $t_1, t_2, \ldots, t_N$, respectively, we seek the line

$$y = a + bt \tag{6.7}$$

which minimizes the sum of squares of errors. We therefore require the value of *two* variables, the slope, b, and the intercept, a.

The error at each point is again defined as the difference between the measured value and the value predicted by the straight line

$$e_k = y_k - a - bt_k$$

Then the sum-of-squares error can be written

$$e^2 = \sum_{k=1}^{N} e_k{}^2 = \sum_{k=1}^{N} [y_k - a - bt_k]^2$$

$$= \sum_{k=1}^{N} [y_k{}^2 - 2ay_k + a^2 + 2abt_k - 2by_kt_k + b^2t_k{}^2]$$

$$= \sum_{k=1}^{N} y_k{}^2 - 2a \sum_{k=1}^{N} y_k + Na^2 + 2ab \sum_{k=1}^{N} t_k - 2b \sum_{k=1}^{N} y_kt_k + b^2 \sum_{k=1}^{N} t_k{}^2$$

$$(6.8)$$

(The term Na^2 results from $\sum_{k=1}^{N} a^2 = a^2 + a^2 + \cdots + a^2 = Na^2$.) The minimum of e^2 with respect to a and b occurs when the partial derivatives with respect to both a and b vanish simultaneously. Considering first the partial derivative with respect to a we have

$$\frac{\partial}{\partial a} e^2 = \frac{\partial}{\partial a}\left[\sum_{k=1}^{N} y_k{}^2\right] - \frac{\partial}{\partial a}\left[2a \sum_{k=1}^{N} y_k\right] + \frac{\partial}{\partial a}[Na^2] + \frac{\partial}{\partial a}\left[2ab \sum_{k=1}^{N} t_k\right] - \frac{\partial}{\partial a}\left[2b \sum_{k=1}^{N} y_kt_k\right]$$

$$\downarrow \qquad\qquad \downarrow \qquad\qquad \downarrow \qquad\qquad \downarrow \qquad\qquad \downarrow$$

$$= \qquad 0 \quad - \quad 2b\sum_{k=1}^{N} y_k \quad + \quad 2Na \quad + \quad 2b\sum_{k=1}^{N} t_k \quad - \qquad 0$$

$$+ \frac{\partial}{\partial a}\left[b^2 \sum_{k=1}^{N} t_k{}^2\right]$$

$$\downarrow$$

$$+ \qquad 0$$

In a similar way the partial derivative with respect to b is

$$\frac{\partial}{\partial b} e^2 = 2a \sum_{k=1}^{N} t_k - 2 \sum_{k=1}^{N} y_kt_k + 2b \sum_{k=1}^{N} t_k{}^2$$

Setting each of the partial derivatives to zero we then obtain the two equations

$$Na + \left[\sum_{k=1}^{N} t_k\right] b = \sum_{k=1}^{N} y_k$$

$$\left[\sum_{k=1}^{N} t_k\right] a + \left[\sum_{k=1}^{N} t_k{}^2\right] b = \sum_{k=1}^{N} y_kt_k$$

For computation it is somewhat more convenient to write these equations in a form in which the coefficients of b will be of similar size, as will the right-hand sides, by dividing by the coefficient of a:

$$a + \left[\frac{1}{N}\sum_{k=1}^{N} t_k\right] b = \frac{1}{N}\sum_{k=1}^{N} y_k \tag{6.9a}$$

$$a + \left[\frac{\displaystyle\sum_{k=1}^{N} t_k^2}{\displaystyle\sum_{k=1}^{N} t_k}\right] b = \frac{\displaystyle\sum_{k=1}^{N} y_k t_k}{\displaystyle\sum_{k=1}^{N} t_k} \tag{6.9b}$$

The best values of a and b are then found by simultaneous solution of Equations 6.9.

Example 6.3

As an example, consider the problem of representing the density of a sodium chloride-water solution as a linear function of the concentration of sodium chloride. The first two columns in Table 6.5 show data for the specific gravity

TABLE 6.5 Specific gravity (density relative to density of water at 4°C) of aqueous NaCl at 20°C. Computation of the best straight line using Equation 6.9

$y_k = \rho/\rho_{H_2O}$	c_k (lb$_m$/ft^3)	c_k^2	$y_k c_k$
1.0005	0.63	0.4	0.63
1.013	1.26	1.6	1.28
1.027	2.56	6.6	2.63
1.041	3.90	15.2	4.06
1.056	5.27	27.8	5.57
1.071	6.68	44.6	7.15
1.086	8.13	66.1	8.83
1.100	9.62	92.5	10.58
1.116	11.15	124.3	12.44
1.132	12.72	161.8	14.40
1.148	14.33	205.3	16.45
1.164	15.99	255.7	18.61
1.180	17.69	312.9	20.87
1.197	19.43	377.5	23.26
$\sum y_k = 15.332$	$\sum c_k = 129.36$	$\sum c_k^2 = 1692.3$	$\sum y_k c_k = 146.76$
$\frac{1}{N}\sum y_k = 1.095$	$\frac{1}{N}\sum c_k = 9.24$	$\dfrac{\sum c_k^2}{\sum c_k} = 13.08$	$\dfrac{\sum y_k c_k}{\sum c_k} = 1.135$

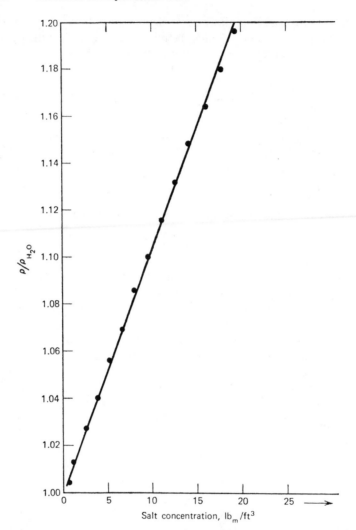

FIGURE 6.3 Specific gravity of sodium chloride-water solutions at 20°C versus sodium chloride concentration. $\rho_{H_2O} = 62.43 \ lb_m/ft^3$. The best straight line is computed from Equations 6.9.

of aqueous NaCl solutions at 20°C. (Specific gravity is the ratio of density to the density of pure water at 4°C. The density of pure water at 4°C is 1.000 gm/cm³, or 62.43 lb_m/ft^3, so the density of the solution in lb_m/ft^3 is obtained by multiplying the value in the first column by 62.43). The data are plotted in Figure 6.3. We denote the specific gravity, ρ/ρ_{H2O}, as y and replace the independent variable t by c, the concentration. According to Equations

6.9 the values of concentration squared and the product of concentration and specific gravity are needed. These are shown in the third and fourth columns of Table 6.5. Calculating the sums as shown in the table we then obtain the two equations

$$a + 9.24b = 1.095$$

$$a + 13.08b = 1.135$$

with solution $a = 0.999$, $b = 0.0104$. Thus the data are best fit by the equation

$$\rho/\rho_{H_2O} = 0.999 + 0.0104c$$

This is the line shown in Figure 6.3. Multiplying by the density of water at 4°C, 62.43 lb_m/ft^3, we obtain the density at 20°C in lb_m/ft^3

$$\rho = 62.37 + 0.649c$$

It should be noted that although the straight line is a good representation of the data, there is some curvature in the data and the deviation from the straight line is systematic, not random. Hence a function with curvature would do a better job on the data, though the systematic deviation here is not sufficient to warrant extra effort.

6.6 PHYSICAL PROPERTY CORRELATION

This is a useful point at which to degress briefly to discuss physical property correlations. We are frequently interested in retaining large amounts of information about physical properties of similar materials. It is sometimes possible to exploit some feature of the chemical constituents to represent data for similar materials by an equation. When this is possible the storage problem is reduced from large amounts of individual information to the few parameters in the equation.

As an example, consider the following list of boiling points of the first eight normal paraffin hydrocarbons.

Compound	Number of Carbon Atoms	Boiling Point, °C
Methane	1	−162
Ethane	2	−88
Propane	3	−42
n-Butane	4	1
n-Pentane	5	36
n-Hexane	6	69
n-Heptane	7	98
n-Octane	8	126

TABLE 6.6 Boiling point (BP) in °C of paraffin hydrocarbons with eight carbon atoms or fewer ($n \leq 8$). Computation of the best straight line using Equation 6.9 and the best quadratic using Equation 6.12

Compound	n	BP	n^2	$n \cdot BP$	$n^2 \cdot BP$	n^3	n^4
5584	1	−162	1	−162	−162	1	1
3488	2	−88	4	−176	−352	8	16
7073	3	−42	9	−126	−378	27	81
4950	4	−10	16	−40	−160	64	256
1998	4	1	16	4	16	64	256
2028	5	28	25	140	700	125	625
6481	5	36	25	180	900	125	625
7101	5	10	25	50	250	125	625
2018	6	50	36	300	1800	216	1296
2019	6	58	36	348	2088	216	1296
4556	6	69	36	414	2484	216	1296
6503	6	60	36	360	2160	216	1296
6504	6	64	36	384	2304	216	1296
2041	7	81	49	567	3969	343	2401
4458	7	98	49	686	4802	343	2401
4571	7	90	49	630	4410	343	2401
4572	7	92	49	644	4508	343	2401
6493	7	79	49	553	3871	343	2401
6494	7	90	49	630	4410	343	2401
6495	7	81	49	567	3969	343	2401
6496	7	86	49	602	4214	343	2401
6498	7	94	49	658	4606	343	2401
2040	8	107	64	856	6848	512	4096
4467	8	116	64	928	7424	512	4096
4468	8	122	64	976	7808	512	4096
4469	8	118	64	944	7552	512	4096
4564	8	114	64	912	7296	512	4096
4565	8	110	64	880	7040	512	4096
4566	8	108	64	864	6912	512	4096
4567	8	117	64	936	7488	512	4096
4569	8	119	64	952	7616	512	4096
6245	8	126	64	1008	8064	512	4096
6499	8	114	64	912	7296	512	4096
6500	8	119	64	952	7616	512	4096
6511	8	99	64	792	6336	512	4096

$$\Sigma\, n_k = 226 \quad \Sigma\, BP_k = 2354 \quad \Sigma\, n_k^2 = 1574 \quad \Sigma\, n_k BP_k = 19125 \quad \Sigma\, n_k^2 BP_k = 145705 \quad \Sigma\, n_k^2 = 11362 \quad \Sigma\, n_k^4 = 83822$$

$$\frac{1}{N}\Sigma\, n_k = 6.46 \qquad \frac{1}{N}\Sigma\, BP_k = 67.3 \qquad \frac{1}{N}\Sigma\, n_k^2 = 45.0$$

$$\frac{\Sigma\, n_k^2}{\Sigma\, n_k} = 6.96 \qquad \frac{\Sigma\, n_k BP_k}{\Sigma\, n_k} = 84.6 \qquad \frac{\Sigma\, n_k^3}{\Sigma\, n_k} = 50.3$$

$$\frac{\Sigma\, n_k^3}{\Sigma\, n_k^2} = 7.22 \qquad \frac{\Sigma\, n_k^2 BP_k}{\Sigma\, n^2} = 92.6 \qquad \frac{\Sigma\, n_k^4}{\Sigma\, n_k^2} = 53.3$$

We might expect to be able to develop an equation for boiling point in terms of the number of carbon atoms or, equivalently, molecular weight, which would be of sufficient accuracy that there would be no need to store all of the data. Furthermore, the spread between boiling points of the normal isomers exceeds the range of boiling points for isomers of the same molecular weight. As an example, consider the paraffins with six carbon atoms.

Compound	Boiling Point, °C
n-Hexane	69
3 Methyl pentane	64
2 Methyl pentane	60
2, 3 Dimethyl butane	58
2, 2 Dimethyl butane	50

Thus, with some moderate accuracy we might seek an equation relating boiling point to number of carbon atoms for all of the paraffins.

The boiling points for the 35 paraffins with 8 or fewer carbon atoms are listed in Table 6.6. The compound number refers to the location in the listing of organic compounds in the *Handbook of Chemistry and Physics*. n refers to the number of carbons and BP to the boiling point in degrees centigrade. We seek an equation of the form

$$BP = a + bn$$

Thus n plays the role of t and BP of y in the previous analysis. The sums needed for computation of a and b are evaluated in the second to fifth columns of Table 6.6, leading to the two equations

$$a + 6.46b = 67.3$$

$$a + 6.96b = 84.6$$

The solution is $a = -156.2$, $b = 34.6$ or

$$BP = -156.2 + 34.6n$$

The computed straight line is shown with the data on Figure 6.4. The correlation is a fairly good one, although it is evident that elimination of methane, whose chemical behavior is somewhat unique, would lead to improvement. The trend in the data clearly involves some curvature and we shall show subsequently how a quadratic might be fit to these data. An empirical correlation of the type developed here is common, for if we can

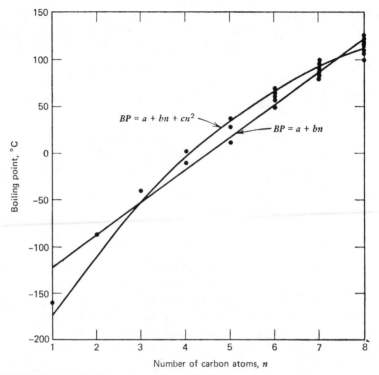

FIGURE 6.4   Boiling points of paraffin hydrocarbons as a function of the number of carbon atoms. The best linear and quadratic functions are computed using Equations 6.9 and 6.12, respectively.

tolerate the inaccuracies inherent in lumping the isomers together we can replace 35 data points by the two parameters in an equation.

6.7   FITTING A QUADRATIC

Before embarking on complete generality it is helpful to consider the problem of fitting a quadratic curve to data:

$$y = a + bt + ct^2 \qquad (6.10)$$

For given data $y_1, y_2, \ldots, y_N$ at points $t_1, t_2, \ldots, t_N$ we now seek the *three* parameters, a, b, and c, which minimize the sum of squares of errors.

The error at each point is the difference between the measured value and the value predicted by the line

$$e_k = y_k - a - bt_k - ct_k^2$$

The sum-of-squares error is

$$e^2 = \sum_{k=1}^{N} e_k^2 = \sum_{k=1}^{N} [y_k - a - bt_k - ct_k^2]^2 \qquad (6.11)$$

We minimize e^2 by taking partial derivatives with respect to each of the variables, a, b, and c, and setting the derivatives to zero:

$$\frac{\partial}{\partial a} e^2 = \frac{\partial}{\partial a} \sum_{k=1}^{N} [y_k - a - bt_k - ct_k^2]^2$$

$$= \sum_{k=1}^{N} \frac{\partial}{\partial a} [y_k - a - bt_k - ct_k^2]^2$$

$$= \sum_{k=1}^{N} 2[y_k - a - bt_k - ct_k^2] \frac{\partial}{\partial a} [y_k - a - bt_k - ct_k^2]$$

$$= \sum_{k=1}^{N} 2[y_k - a - bt_k - ct_k^2][-1]$$

$$= -2\left[\sum_{k=1}^{N} y_k - Na - b\sum_{k=1}^{N} t_k - c\sum_{k=1}^{N} t_k^2 \right] = 0$$

Similarly

$$\frac{\partial}{\partial b} e^2 = -2\left[\sum_{k=1}^{N} y_k t_k - a\sum_{k=1}^{N} t_k - b\sum_{k=1}^{N} t_k^2 - c\sum_{k=1}^{N} t_k^3 \right] = 0$$

$$\frac{\partial}{\partial c} e^2 = -2\left[\sum_{k=1}^{N} y_k t_k^2 - a\sum_{k=1}^{N} t_k^2 - b\sum_{k=1}^{N} t_k^3 - c\sum_{k=1}^{N} t_k^4 \right] = 0$$

or, rearranging to a form convenient for computation,

$$a + \left[\frac{1}{N} \sum_{k=1}^{N} t_k \right] b + \left[\frac{1}{N} \sum_{k=1}^{N} t_k^2 \right] c = \frac{1}{N} \sum_{k=1}^{N} y_k \tag{6.12a}$$

$$a + \left[\frac{\sum_{k=1}^{N} t_k^2}{\sum_{k=1}^{N} t_k} \right] b + \left[\frac{\sum_{k=1}^{N} t_k^3}{\sum_{k=1}^{N} t_k} \right] c = \frac{\sum_{k=1}^{N} y_k t_k}{\sum_{k=1}^{N} t_k} \tag{6.12b}$$

$$a + \left[\frac{\sum_{k=1}^{N} t_k^3}{\sum_{k=1}^{N} t_k^2} \right] b + \left[\frac{\sum_{k=1}^{N} t_k^4}{\sum_{k=1}^{N} t_k^2} \right] c = \frac{\sum_{k=1}^{N} y_k t_k^2}{\sum_{k=1}^{N} t_k^2} \tag{6.12c}$$

These three equations can now be solved for the three unknowns, a, b, and c. Comparison with Equations 6.9 indicates how the equations develop for additional terms in the polynomial.

Example 6.4

As an example of the calculation of coefficients in a quadratic we return to the boiling point correlation of the preceding section. Here we substitute BP for y and n for t. The necessary additional terms are shown in the final three columns of Table 6.6 and the sums are recorded at the bottom of the table. Equations 6.12 then become

$$a + 6.46b + 45.0c = 67.3$$

$$a + 6.96b + 50.3c = 84.6$$

$$a + 7.22b + 53.3c = 92.6$$

with solutions $a = -252.3$, $b = 77.9$, $c = -4.08$. The boiling point correlation is then

$$BP = -252.3 + 77.9n - 4.08n^2$$

This equation is the curved line plotted in Figure 6.4. That the quadratic represents some improvement over the linear function is verified by computing e from Equations 6.8 and 6.11 for the linear and quadratic fits, respectively. The results are

$$\text{linear:} \quad e^2 = \sum_{k=1}^{N} [y_k - a - bt_k]^2 = 4746$$

$$\text{quadratic:} \quad e^2 = \sum_{k=1}^{N} [y_k - a - bt_k - ct_k^2] = 3829$$

In this case the improvement is not particularly large. The question of when an improvement is adequate to justify use of the additional term requires application of some elementary statistical tests which are a consequence of the analytical properties of the sum-of-squares error. We shall not go into these here. For this example we might anticipate that the extra complexity associated with the quadratic function is not justified, since the difference between the linear and quadratic functions at each value of n is in most cases smaller than the total spread in the data.

One reason that we might wish such a correlation is to estimate unavailable physical properties. For example, we could now attempt to predict the boiling points of nine carbon paraffins. For $n = 9$ we obtain $BP = 118$ from the quadratic fit and $BP = 155$ from the linear fit compared to an experimental value for n-nonane, the highest boiler, of 150. Lest we be tempted to try to use an empirical correlation of this type too far from the range of data for which it was developed, however, we should note that the quadratic boiling point equation does underestimate BP for $n > 8$, predicts a maximum boiling point, and ultimately predicts boiling points below absolute zero.

6.8 LINEAR REGRESSION

The curve-fitting procedures that we have studied thus far are special cases of what is commonly known as *linear regression*. This name sometimes causes confusion, for clearly the quadratic applied in the previous section is not a linear function. Notice, though, that the *parameters a, b,* and *c* which we were choosing do appear linearly in that there are no products or powers of the parameters in the expression, and the parameters are found by solving a set of linear algebraic equations.

We can generalize somewhat and still retain the essential simplicity of the previous examples. First let us suppose that we wish to fit the data with an equation of the form

$$y = af(t) + bg(t) \tag{6.13}$$

where $f(t)$ and $g(t)$ are specified functions of t, such as powers, sines or cosines, exponentials, etc. This expression is also linear in a and b! Then for each data point

$$e_k{}^2 = [y_k - af(t_k) - bg(t_k)]^2$$

and the total error is

$$e^2 = \sum_{k=1}^{N} [y_k - af(t_k) - bg(t_k)]^2 \tag{6.14}$$

Then the minimum occurs at

$$\frac{\partial e^2}{\partial a} = -2 \sum_{k=1}^{N} f(t_k)[y_k - af(t_k) - bg(t_k)] = 0$$

$$\frac{\partial e^2}{\partial b} = -2 \sum_{k=1}^{N} g(t_k)[y_k - af(t_k) - bg(t_k)] = 0$$

or, rearranging,

$$a + \left[\frac{\sum_{k=1}^{N} f(t_k)g(t_k)}{\sum_{k=1}^{N} f(t_k)^2} \right] b = \frac{\sum_{k=1}^{N} f(t_k)y_k}{\sum_{k=1}^{N} f(t_k)^2} \tag{6.15a}$$

$$a + \left[\frac{\sum_{k=1}^{N} g(t_k)^2}{\sum_{k=1}^{N} f(t_k)g(t_k)} \right] b = \frac{\sum_{k=1}^{N} g(t_k)y_k}{\sum_{k=1}^{N} f(t_k)g(t_k)} \tag{6.15b}$$

Equation 6.6 is the special case obtained when $f(t) = 0$, $g(t) = t$, and Equations 6.9 are the special case in which $f(t) = 1\left(\sum_{k=1}^{n} 1 = N\right)$ and $g(t) = t$. Notice that we still obtain two linear algebraic equations for the two constants.

In general, we seek to find M constants for an equation of the form

$$y = a_1 f_1(t) + a_2 f_2(t) + \cdots + a_M f_M(t) \tag{6.16}$$

where $f_1(t), f_2(t), \ldots, f_M(t)$ are specified functions. Then:

$$e^2 = \sum_{k=1}^{N} [y_k - a_1 f_1(t_k) - a_2 f_2(t_k) - \cdots - a_M f_M(t_k)]^2 \tag{6.17}$$

$$\frac{\partial e^2}{\partial a_1} = -2 \sum_{k=1}^{N} f_1(t_k)[y_k - a_1 f_1(t_k) - a_2 f_2(t_k) - \cdots - a_M f_M(t_k)]$$

$$\frac{\partial e^2}{\partial a_2} = -2 \sum_{k=1}^{N} f_2(t_k)[y_k - a_1 f_1(t_k) - a_2 f_2(t_k) - \cdots - a_M f_M(t_k)]$$

etc.

In general, for any a_j, $j = 1, 2, \ldots, M$,

$$\frac{\partial e^2}{\partial a_j} = -2 \sum_{k=1}^{N} f_j(t_k) \, [y_k - a_1 f_1(t_k) - a_2 f_2(t_k) - \cdots - a_M f_M(t_k)]$$

$$= 0, \quad j = 1, 2, 3, \ldots, M$$

or

$$a_1 \left[\sum_{k=1}^{N} f_j(t_k) f_1(t_k) \right] + \cdots + a_M \left[\sum_{k=1}^{N} f_j(t_k) f_M(t_k) \right] = \sum_{k=1}^{N} f_j(t_k) y_k$$

$$j = 1, 2, \ldots, M$$

That is, defining M^2 quantities A_{ij}; $i = 1, 2, \ldots, M$; $j = 1, 2, \ldots, M$; by

$$A_{ij} = \sum_{k=1}^{N} f_i(t_k) f_j(t_k) \qquad \begin{matrix} i = 1, 2, \ldots, M \\ j = 1, 2, \ldots, M \end{matrix}$$

and M quantities b_i, $i = 1, 2, \ldots, M$ by

$$b_i = \sum_{k=1}^{N} f_i(t_k) y_k$$

we may write the equations defining $a_1, a_2, \ldots, a_M$ as

$$A_{11} a_1 + A_{12} a_2 + \cdots + A_{1M} a_M = b_1$$
$$A_{21} a_1 + A_{22} a_2 + \cdots + A_{2M} a_M = b_2$$
$$\vdots \qquad\qquad \vdots \qquad\qquad \vdots$$
$$A_{M1} a_1 + A_{M2} a_2 + \cdots + A_{MM} a_M = b_M$$

or, in compact notation,

$$\sum_{j=1}^{M} A_{ij}a_j = b_i, \qquad i = 1, 2, \ldots, M \tag{6.18}$$

Thus, in order to solve linear regression problems to obtain more than two or three coefficients it is necessary for us to have the facility to solve large numbers of simultaneous linear algebraic equations. This computational problem is considered in Section 16.2.

6.9 GRAPH PAPER

In analyzing data we have, on several occasions, plotted the logarithm of one quantity versus t. Special graph paper is available to facilitate this operation, known as "semilogarithmic" paper. Here, one axis is marked off in the logarithm (to the base ten) instead of in an arithmetic scale. An example is shown in Figure 6.5, where the 5 on the ordinate should be interpreted as *log 5*, or 0.697. Thus, the intermediate step of taking logarithms is avoided, and data may be plotted directly.

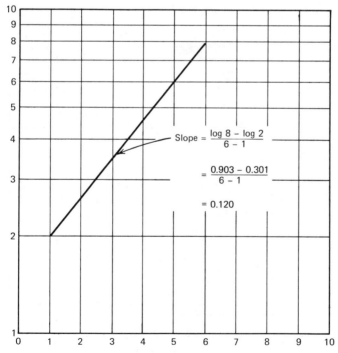

FIGURE 6.5 Semilogarithmic graph paper.

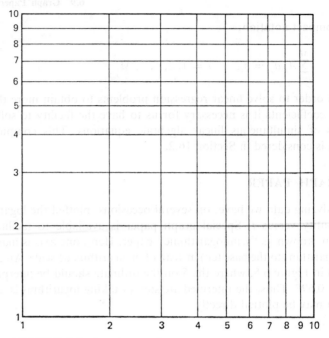

FIGURE 6.6   Log-log graph paper.

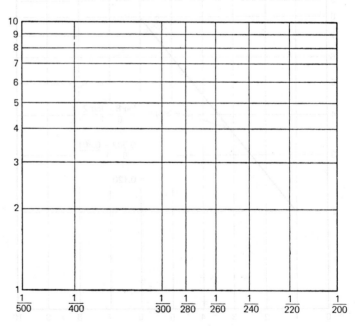

FIGURE 6.7   Log-inverse graph paper.

170

A word of warning is important! If a parameter is to be evaluated from the slope on a plot of log c versus t the difference in logarithms, not scale markings, must be used. This is demonstrated in Figure 6.5.

Several other types of paper are also available. For example, Figures 6.6 and 6.7 show, respectively, logarithm versus logarithm and logarithm versus inverse.

6.10 CONCLUDING REMARKS

The primary aim of this chapter is the development of a rational procedure for representing data by a function. It should be noted that the analytical techniques discussed here can be applied only when the function is linear in the parameters whose values are being calculated.

Correlations which represent physical property data in terms of a small number of parameters are widely used. The boiling point correlation used here as an example was chosen for convenience only and is by no means unique.

Statistical analysis plays an important role in determining how good a fit to the data the minimum squares line really is. This is discussed in texts such as

6.1 A. Hald, *Statistical Theory with Engineering Applications*, Wiley, New York, 1951.

6.2 D. M. Himmelblau, *Process Analysis by Statistical Methods*, Wiley, New York, 1970.

There is a brief summary in Section 2 of the Chemical Engineers' Handbook.

6.3 J. H. Perry, *Chemical Engineers' Handbook*, 4th ed., McGraw-Hill, New York, 1963.

6.11 PROBLEMS

6.1 You are given the following data:

x	1	2	3	4	5
y	5.5	9.0	17.0	17.0	25.5

Find the slope m of the straight line

$$y = mx$$

that is *best* in each of the following ways:
(a) minimizes the sum of squares of deviations
(b) minimizes the sum of absolute values of deviations

(c) minimizes the maximum absolute deviation (i.e., for each slope m one of the data points has the largest absolute deviation. Find the slope that makes this largest absolute deviation as small as possible.)

6.2 The following data for the density of ethanol-water mixtures at 20°C are taken from the *Chemical Engineers' Handbook*, 4th ed., p. 3–84:

% C_2H_5OH, by weight	0	5	10	15
Density, g/cm³	0.99823	0.98938	0.98187	0.97514

% C_2H_5OH, by weight	20	25	30	35
Density, g/cm³	0.96864	0.96168	0.95382	0.94494

Using the criterion of minimum sum of squares find the best straight line through the data in the form of Equation 4.24

$$\rho = \rho_0 + bc$$

where c is concentration of alcohol in moles/cm³
(a) when ρ_0 is taken as the density of pure water at 20°C, $\rho_0 = 0.99823$
(b) when the best values of both ρ_0 and b are sought and ρ_0 is simply an arbitrary intercept with no physical significance.
(Hint: the data must be manipulated, since density is given in terms of weight percent, not concentration.)

6.3 In Problem 5.6 data are given at 160°C for the rate constant k_1 for the reaction of cellulose to glucose as a function of HCl concentration, c_{HCl}. Find the best (minimum squares) values of k_1' and n in the relation

$$k_1 = k_1' c_{HCl}^n$$

6.4 Using the data in Table 2.2, obtain the best (minimum squares) value for the exponent, n, in the orifice equation $q = kh^n$. Take the value of $\ln h$ corresponding to a given value of $\ln [-\Delta h/\Delta t]$ as the average of the two end-point values.

6.5 Using the data in Table 5.3, discussed in Example 5.1, obtain the best (minimum squares) value for the order, n, of the reaction between sulfuric acid and diethyl sulfate in aqueous solution at 22.9°C. Take the value of $\ln c_A$ corresponding to a given value of $\ln [-\Delta c_A/\Delta t]$ as the average of the two end-point values.

6.6 Using the data in Table 5.4, discussed in Example 5.2, calculate the best (minimum squares) value of the rate constant, k, for the reaction between sodium ethoxide and ethyl dimethyl sulfonium iodide in solution in absolute ethanol.

6.7 Find the best (minimum squares) constants a, b, and c in the equation

$$\rho = a + bT + cT^2$$

for the following data for the temperature dependence of the density of 4 percent by weight $MgSO_4$ in water:

$T, {}^{\circ}C$	0	20	30	40	50	60	80
$\rho, g/cm^3$	1.0423	1.0392	1.0362	1.0326	1.0283	1.0234	1.0118

6.8 Heat capacity ($\underline{c}_p$) is sometimes correlated with absolute temperature using an equation

$$\underline{c}_p = a + bT + cT^{-2}$$

(a) For a given set of data, $\{\underline{c}_{pi}\}$, taken at temperatures $\{T_i\}$, obtain the equations for computing the best minimum squares values of a, b, and c.

(b) Apply the result to the following data for aluminum:

$T, {}^{\circ}C$	-200	-150	-100	-50	0
$\underline{c}_p$, cal/g${}^{\circ}C$	0.076	0.137	0.168	0.191	0.208

$T, {}^{\circ}C$	20	100	300	600
$\underline{c}_p$, cal/g${}^{\circ}C$	0.214	0.225	0.248	0.277

Notice that the data are given in ${}^{\circ}C$ and need to be converted to ${}^{\circ}K$.

6.9 Sometimes it is helpful to approximate a function $f(t)$, $0 < t < \pi$, by a *Fourier sine series*,

$$f(t) \simeq c_1 \sin t + c_2 \sin 2t + c_3 \sin 3t + \cdots$$

In the minimum square criterion the sum is over all values of t and becomes an integral,

$$e^2 = \int_0^{\pi} [f(t) - c_1 \sin t - c_2 \sin 2t - c_3 \sin 3t - \cdots]^2 \, dt$$

(a) Obtain the coefficients $c_1, c_2, c_3, \ldots$ which minimize e^2. Notice that a general relation for c_n can be derived.

(b) Calculate the first four terms for the function $f(t) = 1$ and plot. How good an approximation is obtained?

(c) Calculate and plot the first four terms for the function

$$f(t) = \begin{cases} t, & 0 < t < \pi/2 \\ \pi - t, & \pi/2 < t < \pi \end{cases}$$

CHAPTER 7

Design of Simple Reacting Liquid Systems

7.1 INTRODUCTION

In Chapter 5 we discussed rate expressions, their role in the model development step, and the batch reactor experiments needed to obtain a constitutive relationship between rate and concentrations for certain simple reactions. We found that it was necessary to formalize our techniques for comparing data with theoretical predictions and Chapter 6 was devoted to a discussion of this problem. We are now prepared to consider the final step in the analysis process, that of using tested mathematical descriptions for design and predictive purposes.

As we examine various problems in reactor design we will increase our skill in both the model development and model behavior step of the analysis process. In addition we gain some introductory experience in simple engineering design with problems that are fundamental to chemical engineering practice. We shall confine our attention to isothermal liquid systems at this stage to simplify the discussion. Such systems do exist and are of importance, both in the laboratory and the chemical processing industries. We shall also see as we progress to more complex problems in later chapters that the model equations for the simple systems considered here are always part of more complicated mathematical descriptions of reacting systems.

Reactors can be classified into two broad categories, tank type and tubular. We derived the general mathematical description for the tank-type reactor at the beginning of Chapter 5 for the single reaction $A + B \rightarrow nD$, the resulting model equations being Equations 5.7 to 5.10. The model equations for a tubular reactor will be developed later in this chapter.

7.2 MODEL EQUATIONS FOR TANK-TYPE REACTORS

In a tank-type reactor the vessel contents are always assumed to be well mixed. Model Equations 5.7 to 5.10 describe the general physical situation for the reaction $A + B \rightarrow nD$. We further classify the tank-type reactors by the way in which material flows to and from the reactor.

Batch Reactors. In a batch reactor [Figure 7.1(a)] there is no flow to or from the reactor. The raw materials are brought together at time $t = 0$ and the mixture is kept in the reactor until the desired conversion of reactants is achieved. The model equations for the batch reactor are presented as Equations 5.15 to 5.17 and we have shown, for various kinetic schemes, how concentrations vary with time. Batch reactors are used both for experimental study and for commercial production.

Semibatch Reactors. In a semibatch reactor [Figure 7.1(b)] one raw material is put in the reactor and the other reactant or reactants are charged to the reactor over some period of time. There is no flow from the reactor.

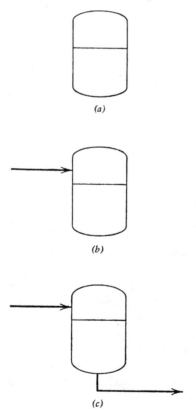

(a)

(b)

(c)

FIGURE 7.1(a) Batch reactor. No flow in or out. (b) Semibatch reactor. Flow in, no flow out. (c) Continuous flow stirred tank reactor.

The volume of the material in the reactor thus varies with time, and concentration changes with time because of reaction, volume variation, and flow to the reactor. The model equations for a semibatch reactor are developed by modifying the general mathematical description (Equations 5.7 to 5.10) for the particular situation. If, for example, we examine the problem in which an initial volume V_0 of material of concentration c_{B0} is put in the reactor and a stream of flow rate q_{Af} containing A at concentration c_{Af} is fed to the reactor starting at time $t = 0$ then $q = q_{Bf} = 0$ and the model equations are

$$\frac{dV}{dt} = q_{Af} \tag{7.1}$$

$$\frac{dVc_A}{dt} = q_{Af}c_{Af} - rV \tag{7.2}$$

$$\frac{dVc_B}{dt} = -rV \tag{7.3}$$

$$\frac{dVc_D}{dt} = +nrV \tag{7.4}$$

Continuous Flow Tank Reactors.   A continuous flow tank reactor [Figure 7.1(c)] is the only tank-type reactor that can be used in a continuous processing unit. These reactors are operated so that the flow to the reactor is equal to the flow from the reactor. The conversion of reactant to product is controlled by the average length of time that an element of fluid remains in the reactor. The model equations for the most general case are Equation 5.7 to 5.10. If the reactor is operating at the *steady state*, the time derivatives in these equations are equal to 0, $q_{Af} + q_{Bf} = q$, and the model equations are

$$0 = q_{Af} + q_{Bf} - q \tag{7.5}$$

$$0 = c_{Af}q_{Af} - c_A q - rV \tag{7.6}$$

$$0 = c_{Bf}q_{Bf} - c_B q - rV \tag{7.7}$$

$$0 = -c_D q + nrV \tag{7.8}$$

7.3   SEMIBATCH REACTORS

In both batch and semibatch reactors the concentration of any species is a function of time. We have seen in Chapter 5 how the behavior of the model equation is determined for batch systems. We now briefly consider the same problem for semibatch reactors. To initiate our discussion we again consider the following reaction and rate expression

$$A + B \rightarrow nD$$

$$r_{A-} = r_{B-} = \frac{1}{n} r_{D+} \equiv r = kc_A c_B$$

In a batch reactor we fill the reactor with the required amounts of A and B and allow the reaction to proceed to some desired degree of conversion of one of the raw materials. If the initial concentrations of A and B are equal, we found that the concentration of A at any time could be expressed by Equation 5.22

$$\frac{1}{c_A} = \frac{1}{c_{A0}} + kt \tag{5.22}$$

Equation 5.22 was obtained by integrating the component mass balance equation for A, after we used the chemical equation and the component mass balance for A and B to relate c_A and c_B. In the batch system we found that $c_A = c_B$ if c_{A0} and c_{B0} are the same (Equation 5.18). In a semibatch reactor the model behavior is not so readily obtained. If all of material B is put into the reactor and if the stream containing A is fed at a constant rate q_{Af} to the reactor over a period of time, model Equations 7.1 to 7.4 apply. We will assume as before, that the densities are constant.

Equation 7.1 can be integrated directly to yield V as a function of time

$$V = V_0 + q_{Af}t \tag{7.9}$$

Equations 7.2, 7.3, and 7.4, written with the assumed form for the reaction rate, become

$$\frac{dVc_A}{dt} = q_{Af}c_{Af} - Vkc_A c_B \tag{7.10}$$

$$\frac{dVc_B}{dt} = -Vkc_A c_B \tag{7.11}$$

$$\frac{dVc_D}{dt} = nVkc_A c_B \tag{7.12}$$

There are two major differences between these equations and the corresponding equations for a batch system. V is not a constant and must be retained inside the derivative on the left-hand side of the equations. The component mass balance equation for A (7.10) has an additional term accounting for the convective flow of A into the system.

Combination of Equations 7.2 and 7.3 or 7.10 and 7.11 leads to the relation

$$\frac{dVc_A}{dt} = \frac{dVc_B}{dt} + q_{Af}c_{Af}$$

which can be integrated for constant $q_{Af}c_{Af}$ to yield

$$Vc_A - V_0c_{A0} = Vc_B - V_0c_{B0} + q_{Af}c_{Af}t \qquad (7.13)$$

Here, V_0 is the volume at $t = 0$. Notice the contrast with Equation 5.18 for the batch reactor. Similarly from Equations 7.3 and 7.4 or 7.11 and 7.12,

$$V_0c_{D0} - Vc_D = nVc_B - nV_0c_{B0} \qquad (7.14)$$

The system behavior can then be expressed in terms of a single equation for c_B by using Equations 7.13 and 7.11:

$$\frac{dVc_B}{dt} = -kc_B[Vc_B + M + q_{Af}c_{Af}t] \qquad (7.15)$$

where V is defined by Equation 7.9 and $M = V_0[c_{A0} - c_{B0}]$.

Equation 7.15 can be solved by a trick similar to the one used to solve Equation 5.53 in Section 5.7.1. By setting $Vc_B = 1/y$ we obtain the equation

$$\frac{dy}{dt} = \frac{k[M + q_{Af}c_{Af}t]y}{V_0 + q_{Af}t} + \frac{k}{V_0 + q_{Af}t}$$

This is again of the form of Equation 15.25, the linear first-order equation, and an analytical solution is obtained which, following some algebra, can be expressed as

$$\frac{Vc_B}{V_0c_{B0}} = \left[\frac{V_0}{V}\right]^{\alpha-1} e^{-kc_{Af}t}\left\{1 + \frac{kc_{B0}V_0}{q_{Af}}\int_0^{[V/V_0]-1} \frac{e^{-[kc_{Af}V_0/q_{Af}]\xi}}{[1+\xi]^\alpha}d\xi\right\}^{-1} \qquad (7.16)$$

$$\alpha = \frac{k}{q_{Af}}[M - V_0c_{Af}] + 1$$

Equation 7.16 expresses the ratio of moles of B which are unreacted to the initial number of moles. We shall not discuss detailed behavior of this equation, but note that the conversion-time relation depends only on a number of parameters with units of inverse time, kc_{Af}, kc_{B0}, kc_{A0} (probably zero), and q/V_0. We can therefore design the system for given production requirements.

Semibatch reactors are used commercially only when there is some particular advantage to be gained by adding one reactant continuously to the other.

This can occur when there are several reactions taking place and the distribution of products can be regulated by adding reactant over a period of time or when there are severe thermal effects that can be regulated in this manner. The latter case requires consideration of the principle of energy conservation, but the analysis leading to Equation 7.16 forms a part of the total study.

7.4 CONTINUOUS FLOW STIRRED TANK REACTOR (CFSTR)

The continuous flow stirred tank reactor represents an extremely important class of reactor because it can be used in a continuous process. For the illustrative problem of the reaction $A + B \rightarrow nD$ the model equations for the tank reactor were obtained in Section 5.2 as

$$\frac{dV}{dt} = q_{Af} + q_{Bf} - q \tag{5.10}$$

$$\frac{dc_A V}{dt} = q_{Af}c_{Af} - qc_A - rV \tag{5.7}$$

$$\frac{dc_B V}{dt} = q_{Bf}c_{Bf} - qc_B - rV \tag{5.8}$$

$$\frac{dc_D V}{dt} = -qc_D + nrV \tag{5.9}$$

We need the complete model if we are concerned with transient behavior, as we are during start-up and shutdown of the process, or when considering control of the reactor. The design of the reactor, however, is carried out for *steady state* operation. At steady state, flows to and from the reactor are equal and there are no changes in level or concentration with time. Thus, all time derivatives are zero and Equations 5.7 to 5.10 reduce to Equations 7.5 to 7.8 given earlier:

$$0 = q_{Af} + q_{Bf} - q \tag{7.5}$$

$$0 = q_{Af}c_{Af} - qc_A - rV \tag{7.6}$$

$$0 = q_{Bf}c_{Bf} - qc_B - rV \tag{7.7}$$

$$0 = -qc_D + nrV \tag{7.8}$$

For definiteness we assume a specific form for the rate expression

$$r = kc_A c_B \tag{7.17}$$

Substituting the rate into Equations 7.6 to 7.9 and solving for q from Equation 7.5 we obtain the final working equations

$$0 = q_{Af}c_{Af} - [q_{Af} + q_{Bf}]c_A - kc_Ac_BV \tag{7.18}$$

$$0 = q_{Bf}c_{Bf} - [q_{Af} + q_{Bf}]c_B - kc_Ac_BV \tag{7.19}$$

$$0 = \qquad - [q_{Af} + q_{Bf}]c_D + nkc_Ac_BV \tag{7.20}$$

Taking n and k as given from other information, these three equations relate eight unknowns, c_{Af}, c_{Bf}, q_{Af}, q_{Bf}, V, c_A, c_B, and c_D. Thus, five values must be specified by other means, with the remaining three quantities calculated from Equations 7.18 to 7.20. The particular five quantities that are specified will depend on the problem to be solved. We might, for example, be given the reactor volume, flow rates, and feed composition and asked to calculate the composition of the exit stream. Or, we might be given feed and effluent compositions and be asked to calculate reactor volume and flow rates. In the several examples that follow we will use the numbers for the sulfuric acid-diethyl sulfate reaction from Examples 5.1 and 5.4. Thus, $n = 2$ and, assuming the reaction to be irreversible, $k = 6.05 \times 10^{-4}$ liter/g-mole min. For some calculations it is helpful to note that Equations 7.18 to 7.20 can be rearranged to yield the equivalent set of three equations consisting of Equation 7.18 and the two following:

$$c_A = c_B + \frac{q_{Af}c_{Af}}{q_{Af} + q_{Bf}} - \frac{q_{Bf}c_{Bf}}{q_{Af} + q_{Bf}} \tag{7.21}$$

$$c_D = -nc_B + \frac{nq_{Bf}c_{Bf}}{q_{Af} + q_{Bf}} \tag{7.22}$$

Example 7.1

Given $V = 25.4$ liters $\qquad q_{Af} = q_{Bf} = 0.1$ liters/min

$$c_{Af} = c_{Bf} = 11 \text{ g-moles/liter}$$

Then

$$\text{reactor holding time} \equiv \theta = \frac{V}{q_{Af} + q_{Bf}} = \frac{25.4}{0.1 + 0.1} = 127 \text{ min}$$

Equation 7.21: $\quad c_A = c_B + \dfrac{0.1 \times 11}{0.1 + 0.1} - \dfrac{0.1 \times 11}{0.1 + 0.1} = c_B$

Equation 7.22: $\quad c_D = -2c_B + \dfrac{2 \times 0.1 \times 11}{0.1 + 0.1} = 11 - 2c_B$

Thus, Equation 7.18 becomes

$$0 = 0.1 \times 11 - [0.1 + 0.1]c_A - 6.05 \times 10^{-4}c_A^2 \times 25.4$$

The quadratic has a solution

$$c_A = c_B = 4.15 \text{ g-moles/liter}$$

$$c_D = 2.70$$

where the extraneous negative root for the quadratic has been discarded. It is left as an exercise to show that the reaction can really be taken as approximately irreversible under the circumstances of the calculation.

Example 7.2

Given $V = 25.4$ $q_{Af} = q_{Bf} = 0.2$ $c_{Af} = c_{Bf} = 11$

This example is the same as the previous one except that the flow rate is doubled. Then, as above,

$$\theta = \frac{V}{q_{Af} + q_{Bf}} = 63.5 \text{ min}$$

$$c_A = c_B$$

$$c_D = 11 - 2c_B$$

The quadratic for c_A, Equation 7.18, is now

$$0 = 0.2 \times 11 - [0.2 + 0.2]c_A - 6.05 \times 10^{-4}c_A^2 \times 25.4$$

$$c_A = c_B = 4.70 \text{ g-moles/liter}$$

$$c_D = 1.60$$

This calculation can be easily repeated for various flow rates. Table 7.1

TABLE 7.1   Concentration in CFSTR for changing flow rate

$q_{Af} + q_{Bf}$	$\theta = V/q$	c_A	c_D
0.2	127	4.15	2.70
0.3	84.7	4.45	2.10
0.4	63.5	4.70	1.60
0.5	50.9	4.80	1.40
0.6	42.8	4.90	1.20
0.8	31.8	5.00	1.00
1.0	25.4	5.05	0.90

shows how c_A is affected by changing the flow $q_{Af} + q_{Bf}$ for $q_{Af} = q_{Bf}$, $c_{Af} = c_{Bf} = 11$, $V = 25.4$. As we would expect, the amount of sulfuric acid converted to product decreases as the reactor holding time, θ, decreases. Examination of the equations shows that the same effect is obtained if the flow rates are held constant and the feed concentrations of A and B are decreased.

Example 7.3

Given $V = 25.4$ $\qquad q_{Af} = q_{Bf}$ $\qquad c_{Af} = 11.0$ $\qquad c_{Bf} = 5.5$ $\qquad c_A = 4.0$

Here we have specified the desired conversion of sulfuric acid (A) and the feed concentrations. We desire the product concentration, c_D, and the flow rates. Notice that $q_{Af} = q_{Bf}$ is only one relation, since we have not specified the value.

We first note that c_B and c_D can be computed from Equations 7.21 and 7.22 without any knowledge of the total flow as long as the ratio q_{Af}/q_{Bf} is known:

$$\text{Equation 7.21:} \quad c_B = c_A + \frac{c_{Bf}}{1 + q_{Af}/q_{Bf}} - \frac{c_{Af}}{1 + q_{Bf}/q_{Af}}$$

$$= 1.25 \text{ g-moles/liter}$$

$$\text{Equation 7.22:} \quad c_D = -2c_B + \frac{2c_{Bf}}{1 + q_{Af}/q_{Bf}} = 3.0$$

The flow rate, $q_{Af} + q_{Bf}$, is then computed easily from Equation 7.20,

$$q_{Af} + q_{Bf} = \frac{nkc_Ac_BV}{c_D} = 0.0512 \text{ liters/min}$$

$$q_{Af} = q_{Bf} = 0.0256 \text{ liters/min}$$

The holding time is

$$\theta = \frac{V}{q_{Af} + q_{Bf}} = \frac{25.4}{0.0512} = 496 \text{ min}$$

As the examples illustrate, we may solve steady state CFSTR problems by manipulation of the set of algebraic equations that make up the mathematical description. For the system used in the examples the specification of five quantities allows the remaining three to be calculated. We cannot, of course, specify any five quantities arbitrarily, since both the physical situation and the model equations require that each component balance must be satisfied. If, for example, we have the situation where $q_{Af} = q_{Bf}$ and c_{Af}, c_A, and V are specified, we are not free to specify c_D. This quantity is known from Equation

7.20 once the other variables are selected. In order to calculate q_{Af} or q_{Bf} we must have an additional piece of information, which in this case is the value of c_{Bf} or c_B.

7.5 AN "OPTIMAL" DESIGN

A very common design problem is the determination of flow rates and reactor size to produce a certain concentration of product, D. The variables that must be calculated are thus q_{Af}, q_{Bf}, and V. The quantities n, k, c_{Af}, c_{Bf}, c_A, c_B, and c_D must be known to solve the problem. It is not our purpose to consider the process synthesis problem in detail, but we can indicate in a general way how the needed quantities are obtained. For a specific reaction n is known and k is computed from the experimental batch analysis which is performed to verify the rate expression. The inlet concentrations c_{Af} and c_{Bf} are specified by consideration of the raw material sources. c_A, c_B, and c_D are determined by a consideration of the economics of the process, which requires an evaluation of the downstream processing equipment as well as the reactor.

We can illustrate the essential features of this design problem by considering a simpler reacting system that allows us to concentrate on the engineering aspects of the design instead of the algebraic manipulation. Motivated by the discussion in Section 5.5, we will consider a first-order decomposition reaction that has the following chemical equation and experimentally verified rate expression:

$$A \to D$$

$$r_{A-} = r_{D+} = kc_A$$

For purposes of computation we will take $k = 0.005$ min^{-1}.

The mathematical description for this situation is readily derived using Equations 7.5 to 7.8, noting that $q_{Bf} = 0$, $q_{Af} = q$:

$$0 = qc_{Af} - qc_A - kc_A V \tag{7.23}$$

$$0 = -qc_D + kc_A V \tag{7.24}$$

These equations can be rearranged to yield an expression for the reactor volume, V, and the holding time, θ:

$$V = \frac{q[c_{Af} - c_A]}{kc_A} = \frac{qc_D}{kc_A} \tag{7.25}$$

$$\theta = \frac{c_{Af} - c_A}{kc_A} = \frac{c_D}{kc_A} \tag{7.26}$$

c_D and c_A are related by the following expressions

$$qc_D = q[c_{Af} - c_A] \qquad (7.27)$$

$$c_D = c_{Af} - c_A \qquad (7.28)$$

In this system of two equations we have six quantities, c_{Af}, c_A, c_D, k, V, and q. The specification of any four allows the system behavior to be determined. Since k is known for the reaction under consideration we can determine the behavior by specifying three of the remaining quantities. The following examples serve as introductory exercises to the optimal design problem.

Example 7.4

$$c_{Af} = 0.2 \text{ g-moles/liter}$$

$$qc_D = 50 \text{ g-moles/min}$$

In this problem the amount of D needed to meet market demand is known ($qc_D = 50$ g-moles/min) and the raw material is available at a given concentration. We need to determine how the reactor size and conversion vary with flow rate

(a) $q = 100 \text{ liters/min}$

$$c_D = 50/q = 0.5 \text{ g-moles/liter}$$

From Equation 7.28

$$c_D = 0.2 - c_A$$

$$c_A = 0.2 - 0.5 = -0.3 \text{ g-moles/liter}$$

Since c_A is negative we find we cannot meet our product demand at this flow rate.

(b) $q = 1000$

$$c_D = 50/q = 0.05$$

Then

$$c_A = 0.2 - 0.05 = 0.15 \text{ g-moles/liter}$$

$$V = \frac{50}{0.005 \times 0.15} = 65,600 \text{ liters}$$

We can meet our product demand but our conversion of A is low. We can decrease c_A by decreasing q.

(c) $q = 500$

$$c_D = 50/q = 0.10$$

Then

$$c_A = 0.10 \text{ g-moles/liter}$$

$$V = \frac{50}{0.005 \times 0.10} = 100,000 \text{ liters}$$

A series of such calculations yields the results shown in Table 7.2. All of the

TABLE 7.2   Volume and conversion in a CFSTR for a given production requirement

q (liters/minute)	c_D $\left(\dfrac{\text{gram-moles}}{\text{liter}}\right)$	c_A $\left(\dfrac{\text{gram-moles}}{\text{liter}}\right)$	V (liters)	$\theta = V/q$ (minutes)
250	0.200	0	∞	∞
300	0.167	0.033	303,000	1000
400	0.125	0.075	133,000	333
500	0.100	0.100	100,000	200
800	0.0625	0.1375	72,600	90.7
1000	0.0500	0.1500	66,600	65
2000	0.0250	0.1750	58,100	27
4000	0.0125	0.1875	53,200	13.3

flow rates over 250 liters per minute will produce the required amount of D and we need additional information if we are to seek an optimal reactor size. The following example illustrates the basic concepts of an optimal design using this simple reaction.

Example 7.5

The raw material A is available as a saturated solution containing 0.2 g-moles of A per liter. A and D can be separated in a single distillation column that produces a stream containing A overhead and a bottom product of the desired material, D. A simple process flow sheet is shown as Figure 7.2. The cost of recovering A from the overhead stream and reusing it as a raw material is a separate economic problem which can be solved independently of the reactor design problem. If recovery costs are lower than the cost of A from its original source, it is worthwhile to recover it. If recovery costs are in excess of what it costs to purchase A, it is not worthwhile to build and operate a recovery unit unless problems of disposal can be assigned an economic value. The cost of A is \$0.20 per g-mole and 50 g-moles/min of D must be produced to meet market demands for the product.

Qualitatively, the economic problem is an obvious one. If we build a large reactor which gives high conversion of A, the raw material cost is lower than

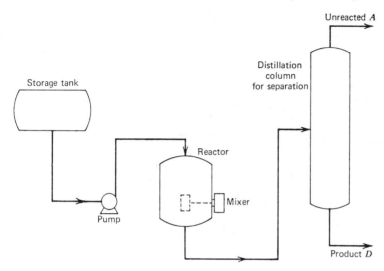

FIGURE 7.2   Process flow diagram for the reaction $A \rightarrow D$.

if we build a smaller reactor giving lower conversion of A. It costs more, however, to build the larger reactor, and we must find some way of selecting the volume that will give us the most economical balance between these competing design considerations. One criterion we can use is the yearly profit, defined as follows:

$\mathscr{P}$ = Yearly Profit = INCOME FROM THE SALE OF D — COST OF THE RAW MATERIAL A — CAPITAL COST OF THE REACTOR DIVIDED BY ITS EXPECTED LIFE IN YEARS (DEPRECIATION) — OPERATING COST OF THE REACTING SYSTEM — CAPITAL COST OF THE DISTILLATION COLUMN AND MISCELLAN- EOUS EQUIPMENT DIVIDED BY ITS EXPECTED LIFE IN YEARS (DEPRECIATION) — OPERATING COST OF THE DISTILLATION COLUMN AND ITS ATTENDANT EQUIP- MENT. (7.29)

In the discussion that follows we will make some overly simple assump- tions about the form of capital and operating costs, depreciation, and so on. More realistic economic equations would not change the approach at all, although the resulting algebra would be much more complicated and the simple qualitative conclusions more difficult to draw. A consequence of the assumptions is that it will be possible to express the profit entirely in terms of one variable, the concentration of A leaving the reactor, c_A, and hence we will find the optimum by setting $d\mathscr{P}/dc_A$ to zero. In fact, many real design problems *do* reduce in practice to consideration of one variable only, though rarely with the simple analytical results obtained here.

We can examine the terms in the profit equation one at a time to see how they depend on c_A. The income from the sale of D is a fixed number, since the production rate of D is specified. The derivative of this term will be zero. The cost of raw material A is determined as follows, assuming a 350-day operating year:

$$\text{cost of } A, \text{\$/min} = 0.20\, qc_{Af}$$

$$\text{cost of } A, \text{\$/year} = 0.20 \times 350 \times 24 \times 60\, qc_{Af} = S_A qc_{Af}$$

where $S_A = 1.008 \times 10^5$ \$ min/mole year. From Equation 7.27

$$q = \frac{qc_D}{c_{Af} - c_A} = \frac{50}{c_{Af} - c_A}$$

Thus

$$\text{cost of } A, \text{\$/year} = \frac{50 S_A c_{Af}}{c_{Af} - c_A} \tag{7.30}$$

The capital costs of the process equipment will consist of two parts, one depending on the required capacity and the other fixed. The derivative of the fixed part will be zero and can be neglected. For the reactor we will take the variable capital cost as proportional to the volume

$$\text{capital cost of reactor} = S_V V$$

where S_V has units \$/liter. This proportional relation is a reasonable first approximation, though it represents a gross simplification of true cost relations. From Equations 7.25

$$\text{capital cost of reactor} = \frac{S_V qc_D}{kc_A} = \frac{50 S_V}{0.005 c_A} = \frac{10^4 S_V}{c_A}$$

For purposes of depreciation we will assume a useful life of ten years and (unrealistically) depreciate the cost evenly over the life, providing the term for the profit equation,

$$\text{reactor depreciation, \$/year} = \frac{\text{capital cost of reactor}}{10} = \frac{10^3 S_V}{c_A}$$

$$\tag{7.31}$$

The variable capital cost of the distillation column and other miscellaneous equipment will depend on the throughput of material. As an approximation we take the cost to be proportional to the flow. Thus

$$\text{capital cost of distillation column} = S_D q = \frac{50 S_D}{c_{Af} - c_A}$$

Here, S_D has units $ min/liter and we have applied Equation 7.24 for q. Again taking a useful life of ten years for depreciation we obtain

$$\text{distillation column depreciation, \$/year} = \frac{\text{capital cost}}{10}$$

$$= \frac{5 S_D}{c_{Af} - c_A} \qquad (7.32)$$

The operating costs for the reactor and its attendant equipment and the distillation column and its attendant equipment are made up of fixed costs and variable costs depending on process throughput. We take the variable operating costs as approximately proportional to the throughput

$$\text{operating costs} = S_0 q = \frac{50 S_0}{c_{Af} - c_A} \qquad (7.33)$$

Here S_0 has units $ min/year liter.

The yearly profit, Equation 7.29, can now be expressed in terms of the results of Equations 7.30 to 7.33,

$$\mathscr{P} = \text{fixed terms} - \frac{50 S_A c_{Af}}{c_{Af} - c_A} - \frac{10^3 S_V}{c_A} - \frac{5 S_D}{c_{Af} - c_A} - \frac{50 S_0}{c_{Af} - c_A}$$

$$= \text{fixed terms} - \frac{50 S_A c_{Af} + 5 S_D + 50 S_0}{c_{Af} - c_A} - \frac{10^3 S_V}{c_A} \qquad (7.34)$$

This depends only on c_A and known quantities. The maximum profit is then found by setting the derivative of $\mathscr{P}$ with respect to c_A to zero,

$$\frac{d\mathscr{P}}{dc_A} = - \frac{50 S_A c_{Af} + 5 S_D + 50 S_0}{[c_{Af} - c_A]^2} + \frac{10^3 S_V}{c_A^2}$$

$$= - \frac{M_1}{[c_{Af} - c_A]^2} + \frac{M_2}{c_A^2} = 0 \qquad (7.35)$$

The constant M_1 refers to costs associated with raw material and throughput, M_2 with reactor capital unit costs. Equation 7.35 can be conveniently rearranged to

$$c_A{}^2 - \frac{2c_{Af}M_2}{M_2 - M_1}\, c_A + \frac{M_2 c_{Af}^2}{M_2 - M_1} = 0 \tag{7.36}$$

M_1 and M_2 contain S_D, S_V, and S_0, which can only be determined by a detailed cost analysis, beyond the scope of this text. We can illustrate the effect of the cost analysis, however, without using the actual numbers. Let $M_2/[M_2 - M_1]$ be denoted as β for convenience, and note that in order for M_1 and M_2 to both be positive it is possible for β to take on all values between $-\infty$ and $+\infty$ *except* values between zero and one. For fixed c_{Af} and fixed unit costs, however, β is a constant for the system in question. Also, to simplify manipulation we can define $x_A = c_A/c_{Af}$, $0 \le x_A \le 1$. x_A is one minus the fractional conversion of A. Then Equation 7.36 becomes

$$x_A{}^2 - 2\beta x_A + \beta = 0 \tag{7.37}$$

with solution

$$x_A = \beta + [\beta^2 - \beta]^{1/2} = \beta\left\{1 + \left[1 - \frac{1}{\beta}\right]^{1/2}\right\} \tag{7.38}$$

TABLE 7.3   Fractional conversion and optimal parameters in a CFSTR for various values of the cost parameter, β

β	x_A	$c_A = 0.2x_A$	V	q	$\theta = V/q$
1.0	1.0	0.200	50,000	∞	0
1.2	0.813	0.163	61,400	1060	58
1.5	0.634	0.127	80,000	685	117
2.0	0.586	0.117	85,000	605	140
5.0	0.527	0.106	94,500	532	177
10.0	0.513	0.103	97,800	516	189
100	0.501	0.100	100,000	500	200
1000	0.500	0.100	100,000	500	200
-1000	0.500	0.100	100,000	500	200
-10.0	0.488	0.098	102,000	486	210
-1.0	0.414	0.083	120,000	424	283
-0.50	0.366	0.073	137,000	396	346
-0.10	0.232	0.046	217,000	327	662
-0.05	0.179	0.036	278,000	303	917
-0.01	0.090	0.018	556,000	275	2020
0	0	0	∞	250	∞

The negative square root is required to obtain a solution between zero and unity for $\beta > 1$ and the positive square root for $\beta < 0$.

The solution to Equation 7.38 and reactor design parameters are shown in Table 7.3 for selected values of β. Notice that the behavior expressed by the table can be anticipated qualitatively. If reactor capital costs greatly exceed raw material and throughput costs ($M_2 \gg M_1$, $\beta \to 1$) then we would meet production requirements by putting large amounts of material through as small a reactor as possible and accepting negligible conversion ($x_A \to 1$). At the other extreme, large raw material and throughput costs relative to reactor capital cost ($M_1 \gg M_2$, $\beta \to 0$), we would use as large a reactor as possible and attempt to operate close to complete conversion ($x_A \to 0$). In the intermediate limiting case of nearly equal costs ($M_2 \simeq M_1$, $\beta \to \infty$) the system is operated at 50 percent conversion ($x_A \to 1/2$, $c_A \to 0.1$).

7.6 ETHYLENE GLYCOL MANUFACTURE

Because the model equations for the steady state continuous flow stirred tank reactor are algebraic, it is possible to deal with quite complex chemical reactions and still obtain needed design information. We consider here a situation in which several reactions occur and the distribution of product plays a key role in the process synthesis and reactor design.

The following chemical equations and rate expressions describe a class of complex reactions that accounts for about two billion pounds per year of chemical production in the United States:

$$A + B \to R$$

$$R + B \to S$$

$$S + B \to T$$

$$r_{A-} = r_{R+} = k_1 c_A c_B \tag{7.39a}$$

$$r_{R-} = r_{S+} = k_2 c_R c_B \tag{7.39b}$$

$$r_{S-} = r_{T+} = k_3 c_S c_B \tag{7.39c}$$

$$r_{B-} = r_{A-} + r_{R-} + r_{S-} = c_B[k_1 c_A + k_2 c_R + k_3 c_S] \tag{7.39d}$$

In Chapter 1 we briefly examined the reaction of water (A) and ethylene oxide (B) to form ethylene glycol (R), diglycol (S), and triglycol (T), and we shall return to this important specific case shortly. A number of other reactions that fall into this general category are listed in Table 7.4.

The model equations for the steady state CFSTR are obtained as outlined in our previous discussions. We will assume that there is no product present in the feed to the reactor and that the raw materials, A and B, are mixed together just before introduction to the reactor and enter in a single stream.

TABLE 7.4 Industrially important complex reactions

Reactants		Products		
A	B	R	S	T
Water	Ethylene oxide*	Ethylene glycol	Diethylene glycol	Triethylene glycol
Ammonia	Ethylene oxide*	Monoethanol-amine	Diethanolamine	Triethanolamine
Methyl, ethyl, or butyl alcohol	Ethylene oxide*	Monoglycol ether	Diglycol ether	Triglycol ether
* The same sets of reactions are also carried out using propylene oxide.				
Benzene	Chlorine	Monochloro-benzene	Dichlorobenzene	Trichlorobenzene
Methane	Chlorine	Methyl chloride	Dichloromethane	Trichloromethane

The steady state equations for conservation of mass of each species are then as follows:

$$A: \quad 0 = q[c_{Af} - c_A] - r_{A-}V \tag{7.40a}$$

$$B: \quad 0 = q[c_{Bf} - c_B] - r_{B-}V \tag{7.41a}$$

$$R: \quad 0 = -qc_R + r_{R+}V - r_{R-}V \tag{7.42a}$$

$$S: \quad 0 = -qc_S + r_{S+}V - r_{S-}V \tag{7.43a}$$

$$T: \quad 0 = -qc_T + r_{T+}V \tag{7.44a}$$

On substitution of Equation 7.39 for the rates these can be rewritten

$$0 = q[c_{Af} - c_A] - k_1 c_A c_B V \tag{7.40b}$$

$$0 = q[c_{Bf} - c_B] - c_B[k_1 c_A + k_2 c_R + k_3 c_S]V \tag{7.41b}$$

$$0 = -qc_R + c_B[k_1 c_A - k_2 c_R]V \tag{7.42b}$$

$$0 = -qc_S + c_B[k_2 c_R - k_3 c_S]V \tag{7.43b}$$

$$0 = -qc_T + k_3 c_S c_B V \tag{7.44b}$$

We have factored out c_B from the rate expressions in order to make some of the subsequent manipulations more apparent. The fact that c_B appears in each rate expression simplifies the analysis.

The concentrations in the effluent can be obtained in terms of one minus the fractional conversion of c_A

$$x_A = \frac{c_A}{c_{Af}}$$

For example, Equation 7.42b can be solved for c_B

$$c_B = \frac{q c_R}{[k_1 c_A - k_2 c_R] V}$$

which yields, on substitution into Equation 7.40b,

$$0 = q[c_{Af} - c_A] - k_1 c_A \frac{q c_R}{[k_1 c_A - k_2 c_R] V} V$$

or, on rearranging,

$$\frac{c_R}{c_{Af}} = \frac{x_A[1 - x_A]}{x_A + [k_2/k_1][1 - x_A]} \tag{7.45}$$

Similarly, Equations 7.43b and 7.44b can be solved for c_B in turn and substituted into Equation 7.40b to yield, respectively,

$$\frac{c_S}{c_{Af}} = \frac{[k_2/k_1][1 - x_A]^2 x_A}{\{x_A + [k_2/k_1][1 - x_A]\}\{x_A + [k_3/k_1][1 - x_A]\}} \tag{7.46}$$

$$\frac{c_T}{c_{Af}} = \frac{[k_2/k_1][k_3/k_1][1 - x_A]^3}{\{x_A + [k_2/k_1][1 - x_A]\}\{x_A + [k_3/k_1][1 - x_A]\}} \tag{7.47}$$

Notice that the ratios c_R/c_{Af}, c_S/c_{Af}, and c_T/c_{Af} are independent of q and V and depend only on fractional conversion of A and the ratios of rate constants, k_2/k_1 and k_3/k_1. Thus, the distribution of product, $c_R:c_S:c_T$, for given rate constants in a particular chemical system depends only on the fractional conversion of A. Furthermore, since the product distribution $c_R:c_S:c_T$ depends on the single quantity x_A, $0 < x_A < 1$, *only a limited class of product distributions is possible.*

Finally, we can obtain the relation between c_B and x_A by rewriting Equation 7.41b as follows:

$$0 = q[c_{Bf} - c_B] - c_B[k_1 c_A - k_2 c_R]V$$

$$- 2c_B[k_2 c_R - k_3 c_S]V - 3k_3 c_S c_B V$$

Substitution of Equations 7.42b to 7.44b then yields

$$c_{Bf} - c_B = c_R + 2c_S + 3c_T$$

or, dividing by c_{Af},

$$\frac{c_{Bf}}{c_{Af}} - \frac{c_B}{c_{Af}} = \frac{c_R}{c_{Af}} + 2\frac{c_S}{c_{Af}} + 3\frac{c_T}{c_{Af}} \qquad (7.48)$$

where the right-hand side is expressed in terms of x_A by means of Equations 7.45 to 7.47.

With the information we have now generated about the system we can prepare a preliminary block flow diagram for the process. It is obvious that we need a reactor and it is equally obvious that without further information restricting our design a system for separating A and B from the products R, S, and T must be included downstream from the reactor. We would probably recycle the recovered A and B back to inlet but, as discussed in the previous example, the decision as to whether we do this depends on the cost of the new materials. As the final step in the process, some means of producing each product must be considered. The block process flow diagram is shown in Figure 7.3.

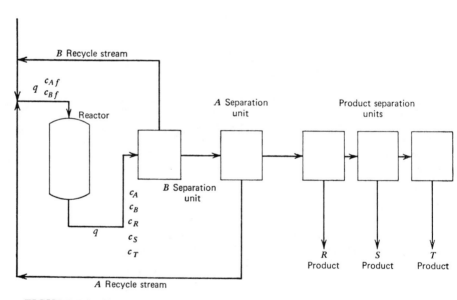

FIGURE 7.3 Simplified block flow diagram for reactions $A + B \rightarrow R$, $R + B \rightarrow S$, $S + B \rightarrow T$.

To illustrate the design aspects of this problem we use as a more specific example the reaction of ethylene oxide and water, which produces the mono-, di-, and triethylene glycols. In our notation A is the water, B is the ethylene oxide, R is the monoethylene glycol, S is the diethylene glycol, and T is the triethylene glycol. This reaction, discussed previously in Chapter 1, is represented in chemical symbols as follows:

$$\underset{O}{\overset{\text{H}\qquad\text{H}}{\text{HC}\!\!-\!\!-\!\!-\!\!\text{CH}}} + \text{HOH} \longrightarrow \text{OH}\!-\!\overset{\text{H}}{\underset{\text{H}}{\text{C}}}\!-\!\overset{\text{H}}{\underset{\text{H}}{\text{C}}}\!-\!\text{OH}$$

$$\text{OH}\!-\!\overset{\text{H}}{\underset{\text{H}}{\text{C}}}\!-\!\overset{\text{H}}{\underset{\text{H}}{\text{C}}}\!-\!\text{OH} + \underset{O}{\overset{\text{H}\qquad\text{H}}{\text{HC}\!-\!-\!-\!\text{CH}}} \longrightarrow \text{O}\!\!\underset{\text{CH}_2\text{CH}_2\text{OH}}{\overset{\text{CH}_2\text{CH}_2\text{OH}}{}}$$

$$\text{O}\!\!\underset{\text{CH}_2\text{CH}_2\text{OH}}{\overset{\text{CH}_2\text{CH}_2\text{OH}}{}} + \underset{O}{\overset{\text{H}\qquad\text{H}}{\text{HC}\!-\!-\!-\!\text{CH}}} \longrightarrow \begin{array}{l}\text{CH}_2\!-\!\text{O}\!-\!\text{CH}_2\text{CH}_2\text{OH}\\ |\\ \text{CH}_2\!-\!\text{O}\!-\!\text{CH}_2\text{CH}_2\text{OH}\end{array}$$

This reaction has been thoroughly investigated in the laboratory and the values of rate constants have been determined by a number of investigators. k_1 is found to be 7.37×10^{-7} liters/g-mole min at 25°C, and within the experimental error the ratios of rate constants are

$$\frac{k_2}{k_1} = \frac{k_3}{k_1} = 2.0$$

The product distribution, expressed as mass fraction of total glycol, is shown

TABLE 7.5　Product distribution of glycols in a CFSTR for various values of the fractional conversion

x_A	R	S	T	$\dfrac{c_{Bf}}{c_{Af}} - \dfrac{c_B}{c_{Af}}$
0.9900	0.9663	0.0327	0.0009	0.0102
0.9850	0.9499	0.0480	0.0021	0.0155
0.9800	0.9338	0.0626	0.0036	0.0208
0.9750	0.9179	0.0766	0.0055	0.0263
0.9700	0.9023	0.0899	0.0079	0.0318
0.9650	0.8869	0.1023	0.0105	0.0375
0.9600	0.8718	0.1146	0.0135	0.0433
0.9550	0.8570	0.1262	0.0168	0.0492
0.9500	0.8424	0.1372	0.0204	0.0552
0.9450	0.8281	0.1476	0.0243	0.0613
0.9000	0.7099	0.2207	0.0694	0.1215
0.8500	0.5993	0.2672	0.1334	0.1993
0.8000	0.5069	0.2888	0.2042	0.2888
0.7500	0.4294	0.2937	0.2769	0.3900

in Table 7.5 for various values of x_A. The mass fraction is computed as follows:

mass fraction of i

$$= \frac{[c_i/c_{Af}] \times \text{mol.wt.}i}{[c_R/c_{Af}] \times \text{mol.wt.}R + [c_S/c_{Af}] \times \text{mol.wt.}S + [c_T/c_{Af}] \times \text{mol.wt.}T}$$

molecular weights: $R = 62$, $S = 106$, $T = 150$

The final column in Table 7.5 is the ratio of moles of B to A in the feed less the ratio of unreacted B to A in the feed.

The product distribution by weight that was required to meet the market demand in the United States for the year 1968 is shown below:

Ethylene Glycol, $R = 90\%$

Diethylene Glycol, $S = 8\%$

Triethylene Glycol, $T = 2\%$

Examining the table we see that a value of one minus the conversion of A, $x_A = c_A/c_{Af}$, between 0.970 and 0.965 will come quite close to meeting the required product distribution. An input mole ratio c_{Bf}/c_{Af} between 0.0318 and 0.0375 is needed if we size the reactor such that c_B/c_{Af} is very small compared to c_{Bf}/c_{Af}.

In Example 7.6 we were concerned with sizing the reactor in such a way that the yearly profit was maximized. We did not, however, have to worry about the product distribution because only one product was produced. In the ethylene oxide-water system the required product distribution must be met as a primary goal. This fixes x_A, the fraction of raw material A used, and specifies the quantity $[c_{Bf} - c_B]/c_{Af}$, the amount of B converted divided by the inlet concentration of A. We can design our system so that almost all the B is converted, in which case $[c_{Bf} - c_B]/c_{Af} \rightarrow c_{Bf}/c_{Af}$, or we can design for partial conversion of B. If we make the first choice we do not need any processing equipment for the removal of B. If we decide on the second course of action, the reactor will be smaller but a B separation unit must be included in the process. The optimal design problem is similar to the one considered in Example 7.5 except that we must also consider the effect of B conversion on the production distribution part of our design.

As can be seen from Table 7.5 and the equations relating the product distribution and $[c_{Bf} - c_B]/c_{Af}$ to x_A, $[c_{Bf} - c_B]/c_{Af}$ is fixed once the desired product distribution is known. As long as the conversion of B has the same value we can pick an infinite number of values for c_{Bf}/c_{Af} and c_B/c_{Af} and still get the desired product distribution. We note, however, that an excess of water is always required, and this must be removed before we can obtain the glycols. The amount of water that must be removed per mole of mixed product produced, $c_A/[c_{Af} - c_A] = x_A/[1 - x_A]$, is always a constant, regardless of the value of $[c_{Bf} - c_B]/c_{Af}$. The conversion of B thus has no effect on the process downstream from the reactor except for its effect on the B separation unit. The B conversion for which we design the reactor can thus be determined without considering the product distribution part of the process design, and the optimal value of c_{Bf} is determined as illustrated in Example 7.6. If we perform the economic analysis for the ethylene oxide-water problem, we find that the capital cost of the separation equipment is so high compared to the saving resulting from the decrease in reactor volume obtained for partial conversion of B that it is never advantageous to design the reactor such that there is a significant amount of B in the effluent. In other words, an optimal design is one in which we do not need a B separation unit. We cannot, of course, design a reactor that will convert all the B, for such a reactor would have an infinite volume. What we mean in an engineering sense by "complete conversion" is a concentration of B in the reactor effluent that can be tolerated in processing equipment downstream. In the ethylene oxide-water system the ethylene oxide (B) unconverted at the reactor exit will be removed with the water and thus either be discarded or recycled. The ethylene oxide will continue to react during any processing step downstream from the reactor and as a result there will be small amounts of product, R, S and T, in any A recycle stream. We could calculate the effect that this would have on the reactor by including in our model equations the terms qc_{Rf}, qc_{Bf}, and qc_{Tf}.

The reactor holding time, $\theta = V/q$, is computed using Equation 7.40b after x_A has been determined by the product distribution analysis

$$\theta = \frac{V}{q} = \frac{1 - x_A}{k_1 x_A c_B} \tag{7.49}$$

The flow rate, q, is determined as before once a decision has been made as to the total amount of product desired. The choice of c_B is made by considering various values of c_B/c_{Af} ranging, say, from $10^{-2}c_{Bf}/c_{Af}$ to $10^{-4}c_{Bf}/c_{Af}$ and computing required volumes. This range of c_B will ensure that the required product distribution is obtained, and the actual volume decision is arrived at by an evaluation of the effect of c_B on the recycle stream. The effect turns out to be negligible for ethylene oxide and water for c_B/c_{Af} equal to about $10^{-2}c_{Bf}/c_{Af}$.

7.7 RATE DETERMINATION

As a final example we consider the use of the steady state continuous flow stirred tank reactor as an experimental device to determine rate expressions. An experiment using a continuous flow reactor is more difficult to perform than an experiment in a batch reactor, since raw material must be supplied continuously over a period of time and product and raw material may have to be separated continuously or discarded. On the other hand, because the system is at steady state, sampling problems are lessened and model behavior is simpler to determine than for batch systems. Thus, the CFSTR is sometimes used to study complex reactions.

To illustrate the application, consider a simple irreversible reaction with the following chemical equation:

$$A \rightarrow D$$

The rate is assumed to be of the form

$$r_{A-} = kc_A{}^n$$

The steady state CFSTR equation for A is

$$0 = \frac{1}{\theta} [c_{Af} - c_A] - kc_A{}^n$$

where $\theta = V/q$. Then we can write

$$\ln \frac{1}{\theta} [c_{Af} - c_A] = \ln k + n \ln c_A \qquad (7.50)$$

and a plot of $\ln c_A$ versus $\ln [1/\theta][c_{Af} - c_A]$ for various values of θ and/or c_{Af} will yield the rate parameters. Notice that this procedure is similar to the one introduced in Section 5.4 for estimating parameters from batch data. The batch procedure is approximate and subject to large errors, however, while the CFSTR result is exact and accurate.

7.8   THE MODEL EQUATIONS FOR TUBULAR REACTORS

There are two types of reaction equipment that can be used in continuous processing units, the continuous flow stirred tank reactor and the tubular reactor. In the tank-type reactor the contents are well mixed while in the tubular reactor there is a steady movement of material from one end of the reactor to the other with no attempt to induce mixing. These two reactor types represent extremes in behavior as far as the gross fluid motions are concerned and their analysis is of immense importance in reaction engineering. We have already analyzed the tank-type reactor and developed descriptions that enabled us to discuss the design problems in a preliminary way. An analysis of the tubular reactor will allow us to compare the performance of these two reactor types and to determine the effect of the gross fluid motions on such important factors as conversion and product distribution. In addition we will develop our modeling skills for a situation in which dependent variables other than time must be considered.

In general, when we refer to tubular reactors we mean a device in which the fluid movement from inlet to outlet is achieved without mixing between fluid elements at any point along the direction of flow. This usually means that the reactor takes the form of a pipe or tube. The assumption of a nonmixing plug fluid motion through the device is in many cases an accurate enough description to assure that predicted and measured performance of reacting systems is the same. In situations in which the assumption is not good the predicted performance provides us with a useful first approximation that may well allow a design decision to be made.

We will conceptually visualize the tubular system to be a circular pipe of length L with constant cross-sectional area, $A = \pi D^2/4$. A schematic sketch is shown in Figure 7.4. We will consider the same reaction as we did in the tank-type system

$$A + B \rightarrow nD$$

The total volumetric flow rate to the reactor will be designated by q and the reactor feed will be assumed to consist of a mixture of the two raw materials A and B with concentrations c_{Af} and c_{Bf}.

As indicated in Figure 7.4, the essential physical assumption in the plug-flow tubular reactor is that at any axial (z) position a cross section of the tube

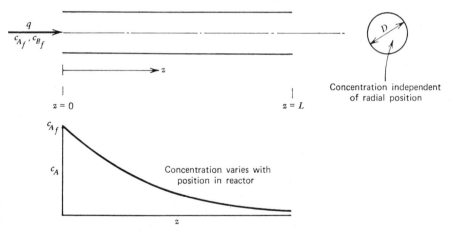

FIGURE 7.4 Schematic diagram of a tubular reactor.

is perfectly mixed, but no mixing whatever takes place in the axial direction between adjacent cross sections. This idealization provides a means of using a prior result to determine the steady state behavior of the tubular reactor. As fluid enters the reactor it travels down the tube undisturbed. Thus, a fluid element is a small batch reactor traveling down the tube. The batch reactor equations, Equations 5.15 to 5.17, are

$$\frac{dc_A}{d\tau} = -r$$

$$\frac{dc_B}{d\tau} = -r$$

$$\frac{dc_D}{d\tau} = +nr$$

where, for reasons that will become obvious subsequently, we have denoted time by τ. Now, the linear velocity in the tube is

$$v = \frac{4q}{\pi D^2} \tag{7.51}$$

so that the time to reach a position z is

$$\tau = \frac{z}{v} = \frac{\pi D^2 z}{4q} \tag{7.52}$$

$$d\tau = \frac{1}{v} dz = \frac{\pi D^2}{4q} dz \tag{7.53}$$

Thus, the variation of concentration with position at the steady state is determined by the equations

$$v \frac{dc_A}{dz} = \frac{4q}{\pi D^2} \frac{dc_A}{dz} = -r \tag{7.54}$$

$$v \frac{dc_B}{dz} = \frac{4q}{\pi D^2} \frac{dc_B}{dz} = -r \tag{7.55}$$

$$v \frac{dc_D}{dz} = \frac{4q}{\pi D^2} \frac{dc_D}{dz} = +nr \tag{7.56}$$

All of the results for the time dependence of the transient batch reactor carry over directly to the steady state position dependence of the tubular reactor by replacing time with position according to Equation 7.52. Thus, we already have available the tools for reactor design. This analogy is valid, however, only under the restrictive assumptions made here about radial and axial mixing. In the next section we shall rederive the equations using the systematic approach developed in the earlier chapters. This approach can be as easily applied to more complicated flow situations.

7.9   DIFFERENTIAL CONTROL VOLUME

If we begin our model development for the tubular reactor by selecting the reactor itself as a control volume, as we did for the tank-type reactors, we find that there is not a unique value for the characterizing variables, c_A, c_B, and c_D. These quantities will vary, at least with distance along the reactor, and in accordance with Figure 3.8, we must select a new control volume. If we select a small incremental length of the reactor, Δz, then the disk with volume $\pi D^2 \Delta z / 4$ shown in Figure 7.5 can be tried as the next simplest control

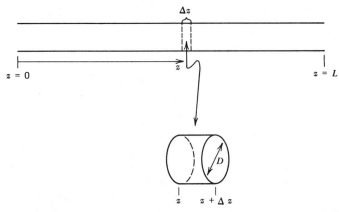

FIGURE 7.5   Control volume for a tubular reactor. Concentrations are approximately constant over the small distance Δz.

volume to the reactor itself. Within the incremental slice of thickness Δz, c_A, c_B, and c_D are considered to have unique values if we can sample at any radial or circumferential position and find no change. Selection of this control volume is equivalent to assuming that the fluid travels through the reactor in plug or piston flow and that the concentration of any species at a particular axial position is uniform throughout the pipe cross section. This uniformity in the radial and circumferential directions is achieved in many situations by random small scale fluid motions which tend to produce a well-mixed fluid within the control volume, $\pi D^2 \Delta z / 4$. The assumption that characterizing variables are essentially constant over the length Δz is consistent with the usual approximations made in applying the calculus and will be valid in the limit as $\Delta z \to 0$, an operation that we will perform subsequently.

The law of conservation of mass can be applied to the control volume $\pi D^2 \Delta z / 4$ in the same manner as we have done with tank-type systems as long as we recognize that conditions at point z in the reactor are different from those at point $z + \Delta z$. The total mass in the control volume is $\rho \pi D^2 \Delta z / 4$, and the overall mass balance can be written as follows:

mass in control   mass in control   mass entered   mass left
volume at time = volume at time + during Δt at $-$ during Δt at
$\quad$ $t + \Delta t$ $\qquad\qquad$ t $\qquad\qquad$ position z $\quad$ position $z + \Delta z$

$$\rho(t + \Delta t)\pi D^2 \Delta z/4 = \rho(t)\pi D^2 \Delta z/4 + \rho(z)q(z)\Delta t$$
$$- \rho(z + \Delta z)q(z + \Delta z)\Delta t \qquad (7.57)$$

As before we are using mean values and ( ) denotes a functional dependence. There are now *two* independent variables, time, t, and position in the reactor, z. If we divide Equation 7.57 by $\Delta z \Delta t$ and rearrange it slightly we obtain

$$\frac{\rho(t + \Delta t)\pi D^2/4 - \rho(t)\pi D^2/4}{\Delta t}$$

$$= - \frac{\rho(z + \Delta z)q(z + \Delta z) - \rho(z)q(z)}{\Delta z}$$

The quantity on the left is the difference approximation to a time derivative where the quantities are evaluated at a mean position in the control volume. In the limit as $\Delta t \to 0$ this becomes the *partial time derivative* as discussed in Section 15.11, the derivative with respect to time at fixed position. Similarly, the quantity on the right is the difference approximation to a derivative with respect to z. In the limit as $\Delta z \to 0$ this becomes the *partial spatial derivative*, differentiation with respect to z at fixed t. Thus, taking the two limits we obtain

$$\frac{\partial \rho \pi D^2/4}{\partial t} = - \frac{\partial \rho q}{\partial z} \qquad (7.58)$$

Using the definition of velocity in Equation 7.51 and assuming that the cross section is constant, so D is independent of z, we obtain

$$\frac{\partial \rho}{\partial t} + \frac{\partial \rho v}{\partial z} = 0 \qquad (7.59)$$

This is often referred to as the *continuity equation* and is most frequently encountered in studies of fluid mechanics. At the steady state ρ does not vary with time and Equation 7.59 reduces to

$$\frac{\partial \rho v}{\partial z} = 0 \qquad (7.60)$$

Equation 7.60 tells us that the quantity ρv is constant and independent of z. Furthermore, if ρ is constant, as it would be for a liquid system, we see that v, the liquid velocity, is a constant. None of these conclusions is very surprising, however, and we see that the overall mass balance is not of primary importance in the analysis of this reacting system.

The component mass balances are readily developed for the control volume and we will go through the derivation in detail for component A. The word statement of the conservation law is expressed in symbols as follows:

$$c_A(t + \Delta t)\pi D^2 \Delta z/4 = c_A(t)\pi D^2 \Delta z/4 + q(z)c_A(z)\Delta t$$
$$- q(z + \Delta z)c_A(z + \Delta z)\Delta t$$
$$- r_{A-}\pi D^2 \Delta z \Delta t/4 \tag{7.61}$$

Dividing each term by $\Delta z \Delta t$ yields

$$\frac{c_A(t + \Delta t)\pi D^2/4 - c_A(t)\pi D^2/4}{\Delta t}$$

$$= - \frac{q(z + \Delta z)c_A(z + \Delta z) - q(z)c_A(z)}{\Delta z} - r_{A-}\frac{\pi D^2}{4}$$

Taking the limit as Δz and Δt approach zero we then obtain

$$\frac{\pi D^2}{4}\frac{\partial c_A}{\partial t} = - \frac{\partial q c_A}{\partial z} - r_{A-}\frac{\pi D^2}{4} \tag{7.62}$$

We are generally concerned with a constant volumetric flow. In that case, using Equation 7.51 to introduce the velocity we obtain

$$\frac{\partial c_A}{\partial t} = - \frac{4q}{\pi D^2}\frac{\partial c_A}{\partial z} - r_{A-}$$

$$\frac{\partial c_A}{\partial t} = -v\frac{\partial c_A}{\partial z} - r_{A-} \tag{7.63}$$

In a similar manner we can derive the component balances for substances B and D:

$$\frac{\partial c_B}{\partial t} = -v\frac{\partial c_B}{\partial z} - r_{B-} \tag{7.64}$$

$$\frac{\partial c_D}{\partial t} = -v\frac{\partial c_D}{\partial z} + r_{D+} \tag{7.65}$$

In order to examine the design problem we need to consider the steady state situation, for which all time derivatives are zero. Equations 7.63 to 7.65

then reduce to

$$v \frac{\partial c_A}{\partial z} = -r_{A-} \tag{7.66}$$

$$v \frac{\partial c_B}{\partial z} = -r_{B-} \tag{7.67}$$

$$v \frac{\partial c_D}{\partial z} = r_{D+} \tag{7.68}$$

Since $r_{A-} = r_{B-} = r_{D+}/n = r$ these are identical to Equations 7.54 to 7.56, which we obtained by analogy with the batch system. Since time is no longer an independent variable we could replace $\partial/\partial z$ by d/dz, as we have done in Equations 7.54 to 7.56.

Despite the mathematical equivalence to the batch reactor there is an essential difference from an engineering viewpoint, namely, that the tubular reactor can be employed as a continuous processing unit. The important comparison is with the continuous flow stirred tank reactor, and we can see the differences by considering the previous examples.

Example 7.6

Using the information given in Example 7.2 we can determine the concentrations of A, B, and D in the exit stream of a tubular reactor of volume 25.4 liters. All the inlet concentrations are as specified in Example 7.1 and the flow is the sum $q_{Af} + q_{Bf}$ as given in the example. The model equations for the tubular reactor are the same as those for the batch reactor, with $\tau = z/v$ substituted for t:

$$\frac{dc_A}{d\tau} = -kc_A c_B$$

$$\frac{dc_B}{d\tau} = -kc_A c_B$$

$$\frac{dc_D}{d\tau} = 2kc_A c_B$$

As was shown in Chapter 5 the following relationships are obtained relating the species concentrations (Equations 5.18, 5.19)

$$c_A - c_{Af} = c_B - c_{Bf}$$

$$c_{Af} - c_A = \tfrac{1}{2}c_D$$

(Here we are denoting the concentration at $\tau = 0$ as c_{Af}, rather than c_{A0}, as for the batch system.) Since c_{Af} and c_{Bf} are equal in this example, $c_A = c_B$

and c_A is readily obtained by integrating the equation

$$\frac{dc_A}{d\tau} = -kc_A^2$$

The result is exactly that given by Equation 5.22 with τ substituted for t:

$$\frac{1}{c_A} = \frac{1}{c_{Af}} + k\tau$$

Here c_{Af} represents the inlet concentration of A in the combined stream containing both A and B. To maintain the same molar feed rate as in Example 7.2 the feed concentrations must be one-half those in the individual feed streams, so

$$c_{Af} = \tfrac{11}{2} = 5.5 \text{ g-moles/liter}$$

$$c_{Bf} = \tfrac{11}{2} = 5.5 \text{ g-moles/liter}$$

The exit of the reactor is at $\tau = \Theta = \pi D^2 L/4q = V/q$. For $q = 0.2$, $\Theta = 127$. Thus, c_A, the concentration of A in the exit stream from the tubular reactor, is

$$\frac{1}{c_A} = \frac{1}{5.5} + 6.05 \times 10^{-4} \times 127$$

$$c_A = 3.84 \text{ g-moles/liter}$$

c_D, the product concentration, is easily computed as

$$c_D = 2[5.5 - 3.84]$$

$$= 3.32 \text{ g-moles/liter.}$$

We see that for the same residence time, $\Theta = 127$ min, we obtain a higher conversion of A in the tubular reactor than in the continuous flow tank reactor. A series of calculations can be performed to yield Table 7.6, which compares the tubular reactor with the continuous flow tank reactor results contained in Table 7.1. It is clear that the tubular configuration is more efficient in that a smaller reactor is always required than that necessary to produce the same conversion in a continuous flow tank reactor. The fact that a smaller reactor is required does not necessarily mean that it will be less expensive to construct, and in fact there are many situations where a CFSTR is less expensive.

TABLE 7.6   Concentration in a tubular reactor compared with a CFSTR with the same residence time

q	$\Theta = V/q$	CFSTR			Tubular		
liters							
min	min	c_A	c_D	qc_D	c_A	c_D	qc_D
0.2	127	4.15	2.70	0.54	3.84	3.32	0.664
0.3	84.7	4.45	2.10	0.63	4.29	2.42	0.726
0.4	63.5	4.70	1.60	0.64	4.54	1.92	0.768
0.5	50.9	4.80	1.40	0.70	4.70	1.60	0.800
0.6	42.8	4.90	1.20	0.72	4.81	1.38	0.828
0.8	31.8	5.00	1.00	0.80	4.97	1.06	0.848
1.0	25.4	5.09	0.88	0.82	5.07	0.86	0.860

Example 7.7

For the reaction $A \to D$ studied in a CFSTR in Example 7.4 find the conversion and production rate in a tubular reactor with the same volume and flow rate.

The conversion of A is described by the equation

$$\frac{dc_A}{d\tau} = -kc_A$$

$$c_A(\tau) = c_{Af}e^{-k\tau}$$

$$c_D(\tau) = c_{Af} - c_A(\tau)$$

At the reactor exit, $\tau = \Theta$, we have, using $c_{Af} = 0.2$, $k = 5 \times 10^{-3}$,

$$c_A = 0.2e^{-0.005\Theta}$$

$$c_D = 0.2[1 - e^{-0.005\Theta}]$$

TABLE 7.7   Volume and conversion in a tubular reactor for a given production requirement compared with a CFSTR

q	V	Θ	CFSTR			Tubular		
liters							
min	liters	min	c_A	c_D	qc_D	c_A	c_D	qc_D
250	∞	∞	0	0.2	50.0	0	0.2	50.0
300	303,000	1000	0.033	0.167	50.0	0.00135	0.1987	59.6
400	133,000	333	0.075	0.125	50.0	0.0378	0.1622	64.9
500	100,000	200	0.10	0.10	50.0	0.0736	0.1264	63.2
800	72,600	90.7	0.1375	0.0625	50.0	0.1271	0.0729	58.3
1000	66,600	65	0.15	0.05	50.0	0.1445	0.0555	55.5
2000	58,100	27	0.175	0.025	50.0	0.1747	0.0253	50.6
4000	53,200	13.3	0.1875	0.0125	50.0	0.1871	0.0129	51.6

The results are shown in Table 7.7 for values of V and q employed in constructing Table 7.2. In all cases the production rate in the tubular reactor exceeds the value in the CFSTR, indicating that a smaller reactor volume is required.

Example 7.8

Consider the irreversible reaction $A + B \rightarrow nD$. Show that for a given conversion the residence time of a tubular reactor will always be smaller than that of a CFSTR.

The tubular and CFSTR equations can be written as follows:

$$\qquad\text{Tubular}\qquad\qquad\qquad\text{CFSTR}$$

$$\frac{dc_A}{d\tau} = -r \qquad\qquad 0 = \frac{1}{\theta}[c_{Af} - c_A] - r$$

$$-\frac{dc_A}{r} = d\tau$$

$$-\int_{c_{Af}}^{c_A} \frac{dc}{r(c)} = \Theta$$

$$\Theta = \int_{c_A}^{c_{Af}} \frac{dc}{r(c)} \qquad\qquad \theta = \frac{c_{Af} - c_A}{r(c_A)}$$

The development assumes that the chemical equation has been used to find r in terms of c_A. Now, r is an increasing function of c_A (e.g., $r = kc_A{}^n$). Thus, $1/r(c)$ is a decreasing function of c as c varies from c_A to c_{Af} ($c_A < c_{Af}$). Figure 7.6 shows a typical plot of $1/r(c)$ versus c, $c_A \leq c \leq c_{Af}$. Θ, the

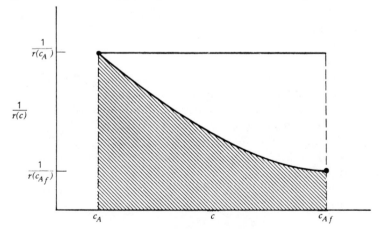

FIGURE 7.6 Reciprocal of rate versus concentration. Residence time of a tubular reactor is the area under the curve. Residence time of a stirred tank reactor is the area of the rectangle.

integral of $1/r$, is the area under that curve. On the other hand, θ is the area of the rectangle shown in the figure with base $c_{Af} - c_A$ and height $1/r(c_A)$. Thus, $\theta > \Theta$.

7.10 CONCLUDING REMARKS

This is perhaps the most important single chapter in the text. Using the principles developed in Chapters 4 and 5 we have arrived at the logical culmination of the analysis process, a practical engineering design. Notice how the equations involving design variables such as reactor volume and flow rate need to be combined with expressions reflecting the economics of the particular situations.

The design examples in this chapter are quite realistic in form. The two limiting cases of perfectly mixed and tubular (completely unmixed) reactors give the limits attainable in practice. Often we are faced with more complex rate constitutive equations and with economic relations that are difficult to express in simple ways. Nothing changes in principle, and frequently the algebra only becomes more difficult. In many cases, however, the effort required to obtain sufficient information to complete the equations is impossible from the point of view of time or money. In these cases we rely on prior experience with similar situations and the structure of the problem which is revealed by a partial analysis in order to arrive at the final design.

There are a number of texts that deal with various aspects of chemical reactor design:

7.1 R. Aris, *Elementary Chemical Reactor Analysis*, Prentice-Hall, Englewood Cliffs, 1969.

7.2 K. G. Denbigh and J. C. R. Turner, *Chemical Reactor Theory*, 2nd ed., Cambridge Univ. Press, Cambridge, 1971.

7.3 O. A. Hougen and K. M. Watson, *Chemical Process Principles*, *Part III: Kinetics and Catalysis*, Wiley, New York, 1947.

7.4 H. Kramers and K. R. Westerterp, *Elements of Chemical Reactor Design and Operation*, Academic, New York, 1963.

7.5 O. Levenspiel, *Chemical Reaction Engineering*, Wiley, New York, 1962.

7.6 J. M. Smith, *Chemical Engineering Kinetics*, 2nd ed., McGraw-Hill, New York, 1970.

7.7 S. M. Walas, *Reaction Kinetics for Chemical Engineers*, McGraw-Hill, New York, 1959.

The sequence Hougen and Watson, Levenspiel, and Denbigh and Turner nicely shows the development of the subject from a chemical engineering viewpoint. Economics of chemical processing is discussed in:

7.8 M. Peters and K. Timmerhaus, *Plant Design and Economics for Chemical Engineers*, 2nd ed., McGraw-Hill, New York, 1968.

7.9 D. F. Rudd and C. C. Watson, *Strategy of Process Engineering*, Wiley, New York, 1968.

Relevant material also appears frequently in the periodical literature. You will want to become acquainted with the major English language chemical engineering journals, *AIChE Journal; Canadian Journal of Chemical Engineering; Chemical Engineering; The Chemical Engineering Journal; Chemical Engineering Progress and Chemical Engineering Progress Symposium Series; Chemical Engineering Science; Chemical Technology; Hydrocarbon Processing and Petroleum Refiner; Industrial and Engineering Chemistry* (no longer published) and three Industrial and Engineering Chemistry quarterlies: *IEC Fundamentals, IEC Process Design and Development*, and *IEC Product Research and Development;* and the *Transactions of the Institution of Chemical Engineers*. You will note that each journal has its own flavor.

7.11 PROBLEMS

7.1 Modify the semibatch reactor development in Section 7.3 for the case in which reactant B is in great excess and c_B may be assumed to be essentially constant. Obtain an equation for the conversion of A analogous to Equation 7.16.

7.2 Using the results in Example 5.4 rederive Equations 7.18 to 7.20 so they can be used in Examples 7.1 to 7.3.

7.3 For the system used in Examples 7.1 to 7.3 compute c_A, c_B, and c_D for $V = 25.4$ liters, $q_{Af} = q_{Bf} = 0.2$ liters/min, and the following feed conditions:

(a) $c_{Af} = c_{Bf} = 10.0$ g-moles/liter
(b) $c_{Af} = c_{Bf} = 8.0$
(c) $c_{Af} = c_{Bf} = 6.0$
(d) $c_{Af} = c_{Bf} = 4.0$

What conclusions can you draw about the dependence of conversion on feed concentration?

7.4 For the systems used in Examples 7.1 to 7.3 compute c_A, c_B, and θ for $q_{Af} = q_{Bf} = 0.2$, $c_{Af} = c_{Bf} = 11$, and the following volumes:

(a) $V = 10$ (b) 20   (c) 30   (d) 40   (e) 50

What general conclusion can you draw?

7.5 Let $c_{Af} = c_{Bf}$ and define $\lambda = q_{Af}/[q_{Af} + q_{Bf}]$. Solve Equation 7.18 for c_A in terms of $k\theta$, λ, and c_{Af}. Use the Taylor series $\sqrt{1 + x} = 1 + x/2 - x^2/8 + \cdots$ to obtain a simple expression for c_A and check against your results for Problems 7.3 and 7.4.

7.6 Repeat the calculation of c_D in Problems 7.3 and 7.4 by assuming that the term kc_Ac_BV can be neglected in Equations 7.18 and 7.19, but not Equation 7.20. Under what conditions is this approximation useful? Explain physically. Explain in light of your answer to Problem 7.5.

7.7 Repeat the calculation in Example 7.3 taking the reverse reaction into account.

7.8 The decomposition of dibromosuccinic acid (A) was shown in Example 5.3 to have the rate expression

$$r_{A-} = 0.031 \text{ min}^{-1} c_A$$

(a) Find the time required in an isothermal batch reactor to reduce the concentration of DBS from 5 g/liter to 1.5 g/liter.

(b) Find the size of an isothermal continuous flow stirred tank reactor to achieve the same conversion of DBS if the steady state flow rate is 50 liters/min. What holding time θ is required?

7.9 Two identical continuous flow stirred tank reactors are to be connected so that the effluent from one is the feed to the other. What size should each be if the same conversion of DBS acid is to be achieved as that specified in Problem 7.8(b)? If these two reactors were operated in parallel with each receiving half the total raw material, what exit concentration of A would be achieved in each?

7.10 N identical isothermal continuous flow stirred tank reactors are connected in series so that the effluent from one is the feed to the next. Let c_{AN} denote the concentration in the effluent from the last reactor and c_{Af} the feed to the first reactor.

(a) For the reaction $A \rightarrow R$, $r_{A-} = kc_A$, develop an expression for c_{AN} in terms of N, c_{Af}, and θ. θ is the (identical) holding time of each reactor.

(b) For the data in Problem 7.8 how many reactors are needed to achieve the conversion required if the total holding time $N\theta$ is two minutes more than the time required in a batch reactor?

7.11 For the reactions

$$A \rightarrow R \rightarrow S$$

$$r_{A-} = r_{R+} = k_1c_A \qquad r_{R-} = r_{S+} = k_2c_R$$

derive the equations for a steady state continuous flow stirred tank reactor.

7.12 For the system in Problem 7.11 assume $c_{Rf} = c_{Sf} = 0$ and obtain an explicit expression for c_R in terms of c_{Af}, k_1, k_2, and θ. Find the holding time θ that will give the maximum concentration of R in the reactor effluent. Develop an expression for this maximum concentration in terms of the ratio k_1/k_2.

7.13 Review Problems 5.6 and 5.14. Compute the maximum glucose concentration at 190°C in a continuous flow stirred tank reactor and plot the maximum concentration and the corresponding holding time versus HCl concentration up to $c_{HCl} = 0.88$. (Extrapolate the data for k_1 to $c_{HCl} = 0.88$ using the rate expression derived in Problem 5.6.)

Find a way of formulating the problem so that the result is valid for any feed concentration c_{Af}. Comment on the significance of the result.

7.14 Does the solution to Problem 7.13 contain enough information to propose an optimal design? What are the economic trade-offs? Reflect upon Section 7.5 and discuss the optimal design of this glucose process in one or two paragraphs.

7.15 For the complex reaction system discussed in Section 7.6 suppose that some product R is recycled so that the feed stream contains a concentration $c_{Rf} \neq 0$.

(a) Redevelop the model equations.
(b) Manipulate the description to obtain expressions for c_R/c_{Af}, c_S/c_{Af}, and c_T/c_{Af}. What effect has the presence of R in the feed on the product distribution?
(c) Find the weight percentage distribution of R, S, and T (as in Table 7.5) for the following conditions:

	$x_A = c_A/c_{Af}$	c_{Rf}/c_{Af}
(i)	0.980	0.05
(ii)	0.980	0.10
(iii)	0.900	0.05
(iv)	0.800	0.05

7.16 The accompanying table shows the product distribution of mono- (R), di- (S), and triglycol (T) by weight percentage in the effluent stream of a *tubular* plug flow reactor. What value of x_A and c_{Bf}/c_{Af} will come closest to meeting the product distribution given in Section 7.6? Why is the tubular reactor superior to the CFSTR if this product distribution must be made? Make your discussion quantitative by considering the amount of water that would have to be removed in each case to make 90 million pounds of R, 8 million pounds of S, and 2 million pounds of T.

TABLE PROBLEM 7.16   Tubular plug flow product distribution

x_A	R	S	T	$\dfrac{c_{Bf}}{c_{Af}} - \dfrac{c_B}{c_{Af}}$
0.9900	0.9830	0.0168	0.0002	0.0101
0.9850	0.9745	0.0251	0.0004	0.0152
0.9800	0.9661	0.0333	0.0006	0.0204
0.9750	0.9577	0.0413	0.0010	0.0256
0.9700	0.9494	0.0492	0.0014	0.0309
0.9650	0.9411	0.0570	0.0019	0.0363
0.9600	0.9328	0.0647	0.0025	0.0416
0.9550	0.9246	0.0722	0.0032	0.0471
0.9500	0.9164	0.0780	0.0039	0.0526
0.9450	0.9082	0.0870	0.0048	0.0581
0.9000	0.8366	0.1480	0.0154	0.1107
0.8500	0.7606	0.2056	0.0337	0.1748
0.8000	0.6885	0.2529	0.0585	0.2456
0.7500	0.6203	0.2905	0.0891	0.3236
0.7000	0.5560	0.3190	0.1251	0.4095

7.17 What size continuous flow stirred tank reactor is required to meet the product demand specified in Problem 7.16 if the reactor is operated at 25°C? (Hint: Follow the discussion at the end of Section 7.6.). Would it be feasible to operate the reactor at this temperature?

7.18 Develop the model equations for a plug flow tubular reactor for the general reaction scheme presented in Section 7.6. Assume that the rate expressions given by Equation 7.39 are valid.

7.19 Algebraic expressions relating c_R/c_{Af}, c_S/c_{Af}, and c_T/c_{Af} similar to Equations 7.45, 7.46, and 7.47 may be developed from the mathematical description derived in Problem 7.18. To do this the conservation equations for R and S are divided by the conservation equation for A. The resulting first-order equation in c_R and c_A is solved using the methods of Section 15.10. With this result in hand the equation in c_S and c_A can be solved using the techniques of Section 15.10. Show that the following expression for c_R/c_{Af} is correct for the glycol reaction:

$$\frac{c_R}{c_{Af}} = x_A - x_A^2 + \frac{c_{Rf}x_A^2}{c_{Af}}$$

7.20 The table for Problem 7.16 was prepared using the following equations with $c_{Rf} = c_{Sf} = 0$

$$\frac{c_R}{c_{Af}} = x_A - x_A^2 + \frac{c_{Rf}}{c_{Af}} x_A^2$$

$$\frac{c_S}{c_{Af}} = 2 \left[x_A - x_A^2 + x_A^2 \ln x_A - \frac{c_{Rf}}{c_{Af}} x_A^2 \ln x_A \right] + \frac{c_{Sf}}{c_{Af}} x_A^2$$

Develop an expression analogous to Equation 7.47 for c_T/c_{Af}.

7.21 To get a quick and fairly reasonable estimate of the size of an isothermal tubular reactor to meet the product demand specified in Problem 7.16 the mass balance equation for the ethylene oxide can be integrated by assuming everything is constant except c_B. Perform this computation and compare the reactor size with that computed in Problem 7.17 for the CFSTR. What major assumptions have been made? How valid are they? Is $25°$ C a reasonable operating temperature?

7.22 To calculate the volume of the tubular reactor for the glycol reaction exactly the mass balance equation for A must be integrated. Show how this can be done graphically.

7.23 The chemical reaction

$$A \rightarrow \text{Products}$$

is known to have a rate proportional to the nth power of the concentration of $A (r_{A-} = k c_A^n)$. An experiment is performed in which the reaction is carried out in two continuous flow stirred tank reactors operating at steady state. The composition of A in the feed to the first reactor is $c_{Af} = 1$ g-mole/liter. The effluent from the first reactor feeds the second. The residence time (V_1/q) of the first reactor is 96 sec, and the volume of the second reactor is *twice* that of the first. It is found experimentally that the concentration of A, c_{A1}, in the first reactor is 0.5 g-mole/liter and in the second reactor the concentration of A, c_{A2}, is 0.25 g-mole/liter. In order to design a tubular reactor we must have a verified rate expression. Develop this from the experimental information.

7.24 The reaction in Problem 7.23 is carried out in a tubular reactor of 2-in. diameter at a volumetric flow rate of 1/3 cu ft/min. If the feed has a concentration $c_{Af} = 1$ g-mole/liter, how long a reactor is needed to achieve 99 percent conversion?

7.25 A method of separating cyclopentadiene from a stream containing other C_5 hydrocarbons has been proposed in which the cyclopentadiene is dimerized to dicyclopentadiene and the C_{10}'s separated from the C_5's by distillation. The kinetics for this reaction can be represented as follows:

$$2A \underset{k_2}{\overset{k_1}{\rightleftarrows}} R$$

$$A + B \xrightarrow{k_3} D$$

A = cyclopentadiene
R = dicyclopentadiene
B = t-1, 3, pentadiene
D = codimer
$r_{A-} = 2k_1 c_A{}^2 - k_2 c_R + k_3 c_A c_B$
$r_{R-} = k_3 c_A c_B$

Develop the steady state mass balance equations for this system for
(a) continuous flow stirred tank reactor
(b) tubular plug flow reactor.

7.26 Develop the steady state tubular plug flow reactor model equations for the reaction described in Problem 7.11. Solve for the effluent concentration of R using the methods in Section 15.10.

CHAPTER 8

Isothermal Two-Phase Systems and the Rate of Mass Transfer

8.1 INTRODUCTION

In Chapter 5 the concept of a reaction rate was introduced and the experimental means of determining it was discussed for some simple systems. The use of the verified rate expression in some introductory design problems was illustrated in Chapter 7. In this chapter we will direct our attention to the analysis of some simple two-phase systems, extending some of the concepts of modeling initiated in Chapter 4 and paralleling the general development in Chapters 5 and 7. Our major objective in this chapter is to extend and sharpen our ability to develop mathematical descriptions. Isothermal two-phase systems represent the next stage in complexity over isothermal single-phase systems and thus provide us with an opportunity to deal with quite complicated modeling situations while still basing our analysis solely on the law of conservation of mass and the appropriate constitutive equations. Two-phase systems are encountered in almost every aspect of chemical engineering and a secondary objective of this chapter is to gain a preliminary understanding of this important class of problems. We will deal specifically with two basic ideas, the rate of transfer between phases and the equilibrium relationship between the concentration of a species in one phase with its concentration in another.

We will begin with an analysis of batch systems to gain an appreciation of the experimental techniques and then extend the study to tank-type two-phase continuous flow systems. Before any of the mathematical descriptions are developed, however, it is convenient first to classify the types of two-phase systems that may be encountered and to discuss two-phase mass transfer operations that are of importance. We do this to show where the analysis developed in this chapter fits into the total picture.

8.2 CLASSIFICATION OF TWO-PHASE SYSTEMS

Substances can occur in processing applications in any of three states, solid, liquid, or gas. For engineering purposes we define a *phase* as a macroscopic portion of a system which is composed entirely of material in one state and which has an identifiable interface with the other phase or phases of the system. Thus, we might expect to encounter two-phase systems which are solid-liquid, solid-gas, or liquid-gas. Immiscible liquids will form distinct phases with a well-defined interface, and liquid-liquid two-phase systems are common in processing applications. Solid-solid systems are not often encountered in the applications with which we will be concerned, and gas-gas systems cannot exist, since mixtures of gases do not form distinct phases.

A two-phase system may be the result of a processing operation in which the raw materials exist in two phases, it may result from the deliberate contacting of one phase with another, or it may be the consequence of a second phase being produced as a result of a processing operation. For classification purposes and for simplicity of analysis it is convenient to draw a distinction between processes where a significant phase change occurs and those where it does not. We will not deal with the latter situation in this text. In this chapter and the next we shall consider processes where two phases are contacted for the purpose of transfer of mass between the phases.

Most mass transfer processes are quite similar in concept, implementation, and quantitative description. As technology has developed various names have been given to these processes, commonly referred to as *unit operations*. These operations are briefly described below.

Solid-Liquid Systems. Solid-liquid mass-transfer systems are of two types. In *adsorption* dissolved or suspended material is transferred from the liquid phase to the surface of the solid. Activated charcoal, silica gel, and magnesium oxide are typical adsorbents for applications such as the removal of sulphur compounds from gasoline, water from hydrocarbons, and various impurities from water. *Ion exchange* is like adsorption except that the ion removed to the solid surface is replaced in the liquid by an ion of the same charge removed from the solid. Typical ion exchange solids are certain synthetic resins and zeolites. Water may be demineralized by ion exchange.

Solvent extraction is the transfer of a soluble material from a solid phase to a liquid solvent. The term *washing* is often used when water is the solvent and the solid to be dissolved is adhering to the surface of an insoluble solid. More complex separations are designated as *leaching*. Typical applications are the recovery of copper from low-grade copper oxide ores by extraction with sulfuric acid and the leaching of sugar from beet pulp with water.

Solid-Gas Systems. The transfer of material from a gas to a solid surface is also classified as *adsorption*. A typical example is the use of silica gel for removal of SO_2 and water from gas streams. Adsorption is involved also in the use of solid catalysts for chemical reaction of gaseous species. *Drying*, or *desorption*, is the transfer of a volatile substance bound in the solid phase to a gas stream.

Liquid-Gas Systems. Transfer of a component from a gas phase to a liquid is denoted as *absorption*, or sometimes as *scrubbing*. Removal of SO_2 from air by absorption into aqueous sodium carbonate and removal of H_2S from natural gas with monoethanolamine solution are typical applications. Absorption is also the way in which a gaseous raw material is brought into contact with a liquid phase reactant, as in the secondary treatment process for liquid wastes. The oxygen necessary for the bio-oxidation reactions that break down the organic matter is supplied by bubbling air through the liquid sewage. *Desorption* or *stripping* is the term used when a component of the liquid is transferred to the gas, as in the removal of CS_2 from an oil phase with steam.

Liquid-Liquid Systems. Liquid *solvent extraction* is the transfer of a solute from one liquid phase to another where the two solvents are themselves nearly immiscible. Because both phases are liquids, the intimate contact step in which the mass transfer is effected must be followed by a step in which the two liquid phases are separated, usually by settling through density differences. Typical uses of extraction are in the recovery of penicillin with cyclohexane or chloroform and the dewaxing of lubricating oils using ketones or liquid propane.

All of the unit operations described here involve intimate contact between two phases in order to transfer material from one phase to the other. All can be carried out isothermally. Although the different materials involved in the various unit operations necessarily means that the processing equipment for effecting the mass transfer must differ in each case, the basic principles are essentially the same, and it will be possible to consider the mathematical description of the mass transfer for all situations simultaneously.

Operations involving a change of phase require further consideration using the principle of conservation of energy and are left for subsequent study. The most common phase change operation is *distillation*, in which a liquid mixture is boiled to produce a vapor of different composition. Repeated systematic application leads to purification. In *crystallization* a liquid mixture is cooled to produce a solid phase of different composition. Despite the different physical basis these phase-change operations can be described with mathematical models similar to those needed for the mass transfer unit operations described above, and the techniques developed in Chapter 9 will apply to them as well.

8.3   BATCH TWO-PHASE SYSTEMS

8.3.1   Basic Model Equations

A batch process is one in which there is no flow to or from the system during the operation. As we saw in Chapter 4, it is a convenient and useful system for obtaining experimental information. A batch two-phase system is one in which both phases are charged at time $t = 0$ and allowed to interact for a specified period of time so that data may be collected. The physical situation is sketched in Figure 8.1. We will assume that the vessel contents are well mixed and that the two phases are in intimate contact, with one phase

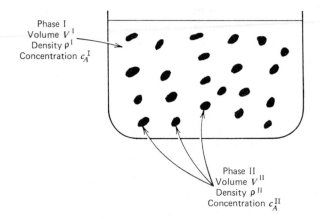

Phase I
Volume V^I
Density ρ^I
Concentration c_A^I

Phase II
Volume V^{II}
Density ρ^{II}
Concentration c_A^{II}

FIGURE 8.1   Two-phase system. Each phase must be taken as a separate control volume.

uniformly distributed throughout the other. This is fairly easy to achieve in practice and it is essential that the batch two-phase system have this characteristic if the analysis is to be done easily and hence is to be useful in giving us some insight into the physical behavior. Proper mixing or agitation also assures that the concentration of all species is spatially independent. The consequences of this perfect mixing assumption in single-phase systems have been discussed previously and in a two-phase situation the same conclusions apply for each phase. If we sample a phase at some particular time, the sample chosen from one location in the tank will be identical to that drawn from the same phase at any other location.

Although there are both isothermal and nonisothermal two-phase systems we will restrict our analysis in this chapter to the isothermal case. There are many situations of interest that are isothermal and in those that are not, the isothermal analysis is always part of the more complex problem. With this restriction we are concerned only with mass as our fundamental variable and we must construct our analysis using only the principle of conservation of mass and whatever constitutive relations we can experimentally verify. As before, we will use as characterizing variables V for volume, ρ for density, and c for concentration. Concentration will be measured in mass units (e.g., lb_m/ft^3). Selection of the vessel as our control volume is not useful for a two-phase system. Since we are interested in transfer between the phases, we must work with two control volumes, one for each phase. We will arbitrarily designate one phase as the continuous phase (Phase I) and one as the dispersed (Phase II). The continuous phase will have the concentration of any species designated as c_i^I and is the Swiss cheeselike control volume shown in Figure 8.1. We will designate the volume of the continuous phase as V^I, the density as ρ^I. The dispersed phase control volume, V^{II}, is made up of all the elements of other phase and, although we consider it as a single control volume for modeling purposes, it may consist physically of a number of distinct volumes. Concentrations in the dispersed phase will be designated with a superscript II as c_i^{II}, density as ρ^{II}.

We will develop our mathematical descriptions for the general two-phase case, recognizing that with appropriate identification of continuous and dispersed phases the basic model equations will apply for batch solid-liquid, solid-gas, liquid-liquid, or liquid-gas systems.

To simplify the algebraic manipulation we will develop our model equations for the physical situation in which a single component, A, is transferred between the phases. A has a concentration c_A^I in the continuous phase and c_A^{II} in the dispersed phase and does not react with any component in either phase. It is not conceptually difficult to extend the analysis to any number of species and also to include reaction in one or both phases. We would only increase the number of equations and the amount of manipulation required to determine behavior.

Application of the law of conservation of mass to each control volume yields the following:

$$\rho^I V^I \big|_{t+\Delta t} = \rho^I V^I \big|_t + \begin{Bmatrix} \text{amount of } A \text{ transferred} \\ \text{from phase II to phase} \\ \text{I from } t \text{ to } t + \Delta t \end{Bmatrix}$$

$$- \begin{Bmatrix} \text{amount of } A \text{ transferred from} \\ \text{phase I to phase II from } t \text{ to } t + \Delta t \end{Bmatrix} \quad (8.1\text{I})$$

$$\rho^{II} V^{II} \big|_{t+\Delta t} = \rho^{II} V^{II} \big|_t + \begin{Bmatrix} \text{amount of } A \text{ transferred} \\ \text{from phase I to phase II} \\ \text{from } t \text{ to } t + \Delta t \end{Bmatrix}$$

$$- \begin{Bmatrix} \text{amount of } A \text{ transferred from} \\ \text{phase II to phase I from } t \text{ to } t + \Delta t \end{Bmatrix} \quad (8.1\text{II})$$

The symbol $\big|_t$ means "evaluated at t."

As with chemical reaction, it is convenient to define the *rate of mass transfer*. To distinguish this rate from the rate of reaction we will use the boldface symbol **r**. Thus, the rate at which A accumulates in phase I is designated $\mathbf{r}_{A+}^{I}$, while the rate at which A is depleted from phase I by mass transfer through the interface is $\mathbf{r}_{A-}^{I}$. Similarly, the rate of accumulation of A in phase II because of mass transfer is $\mathbf{r}_{A+}^{II}$, and the rate at which A is lost from phase II is $\mathbf{r}_{A-}^{II}$. *The dimensions of rate of mass transfer are mass per area per time.* The rate is written on a per area basis because, all other things being equal, an increase in the area between phases will lead to a proportionate increase in the mass transferred. Notice that this is the same logic which leads to our writing reaction rate on a per volume basis.

Let a denote the total interfacial area between the two phases. Equations 8.1 can then be written

$$\rho^{I}V^{I}\big|_{t+\Delta t} = \rho^{I}V^{I}\big|_{t} + \int_{t}^{t+\Delta t} a\mathbf{r}_{A+}^{I}\, d\tau - \int_{t}^{t+\Delta t} a\mathbf{r}_{A-}^{I}\, d\tau \qquad (8.2\text{I})$$

$$\rho^{II}V^{II}\big|_{t+\Delta t} = \rho^{II}V^{II}\big|_{t} + \int_{t}^{t+\Delta t} a\mathbf{r}_{A+}^{II}\, d\tau - \int_{t}^{t+\Delta t} a\mathbf{r}_{A-}^{II}\, d\tau \qquad (8.2\text{II})$$

or, after application of the mean value theorem, dividing by Δt, and taking the limit as $\Delta t \to 0$,

$$\frac{d\rho^{I}V^{I}}{dt} = a[\mathbf{r}_{A+}^{I} - \mathbf{r}_{A-}^{I}] \qquad (8.3\text{I})$$

$$\frac{d\rho^{II}V^{II}}{dt} = a[\mathbf{r}_{A+}^{II} - \mathbf{r}_{A-}^{II}] \qquad (8.3\text{II})$$

It is evident for the batch system that the mass which leaves I must go to II, and vice versa. Thus, $\mathbf{r}_{A+}^{I} = \mathbf{r}_{A-}^{II}$ and $\mathbf{r}_{A-}^{I} = \mathbf{r}_{A+}^{II}$. For convenience we define

$$\mathbf{r}_{A} = \mathbf{r}_{A+}^{II} - \mathbf{r}_{A-}^{II} \qquad (8.4)$$

$$\mathbf{r}_{A}\ [=]\ \text{mass/length}^2\ \text{time}$$

Then

$$\frac{d\rho^{I}V^{I}}{dt} = -a\mathbf{r}_{A} \qquad (8.5\text{I})$$

$$\frac{d\rho^{II}V^{II}}{dt} = a\mathbf{r}_{A} \qquad (8.5\text{II})$$

Application of the law of conservation of mass to the transferred species, component A, leads in an identical manner to the component equations

$$\frac{dc_{A}^{I}V^{I}}{dt} = -a\mathbf{r}_{A} \qquad (8.6\text{I})$$

$$\frac{dc_{A}^{II}V^{II}}{dt} = a\mathbf{r}_{A} \qquad (8.6\text{II})$$

8.3.2 Rate Expression

As with the reacting system, it is necessary to establish the constitutive equation relating the rate, r_A, to the other system variables, particularly the concentrations c_A^I and c_A^{II}. Here, too, it is possible to examine the phenomenon of mass transport at a molecular level to obtain expressions for the rate, but we shall not pursue that path in this text, for ultimate experimental verification of the rate expression is required in any event. Instead, as with reacting systems, we shall construct the simplest form possible that is compatible with experimental observation.

In a batch system it is observed that after a sufficient period of time an equilibrium is reached in which there is a definite relationship between the concentration of the transferred substance in the two phases. That is, at equilibrium,

$$c_{Ae}^I = f(c_{Ae}^{II}) \tag{8.7}$$

where the subscript "e" denotes equilibrium. Figure 8.2 shows a typical set of experimental data for the concentration of acetone in water and 1,1,2-trichloroethane. The concentration of acetone in the organic phase is a unique function of the concentration in the aqueous phase. Also at equilibrium the time derivatives in Equations 8.5 and 8.6 must vanish, indicating that ar_A must go to zero. Since a cannot go to zero

$$\text{equilibrium:} \ r_A = 0 \tag{8.8}$$

r_A must have a form such that Equation 8.8 implies Equation 8.7.

A further observation about the form of the rate follows from the way in which we have defined r_A in Equation 8.4 in terms of r_{A+}^{II} and r_{A-}^{II}. If c_A^I is greater than its equilibrium value for a given c_A^{II}, then there must be a net transfer of A from phase I to phase II, so $r_{A+}^{II} > r_{A-}^{II}$ and $r_A > 0$. Thus

$$c_A^I - f(c_A^{II}) > 0 \Rightarrow r_A > 0 \tag{8.9}$$

Similarly

$$c_A^I - f(c_A^{II}) < 0 \Rightarrow r_A < 0 \tag{8.10}$$

The simplest form compatible with Equations 8.7 to 8.10 is

$$r_A = K_m[c_A^I - f(c_A^{II})] \tag{8.11}$$

where the *overall mass transfer coefficient*, K_m, has dimensions of length/time,

$$K_m \ [=] \ \text{length/time}$$

The equilibrium relationship, Equation 8.7, is often written

$$c_{Ae}^I = M c_{Ae}^{II} \tag{8.12}$$

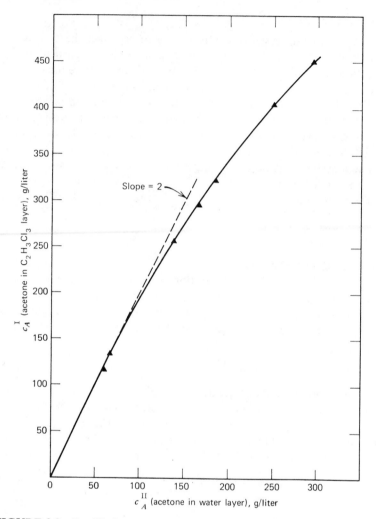

FIGURE 8.2   Equilibrium concentration of acetone in 1,1,2-trichloroethane as a function of equilibrium concentration of acetone in water. [Data of Treybal, Weber, and Daley, *Ind. Eng. Chem.*, **38**, 817 (1946), reproduced by permission.]

where the *distribution coefficient*, M, will not generally be a constant except at low concentrations. (Compare Figure 8.2.) In that case the rate expression will be written

$$\mathbf{r}_A = K_m[c_A{}^I - Mc_A^{II}] \tag{8.13}$$

If K_m and M are both allowed to be functions of concentration, then this form is quite general. For simplicity of illustration we will generally take the coefficients as constants.

It is sometimes helpful to note that Equation 8.13 has the form of a *driving force*, the distance from equilibrium $c_A{}^I - Mc_A^{II}$, divided by a *resistance*, K_m^{-1}, much like Ohm's law in electricity. It is shown in texts on mass transfer that K_m can be expressed in terms of a resistance relation like that in electricity

$$\frac{1}{K_m} = \frac{1}{k^I} + \frac{M}{k^{II}}$$

where k^I and k^{II} are associated with the individual phases.

8.3.3 Equilibrium

The most easily obtained data in a two-phase system are those at equilibrium when sufficient time has passed so that there are no further changes and the concentrations $c_A{}^I$ and c_A^{II} become constant. Everyday experience with two-phase situations indicates that equilibrium is easily obtained for many systems, and we have made use of the familiarity with the concept in developing the form of the rate expression for mass transfer, $\mathbf{r}_A$. An enormous amount of data on equilibrium concentrations in two-phase systems of all types has been obtained and recorded in the technical literature, using a variety of techniques to measure phase concentrations.

Solid-Liquid Systems. The experimental procedure for determining equilibrium concentrations in solid-liquid systems is particularly simple since the phases are easily separated. Equilibrium concentrations of the solid in the liquid are frequently determined by merely weighing the solid before the phases are contacted and again after equilibrium is established. A liquid phase with its equilibrium concentration of dissolved solid is often referred to as *saturated*, and the equilibrium concentration of solute is called the *saturation concentration*. We will denote it by c_{As}^I instead of c_{Ae}^I when dealing with solid-liquid systems. For the pure solid c_A^{II} is always just the solid density, ρ^{II}, so we have at equilibrium

$$c_{As}^I = Mc_{Ae}^{II} = M\rho^{II}$$

Solid density ρ^{II} and saturation concentration c_{As}^I are both constants at a given temperature, so M is a constant,

$$M = \frac{c_{As}^I}{\rho^{II}}$$

Saturation concentrations for several inorganic salts in water are shown in Table 8.1.

The rate expression, Equation 8.13, contains the term Mc_A^{II}. Since in a solid-liquid system $Mc_A^{II} = M\rho_A^{II} = c_{As}^I$, we can conveniently rewrite the rate expression for mass transfer explicitly in terms of the saturation concentration, which is the quantity actually measured experimentally,

$$\mathbf{r}_A = K_m[c_A{}^I - c_{As}^I] \tag{8.14}$$

TABLE 8.1   Saturation concentrations of inorganic compounds in water

Substance	g salt/100 g water		
	0°C	50°C	100°C
Ammonium Iodide NH$_4$I	154.2	199.6	250.3
Barium Nitrate Ba(NO$_3$)$_2$	5.0	17.1	34.2
Magnesium Chloride MgCl$_2$	52.8	59.2	73.0
Sodium Chloride NaCl	35.7	37.0	39.8

Liquid-Liquid Systems.   Most liquid-liquid systems of pragmatic interest are more complicated than solid-liquid systems, where problems can usually be dealt with by knowing the saturation concentration. In liquid-liquid situations it might happen that the component of interest is distributed between two liquids, which are themselves partially miscible over a range of composition. Equilibrium data on many liquid-liquid systems (often called *ternary*, or *three-component systems*) is available in the literature and in technical bulletins issued by solvent manufacturers, though it is frequently necessary to collect one's own data on the system of interest. The simplest liquid-liquid system, which is fairly commonly dealt with, is one in which the two liquids are immiscible. The equilibrium relationship $c_A^I = Mc_A^{II}$ is called Nernst's law when the distribution coefficient M is a constant, and the system is referred to as *ideal*. Few ternary systems are ideal over a wide range of concentrations, though M may be nearly constant over the concentration range encountered in a particular problem.

The experiments to determine the distribution coefficient are easy to carry out when the liquid phases are immiscible. A known amount of solute A is added to the two solvents and the system is thoroughly mixed and allowed to reach equilibrium. If the solvents are of different density, then the phases are readily separated and the amount of A in each is measured by some appropriate analytical method. M is determined by dividing c_{Ae}^I by c_{Ae}^{II}. The experiment is then repeated with a different amount of A. The data shown in Figure 8.2 for the system acetone-water-1,2,2-trichloroethane at 25°C were obtained in this way. The straight line represents the range of ideal behavior with $M = 2$. It is evident that M is not a constant over the full concentration range. In Chapter 9 we will discuss equilibrium in liquid-liquid systems further with an eye toward dealing with situations in which the two liquids are partially miscible.

8.3.4 Rate Experiment

Batch mass transfer experiments to determine K_m are usually exceedingly difficult to carry out because equilibrium is reached too rapidly to obtain reasonable data. We shall discuss this point further subsequently. We have designed an experiment to measure K_m using tablets of table salt (NaCl) where the tablet surface area is sufficiently small to allow mass transfer to take place over several minutes. The aqueous phase is initially distilled water, and dissolved salt concentration is measured using a conductivity probe.

The system is described by Equations 8.5 and 8.6. Since $c_A^{II} = \rho^{II}$ for pure salt, Equation 8.6II is redundant. Thus we have, substituting for the rate,

$$\frac{d\rho^I V^I}{dt} = -a\mathbf{r}_A = -K_m a[c_A^{\ I} - c_{As}^I] \tag{8.15I}$$

$$\frac{d\rho^{II} V^{II}}{dt} = a\mathbf{r}_A = K_m a[c_A^{\ I} - c_{As}^I] \tag{8.15II}$$

$$\frac{dc_A^I V^I}{dt} = -a\mathbf{r}_A = -K_m a[c_A^{\ I} - c_{As}^I] \tag{8.16}$$

The basic data for the experiment are shown in Table 8.2, and they lead to some further simplifications. For example, the total mass of salt is one-third of one percent of the mass of water. Thus, there will never be a significant change in the density or volume of the aqueous phase, so we can disregard the overall mass balance for phase I (Equation 8.15I) and take V^I as a constant, setting $d/dt[c_A^I V^I] = V^I[dc_A^I/dt]$. Furthermore, c_A^I is at all times more than two orders of magnitude less than c_{As}^I, so $c_{As}^I - c_A^I$ is negligibly different from c_{As}^I

$$c_{As}^I - c_A^{\ I} \simeq c_{As}^I$$

Also, ρ^{II} is a constant, so $d/dt[\rho^{II} V^{II}] = \rho^{II}[dV^{II}/dt]$ and the experiment can be described by the two equations

$$\rho^{II} \frac{dV^{II}}{dt} = -K_m a c_{As}^I \tag{8.17}$$

$$V^I \frac{dc_A^I}{dt} = K_m a c_{As}^I \tag{8.18}$$

Equations 8.17 and 8.18 can be added to give

$$\rho^{II} \frac{dV^{II}}{dt} + V^I \frac{dc_A^I}{dt} = 0$$

TABLE 8.2   Concentration of salt dissolved versus time

Number of tablets	$N = 30$
Density of salt	$\rho^{II} = 2.16$ g/cm^3
Total mass of tablets	$= 19.2$ g
Initial volume of tablets	$V_0^{II} = 19.2/2.16 = 8.85$ cm^3
Volume of water	$V^I = 6000$ cm^3
Surface-to-volume factor	$\alpha = 5.32$
Saturation concentration	$c_{As}^I = 0.360$ g/cm^3

Time (seconds)	$1000 c_{As}^I$ (g/cm^3)
0	0
15	0.30
30	0.35
45	0.64
60	0.89
75	1.08
90	1.10
105	1.24
120	1.40
135	1.49
150	1.68
165	1.76
195	2.06
200	2.14
240	2.31
270	2.43

which, on integration, yields

$$\rho^{II} V_0^{II} - \rho^{II} V^{II} = V^I c_A^I \tag{8.19}$$

where V_0^{II} is the initial volume of solid. $\rho^{II} V_0^{II}$ is, of course, the initial mass of solid.

The interfacial area, a, is related to the solid volume, V^{II}, through the solid geometry of the tablets. The N tablets are assumed to be identical, so the volume per tablet is V^{II}/N. The area per tablet is then $\alpha[V^{II}/N]^{2/3}$, where α is a constant which depends on tablet geometry. For a sphere, $\alpha = [36\pi]^{1/3} = 4.84$. For a cube, $\alpha = 6$, and for a square cylinder (height = diameter) $\alpha = 5.50$. The tablets used in these experiments were nearly square cylinders, with $\alpha = 5.32$. The total surface area is then N times the area per tablet

$$a = N\alpha[V^{II}/N]^{2/3} = \alpha N^{1/3}[V^{II}]^{2/3} \tag{8.20}$$

Equation 8.17 is then

$$\rho^{II} \frac{dV^{II}}{dt} = -K_m \alpha N^{1/3} c_{As}^{I} [V^{II}]^{2/3}$$

$$\frac{dV^{II}}{[V^{II}]^{2/3}} = [-K_m \alpha N^{1/3} c_{As}^{I} / \rho^{II}] \, dt$$

which, on integration, yields

$$V^{II} = \left\{ [V_0^{II}]^{1/3} - \frac{K_m \alpha N^{1/3} c_{As}^{I} t}{3\rho^{II}} \right\}^3 \tag{8.21}$$

This is substituted into Equation 8.19 and rearranged to a form convenient for comparison with data

$$[1 - \hat{c}]^{1/3} = 1 - \omega t \tag{8.22}$$

$$\hat{c} = V^I c_A{}^I / \rho^{II} V_0^{II}$$

$$\omega = \frac{K_m \alpha N^{1/3} c_{As}^{I}}{3[V_0^{II}]^{1/3} \rho^{II}}$$

Notice that $\hat{c}$ is simply the fraction of the salt that is in solution.

The data are plotted according to Equation 8.22 in Figure 8.3. They do follow a straight line passing through unity, as required. Using the experimental value $\omega = 1.46 \times 10^{-3}$ sec^{-1} and the data given in Table 8.2 we then

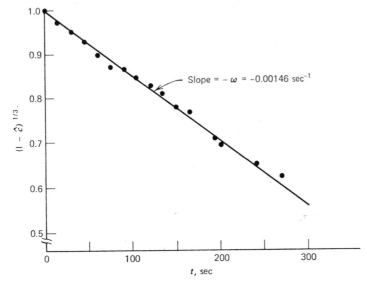

FIGURE 8.3 Rate of solution of salt in water. Data are plotted according to Equation 8.22.

compute $K_m = 3.3 \times 10^{-3}$ cm/sec. Though we shall not show any further experimental data, it is found that K_m in two-phase tank-type systems is nearly always within an order of magnitude of the value computed here. Since the range of K_m is so limited the significant problem is determination of the interfacial area for each situation. The interfacial area varies greatly depending on the degree of agitation, and the development of relationships between power input to the mixer, geometry, physical properties, and interfacial area is an active research field at present.

8.3.5 Approach to Equilibrium

We remarked previously that batch experiments for mass transfer rate data are generally too rapid to obtain useful data. Now that we have a reasonable value for K_m this can be easily demonstrated using the system equations. We will assume that we have two liquid phases which are initially of equal volume and that the total amount of dissolved material A in the system which is transferred between phases is so small that changes in volume are negligible throughout the course of the experiment. The overall equations provide no useful information. Equations 8.6, the component balances for the two phases, can be written

$$V \frac{dc_A^{\,I}}{dt} = -a\mathbf{r}_A = -K_m a[c_A^{\,I} - Mc_A^{\,II}] \tag{8.23I}$$

$$V \frac{dc_A^{\,II}}{dt} = a\mathbf{r}_A = K_m a[c_A^{\,I} - Mc_A^{\,II}] \tag{8.23II}$$

We have not included superscripts I and II for the volumes, since they are equal and constant. M will be taken as a constant.

Addition of Equations 8.23I and II gives

$$V \frac{dc_A^{\,I}}{dt} + V \frac{dc_A^{\,II}}{dt} = 0$$

or

$$c_A^{\,I} - c_{A0}^{\,I} + c_A^{\,II} - c_{A0}^{\,II} = 0 \tag{8.24}$$

$c_{A0}^{\,I}$ and $c_{A0}^{\,II}$ are the values of $c_A^{\,I}$ and $c_A^{\,II}$, respectively, at $t = 0$. Substitution for $c_A^{\,II}$ in Equation 8.23I then gives

$$\frac{dc_A^{\,I}}{dt} = -\frac{K_m a[1 + M]}{V} c_A^{\,I} + \frac{mMK_m a}{V} \tag{8.25}$$

where $m = c_{A0}^{\,I} + c_{A0}^{\,II}$. This is an equation of a form that we have solved many times before, and the solution can be written

$$c_A^{\,I} = c_{Ae}^{\,I} \left\{ 1 - \left[1 - \frac{c_{A0}^{\,I}}{c_{Ae}^{\,I}} \right] e^{-\{K_m a[1+M]t\}/V} \right\} \tag{8.26}$$

where the equilibrium concentration, c_{Ae}^{I}, is computed from the equilibrium relation and Equation 8.24 as

$$c_{Ae}^{I} = \frac{mM}{1 + M} \tag{8.27}$$

According to Equation 8.26, the system will be at equilibrium when the exponential term is negligible. This will occur for an exponent of about -3, so the total time for the experiment will be determined approximately by

$$\frac{K_{m}a[1 + M]t_{e}}{V} \simeq 3 \tag{8.28}$$

For the data shown in Figure 8.2, $M \simeq 2$, and we have seen that K_m can be expected to be of order 3×10^{-3} cm/sec. Thus

$$t_{e} \simeq \frac{10^{3}V}{3a} \tag{8.29}$$

where lengths are in centimeters and time in seconds. Now, if the agitation is such that one phase is effectively dispersed into N droplets then, according to Equation 8.20,

$$a = \alpha N^{1/3}V^{2/3}$$

where α is approximately 5. Thus

$$t_{e} \simeq 10^{2}\left[\frac{V}{N}\right]^{1/3} \tag{8.30}$$

V/N is the volume of a typical droplet. Droplets with a diameter of one millimeter are easily obtained in most low viscosity systems, so $V/N \sim [1/6]\pi[10^{-1}]^{3} \sim 0.5 \times 10^{-3}$, and $t_{e} \sim 10$ seconds. Thus, the system will be more than half way to equilibrium after three seconds and completely there by ten seconds. This is to be compared with the several minutes typically shown for batch reaction systems in Chapter 5.

8.3.6 Further Comments

It is important to note the essential difference in usefulness between the batch experiment for a mass transfer system and the batch reactor experiments described in Chapter 5. We have already observed that the batch mass transfer experiment may be considerably more difficult to carry out than the reaction experiment. This is true not only because of the rapid approach to equilibrium but also because the presence of two phases in intimate contact can often increase the sampling and measurement problems considerably. There is a second extremely important difference that might not be obvious from the discussion thus far. The data from the single-phase batch reactor experiment are sufficient to enable us to compute the design specifications for a continuous processing unit, either a well-stirred continuous flow reactor or a tubular reactor. Such is not the case in the two-phase process. Both the rate and the interfacial area are required, and it is rare that data on interfacial area obtained in a small-scale batch contactor provide meaningful information about interfacial areas in large, continuous flow devices.

8.4 CONTINUOUS FLOW TWO-PHASE SYSTEMS

Continuous flow two-phase systems are ones in which both phases are continuously fed to and removed from the system. They are widely employed for the variety of unit operations described in Section 8.2 and may be carried out in both tank-type and tubular systems. They are also used in the laboratory to collect experimental data for certain systems. Systems in which only one phase flows to and from the system are classified as semiflow continuous systems, and although not commonly used in processing units they are often employed to collect experimental rate data for fluid-fluid mixtures.

Our goal in this section is to gain an appreciation of the problems of two-phase design and operation by examining a simple well-stirred continuous flow tank-type mass transfer system in detail. This will give us an introductory understanding of the more complex problems and some further practice with model development and manipulation for design purposes.

A sketch of the continuous flow system is shown in Figure 8.4. Two pieces of equipment are shown, a tank-type contactor and some device which effects

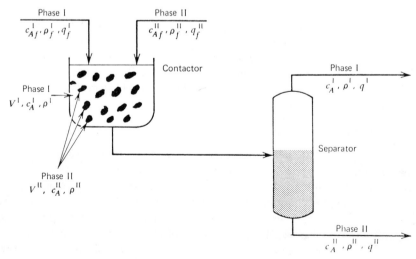

FIGURE 8.4   Continuous flow two-phase process.

a separation between the phases. We assume that the contents of the contactor are well mixed so that the two phases are in intimate contact and there is no spatial dependence of the concentration of a species in either phase. We will assume that all mass transfer takes place in the contactor and that the sole function of the separator is to separate the two phases that are mixed in the contactor. We do not consider the separator operation in detail. It will often be no more than a holding tank large enough to allow separation by gravity. In some gas-liquid systems the separation takes place in the contactor and a distinct separation device is not needed. The concentrations in the exit stream from the separator are assumed to be the same as those that we would find if we sampled the individual phases in the contactor. This is reasonable since, without agitation, there is little mass transfer area available in the separator. The contactor and the separator are frequently referred to as a *stage*.

Our control volumes will be the same as those designated for the batch two-phase systems, and we can develop the model equations by applying the law of conservation of mass in the same manner as we did in Section 8.3. The overall mass balance equations are

$$\frac{d\rho^{I} V^{I}}{dt} = q_f^{I} \rho_f^{I} - q^{I} \rho^{I} - a\mathbf{r}_A \tag{8.31I}$$

$$\frac{d\rho^{II} V^{II}}{dt} = q_f^{II} \rho_f^{II} - q^{II} \rho^{II} + a\mathbf{r}_A \tag{8.31II}$$

Notice that we do not consider mass to have left the control volume until the exit of the separator. This is conceptually the same as if we assumed that separated streams would issue from the contactor. The component balance equations will be written for the case where only one component is being transferred. This is the most commonly encountered situation and allows us to keep the development from becoming too complex. c_{Af}^{I} and c_{Af}^{II} are the inlet stream concentrations. Concentration in the tank and at the separator exit are indicated by c_A^{I} and c_A^{II}:

$$\frac{dV^{I} c_A^{I}}{dt} = q_f^{I} c_{Af}^{I} - q^{I} c_A^{I} - a\mathbf{r}_A \tag{8.32I}$$

$$\frac{dV^{II} c_A^{II}}{dt} = q_f^{II} c_{Af}^{II} - q^{II} c_A^{II} + a\mathbf{r}_A \tag{8.32II}$$

Although the transient behavior is important for problems involved with start-up, shut-down, and control of the system, the basic design is carried out for the steady state case where the time derivatives are equal to zero. Thus,

Equations 8.31 and 8.32 become

$$0 = q_f^I \rho_f^I - q^I \rho^I - a\mathbf{r}_A \tag{8.33I}$$

$$0 = q_f^{II} \rho_f^{II} - q^{II} \rho^{II} + a\mathbf{r}_A \tag{8.33II}$$

$$0 = q_f^I c_{Af}^I - q^I c_A^I - a\mathbf{r}_A \tag{8.34I}$$

$$0 = q_f^{II} c_{Af}^{II} - q^{II} c_A^{II} + a\mathbf{r}_A \tag{8.34II}$$

This set of equations is similar to that developed in Chapter 7 for the continuous flow tank reactor (Equations 7.18 to 7.20). The rate of mass transfer is taken as the expression defined by Equation 8.13

$$\mathbf{r}_A = K_m[c_A^I - Mc_A^{II}] \tag{8.13}$$

A complete solution to this set of equations will require the further constitutive relations between phase densities and compositions. We will return to this more general problem subsequently, but for the moment we shall deal only with the limited, but commonly occurring case in which the total mass of A transferred is not sufficient to change phase volumes. In that case $q_f^I = q^I$, $q_f^{II} = q^{II}$, and only the component equations are needed to describe the system fully. *This approximation leads to negligible error for most applications.* Combining the rate and component equations then gives

$$0 = q^I[c_{Af}^I - c_A^I] - K_m a[c_A^I - Mc_A^{II}] \tag{8.35I}$$

$$0 = q^{II}[c_{Af}^{II} - c_A^{II}] + K_m a[c_A^I - Mc_A^{II}] \tag{8.35II}$$

Addition of Equations 8.35I and 8.35II yields an alternate relation between the concentrations

$$c_A^I = c_{Af}^I + \frac{q^{II}}{q^I}[c_{Af}^{II} - c_A^{II}] \tag{8.36}$$

Equations 8.35 contain nine quantities, $q^I, q^{II}, c_{Af}^I, c_{Af}^{II}, c_A^I, c_A^{II}, M, K_m$, and a. Specification of any seven independent quantities allows us to determine the remaining two by solving the two algebraic equations. The situation is mathematically equivalent to the continuous flow tank reactor problems we discussed in Chapter 7. We can work with Equations 8.35I and 8.35II or, if more convenient, 8.35I and 8.36 or 8.35II and 8.36.

8.4.1 Equilibrium Stage

A typical problem is to determine the effluent concentrations c_A^I and c_A^{II}. Equations 8.35 can be readily combined to obtain c_A^{II} in terms of the remaining parameters

$$c_A^{II} = \frac{c_{Af}^I}{\lambda + M + q^{II}/K_m a} + \frac{c_{Af}^{II}}{1 + \dfrac{M}{\lambda + q^{II}/K_m a}} \tag{8.37}$$

Here $\lambda = q^{II}/q^I$. Notice that the volume, and therefore the holdup in the tank of neither phase appears!

The ratio $q^{II}/K_m a$ can be written

$$\frac{q^{II}}{K_m a} = \frac{V^{II}/K_m a}{V^{II}/q^{II}} = \frac{V^{II}/K_m a}{\theta^{II}}$$

where θ^{II} is the residence time of the dispersed phase. M will be a number of order unity. We demonstrated in Section 8.3.5 that $V^{II}/K_m a$ will be of the order of several seconds provided there is sufficient agitation to produce a dispersed phase with droplets of the order of a millimeter in diameter. If the system is designed for a holdup of, say, the order of minutes, so that $q^{II}/K_m a \ll 1$, then the $q^{II}/K_m a$ terms in Equation 8.37 can be neglected (they can be neglected in the second term only if $q^{II}/K_m a \ll \lambda$) and the effluent concentration can be written

$$c_A^{II} = \frac{c_{Af}^I + \lambda c_{Af}^{II}}{\lambda + M} \tag{8.38}$$

The rate of mass transfer does not appear and the calculations can be carried out without knowledge of $K_m a$! Clearly this is an extraordinarily helpful simplification.

Equation 8.38 is formally equivalent to taking the limit as $K_m a \to \infty$ in Equation 8.37. This result is known as an *equilibrium stage* and can be obtained, with considerably less insight, in a more direct manner. Equation 8.36 is always valid as a relation between the concentrations, independent of the form of the rate expression. It is simply a statement that at steady state the rate at which A enters the tank in all streams, $q^I c_{Af}^I + q^{II} c_{Af}^{II}$, equals the rate at which it leaves in all streams, $q^I c_A^I + q^{II} c_A^{II}$. By assuming that the two phases are in equilibrium and setting $c_A^I = M c_A^{II}$ in Equation 8.36, the result in Equation 8.38 follows directly. Hence the name equilibrium stage. The concept of the equilibrium stage is used extensively in design computations for the unit operations touched on in Section 8.2, and we will return to this important topic in considerably more detail in Chapter 9.

Example 8.1

An aqueous solution containing 200 g/liter of acetone is to be purified by continuous extraction with pure trichloroethane. How much acetone can be removed if the flow rate of both aqueous and organic streams is 10 liters/min?

We call the organic phase I and the aqueous phase II. If we assume that the contactor is ideal and flow rates are constant, then Equation 8.38 applies with $c_{Af}^I = 0$ (pure trichloroethane):

$$c_A^{II} = \frac{\lambda c_{Af}^{II}}{\lambda + M} = \frac{200\lambda}{\lambda + M}$$

$$\lambda = \frac{q_{II}}{q_I} = \frac{10 \text{ liter/min}}{10 \text{ liter/min}} = 1$$

For the acetone-water-trichloroethane system the data in Figure 8.2 show that for $c_A{}^I \leq 150$ the equilibrium can be approximated by

$$c_A{}^I = 2c_A^{II}, \qquad M = 2$$

Thus

$$c_A^{II} = \frac{200}{1 + 2} = 66.7 \text{ g/liter}$$

$$c_A{}^I = 2c_A^{II} = 133.3 \text{ g/liter}$$

That is, the acetone content in the aqueous phase is reduced from 200 to 66.7 g/liter in this equilibrium stage.

Example 8.2

Suppose the flow rate of organic is increased to 20 liters/min with the flow rate of aqueous phase maintained at 10 liters/min? Then

$$\lambda = \frac{q^{II}}{q^I} = \frac{10}{20} = \frac{1}{2}$$

$$c_A^{II} = \frac{\lambda c_{Af}^{II}}{\lambda + M} = \frac{[1/2] \times 200}{[1/2] + 2} = 40 \text{ g/liter}$$

$$c_A{}^I = 2c_A^{II} = 80 \text{ g/liter}$$

That is, by doubling the amount of solvent the acetone content in the aqueous phase changes only from 66.7 to 40 g/liter.

Example 8.3

As an alternate to the increased solvent flow in Example 8.2 and to prepare for the development in Chapter 9, suppose that the aqueous stream leaving the contactor in Example 8.1 is brought into contact in a second unit with another pure trichloroethane stream at 10 liter/min. Then $c_{Af}^{II} = 66.7$ g/liter and $\lambda = 1$

$$c_A^{II} = \frac{\lambda c_{Af}^{II}}{\lambda + M} = \frac{66.7}{1 + 2} = 22.2 \text{ g/liter}$$

$$c_A{}^I = 2c_A^{II} = 44.4 \text{ g/liter}$$

Thus, by using the same amount of solvent as in Example 8.2, but by dividing the total between two consecutive contacting stages, the residual acetone in the water stream is reduced by a factor of nearly two.

8.4.2 Deviation from Equilibrium

Because energy in the form of agitation must be put into two-phase systems to generate adequate interfacial area for mass transfer, it is evident that there will be situations where equilibrium will not be reached in a stage. This has led to attempts to correlate $K_m a$ with factors like the power per cubic foot in order to compute the effluent composition. One quantity often used is the *stage efficiency*

$$\mathscr{E} = \frac{c_A^{II} - c_{Af}^{II}}{c_{Ae}^{II} - c_{Af}^{II}} \tag{8.39}$$

This is nothing more than the fractional approach to equilibrium. Noting that $c_{Ae}^{II} = c_A{}^I/M$ and substituting Equation 8.39 into 8.36 gives an equation for the effluent in terms of this additional parameter,

$$c_A^{II} = \mathscr{E}\,\frac{c_{Af}^{I} + \lambda c_{Af}^{II}}{\lambda\mathscr{E} + M} + \frac{M[1 - \mathscr{E}]c_{Af}^{II}}{\lambda\mathscr{E} + M} \tag{8.40}$$

When $\mathscr{E} \to 1$, Equation 8.40 reduces to Equation 8.38 for the equilibrium stage. As might be expected, the efficiency is simply related to the mass transfer coefficient. By comparison of Equations 8.37 and 8.40 the relationship can be established as

$$\mathscr{E} = \frac{M}{M + q^{II}/K_m a} \tag{8.41}$$

The stage efficiency can often be correlated with design variables. For example, Figure 8.5 shows some data of Flyn and Treybal for transfer of benzoic acid in toluene-water and benzoic acid-water systems. $\mathscr{E}_0$ is the measured efficiency at zero agitator speed (which is high) and $\mathscr{E}_a$ is the residual defined by

$$\mathscr{E} = \mathscr{E}_0 + [1 - \mathscr{E}_0]\mathscr{E}_a \tag{8.42}$$

Clearly, as $\mathscr{E}_a \to 1$, $\mathscr{E} \to 1$. ε is the energy per unit volume

$$\varepsilon = \frac{P}{q^I + q^{II}} \tag{8.43}$$

where P is the power supplied to the agitator. The correlation appears to work for each system for the type of mixer used (six-bladed turbine impellers), and it is evident that beyond approximately 100 ft lb$_f$/ft^3 an equilibrium stage can be assumed.

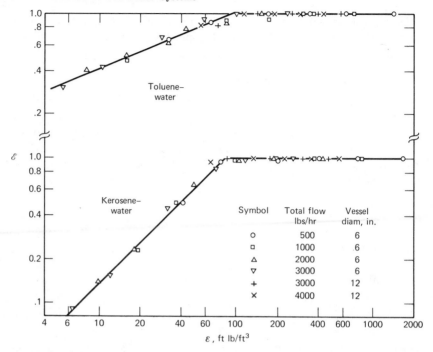

FIGURE 8.5   Residual efficiency in a stirred tank as a function of energy per unit volume. [Data of Flyn and Treybal, *A.I.Ch.E. Journal*, **1**, 324 (1955), reproduced by permission.]

Many other correlations for the efficiency and the quantity $K_m a$ are available in the published literature. A further discussion of this important topic is beyond the scope of this introductory treatment, however.

8.5   TWO-PHASE REACTORS

Many important reactions take place in two-phase reactors, in which a reactant is transferred from one phase in which it is fed to the reactor to the second phase in which it reacts. This is often done to allow the reactants to come into contact with one another uniformly throughout the reactor when high localized concentrations at the entry might lead to undesirable side reactions and, perhaps, fouling or plugging. In other cases, one of the reactants is naturally present in another phase, as in the bioxidation of liquid waste discussed in Example 1.2, where oxygen must be transferred from an air stream to solution in the liquid.

We will consider here only the simplest case of a two-phase reactor. Phase I contains a species, A, which, when transferred to phase II undergoes the pseudo first-order reaction $A \rightarrow D$. We will assume that the total amount of A transferred is sufficiently small that flow rate changes are negligible because of its transfer and reaction, so that the overall balances can be neglected. The component mass balances are then obtained as

$$\frac{dc_A{}^I V^I}{dt} = q^I c_{Af}{}^I - q^I c_A{}^I - a\mathbf{r}_A \tag{8.44I}$$

$$\frac{dc_A{}^{II} V^{II}}{dt} = -q^{II} c_A{}^{II} + a\mathbf{r}_A - V^{II} r_{A-} \tag{8.44II}$$

$\mathbf{r}_A$ is as given previously, and for the pseudo first-order reaction $r_{A-} = kc_A^{II}$. At steady state the time derivatives are zero, and after substitution of $\mathbf{r}_A$ and r_{A-} into Equations 8.44 the resulting algebraic equations are easily solved for c_A^{II}

$$c_A^{II} = \frac{c_{Af}^I}{M + [1 + k\theta^{II}][\lambda + q^{II}/K_m a]} \tag{8.45}$$

Here $\lambda = q^{II}/q^I$ and $\theta^{II} = V^{II}/q^{II}$.

Some interesting limiting cases follow. If agitation is sufficient to ensure that $q^{II}/K_m a \ll \lambda$, then the mass transfer rate term can be neglected and we obtain

$$\frac{q^{II}}{K_m a} \ll \lambda: \quad c_A^{II} \to \frac{c_{Af}^I}{M + \lambda[1 + k\theta^{II}]} \tag{8.46}$$

This case is often referred to as *reaction limited*, since the mass transfer is sufficiently rapid that the reaction rate is the only rate term of importance. This is analogous to the equilibrium stage and reduces to it as $k\theta^{II} \to 0$.

The second limit of interest, which can occur in a highly viscous system where efficient agitation is difficult, is one in which a is sufficiently small that $K_m a/q^{II}$ is small, or $q^{II}/K_m a \gg \lambda$. In that case we would also have $q^{II}/K_m a \gg M$ and both λ and M terms will drop out of Equation 8.45, leading to

$$\frac{q^{II}}{K_m a} \gg \lambda: \quad c_A^{II} \to \frac{K_m a}{q^{II}} \frac{c_{Af}^I}{1 + k\theta^{II}} \tag{8.47}$$

This case is referred to as *mass transfer limited*. A characteristic of such a system, which is often encountered in the production of polymers, is that the conversion depends strongly on the design and intensity of the mixing device even when complete mixing in the sense of spatial uniformity has been achieved.

8.6 CONCLUDING REMARKS

In reviewing this chapter you should return again to the logic diagram in Figure 3.8. Notice the critical role played by the selection of the control volume and the parallel between the development of the rate of mass transfer here and the reaction rate in Chapter 5.

The estimates on the speed at which equilibrium is attained are of particular importance and should be examined carefully. Calculations of this type often lead to substantial simplifications in engineering problems. In the case of mass transfer processes the consequence is the equilibrium stage. Compare Section 8.3.3 with Section 5.6 on reactions going to equilibrium. Nearly all practical design procedures are based on the equations for the equilibrium stage, sometimes taking the stage efficiency $\mathscr{E}$ into account.

A good discussion of two-phase systems can be found in Sections 18–21 of the *Chemical Engineers' Handbook*,

8.1 J. H. Perry, *Chemical Engineers' Handbook*, 4th ed., McGraw-Hill, New York, 1963.

This is well worth consulting to get a feeling for the actual equipment involved, particularly, in the context of this chapter, the material in Section 21 on liquid-liquid systems. Several texts are concerned in part with the question of determination of rates of mass transfer:

8.2 R. B. Bird, W. E. Stewart, and E. N. Lightfoot, *Transport Phenomena*, Wiley, New York, 1960.

8.3 T. K. Sherwood and R. L. Pigford, *Absorption and Extraction*, 2nd ed., McGraw-Hill, New York, 1952.

8.4 R. E. Treybal, *Mass Transfer Operations*, 2nd ed., McGraw-Hill, New York, 1968.

8.5 G. Astarita, *Mass Transfer with Chemical Reaction*, Elsevier, Amsterdam, 1967.

8.7 PROBLEMS

8.1 Develop the basic model equations for an isothermal batch two-phase system in which there are two species, A, and B, which can transfer between phases. You may write your equations in terms of $\mathbf{r}_A$ and $\mathbf{r}_B$. What form would you expect $\mathbf{r}_A$ and $\mathbf{r}_B$ to have?

8.2 An experiment is carried out with the system NaCl and water to determine the saturated concentration, $c_{NaCl_s}^I$, at 25°C. One hundred grams of NaCl is added to 1 liter of water and the system is brought to equilibrium. Can c_s^I be measured? Describe in detail how you would measure c_s^I if all you had available was a 1 liter beaker, water, NaCl, and a balance.

8.3 In the development leading to Equations 8.5 and 8.6 we made use of the relationships

$$\mathbf{r}_{A-}^I = \mathbf{r}_{A+}^{II}, \qquad \mathbf{r}_{A+}^I = \mathbf{r}_{A-}^{II}$$

The rate at which A disappears from phase I and the rate at which it appears in phase II can be postulated in terms of the concentration of A at each side of the phase interface, c_A^{I*}, c_A^{II*}:

$$\mathbf{r}_{A+}^{I} - \mathbf{r}_{A-}^{I} = k^{I}[c_A^{I*} - c_A^{I}]$$

$$\mathbf{r}_{A+}^{II} - \mathbf{r}_{A-}^{II} = k^{II}[c_A^{II*} - c_A^{II}]$$

If $c_A^{I*} = Mc_A^{II*}$ show that Equation 8.13 can be derived and that K_m is related to k^{I} and k^{II} as follows.

$$\frac{1}{K_m} = \frac{1}{k^{I}} + \frac{M}{k^{II}}$$

8.4 Suppose that c_A^{I*} and c_A^{II*} are related as follows.

(a) $c_A^{I*} = a + bc_A^{II*}$

(b) $c_A^{I*} = k[c_A^{II*}]^2$

Repeat the derivation developed in Problem 8.3 and comment on the remarks regarding K_m and M following Equation 8.13.

8.5 In Section 8.3.4, Equations 8.5 and 8.6 were modified to apply to the salt tablet experiment, leading to Equations 8.15 and 8.16. In this problem we wish to develop model equations modified to meet the following conditions:

(a) An experiment in which both V^{I} and V^{II} are constant, and $\mathbf{r}_A$ is given by Equation 8.13. What is the physical significance of the assumptions about V^{I} and V^{II}?

(b) An experiment with the same conditions as (a), but with the added stipulation that c_A^{II} is constant (a solid-liquid system).

8.6 Solve the equations developed in Problem 8.5(a) and compare the result with Equation 8.26.

8.7 Using the model equations developed for Problem 8.5(b) find how c_A^{I} varies with t. Compare your result with the case considered in Section 8.3.5.

8.8 The following data were collected by Rushton, Nagata, and Rooney [*AIChE Journal*, **10**, 298 (1964)] in a batch experiment on the distribution of octanoic acid between an aqueous (corn syrup-water) and organic (xylene) phase. The concentrations c_A^{I} were measured with a calibrated conductivity probe placed in the continuous aqueous phase. Two thousand milliliters of aqueous phase were initially placed in the tank with 200 ml of xylene. The mixer was started and this system was brought to equilibrium with respect to drop size. When this completely mixed state was achieved 250 ml of an aqueous solution of octanoic acid was added and the concentrations recorded.

t seconds	$c_A^{\mathrm{I}} \times 10^4$ Concentration of octanoic acid in water phase, g-moles/liter
0	2.75
10	2.13
20	1.72
30	1.45
40	1.23
60	1.03
80	0.94
120	0.83
∞	0.78

(a) Develop the mathematical description for this particular experiment.

The initial concentration of octanoic acid in the 250 ml charge is 0.0041 g-moles/liter.

(b) How should the data be plotted so that it will lie on a straight line?

(c) Prepare the plot, determine the slope of the line, calculate $K_m a$ and compare your value with that reported, 75cm³/sec.

8.9 Develop the basic model equations for an isothermal batch two-phase system in which a component A can transfer between phases and react with B in phase II. You should assume that B is insoluble in phase I.

8.10 (a) Make the following assumptions with the model developed in Problem 8.9:

(i) The rate of mass transfer is described by Equation 8.13.

(ii) The amount of A transferred is so small that the phase volumes V^{I} and V^{II} are constant.

(iii) The reaction rate is first order in the concentration of A.

(iv) B is in excess in phase II so that the reaction rate is approximately independent of the concentration of B.

(v) At $t = 0$ there is no A in phase II.

Obtain a differential equation for the concentration c_A^{II} in the effluent (refer to Section 17.10 for the method of combining two first-order differential equations into one second-order equation). Solve if you are familiar with the solution of linear second-order differential equations with constant coefficients, Section 17.3.

(b) Making the assumptions outlined in part (a), suppose further that the system is far from equilibrium, so that $M c_A^{\mathrm{II}} \ll c_A^{\mathrm{I}}$. Find

$c_A{}^I$ and c_A^{II} as functions of time. What physical restriction is implied by this additional assumption?

8.11 One step in the preparation of nicotine sulfate from tobacco is the extraction of nicotine from aqueous solution by kerosene in a single equilibrium stage. The following data on the equilibrium distribution of nicotine between water and kerosene were reported by Claffey et al., *Industrial & Engineering Chemistry*, **92**, 166 (1950):

Nicotine in aqueous phase, g/liter	0.62	1.49	2.92	5.70
Nicotine in organic phase, g/liter	0.39	0.96	2.07	4.18

11.8	17.9	24.9	31.5
8.22	12.2	15.3	18.9

(a) Estimate the distribution coefficient using a least-squares fit to the data.

(b) Compute the kerosene/water ratio required to remove 90 and 99 percent of the nicotine from the aqueous solution.

8.12 The following data showing the distribution of picric acid in a benzene-water system at 15°C are reported in Daniels, *Outlines of Physical Chemistry*, Wiley, New York, 1948. Notice that concentrations are in molar units.

$c_A{}^I$ Picric acid in benzene g-moles/liter	c_A^{II} Picric acid in water g-moles/liter
0.000932	0.00208
0.00225	0.00327
0.0101	0.00701
0.0199	0.0101
0.0500	0.0160
0.100	0.0240
0.180	0.0336

(a) Plot this data as $c_A{}^I$ versus c_A^{II} and comment on the suitability of Nernsts "law" for correlating the data. Is there some range over which this "law" could be used with reasonable accuracy?

(b) Using the procedures described in Section 6.7, fit a quadratic expression of the following form to the data

$$c_A{}^I = a + bc_A^{II} + d[c_A^{II}]^2$$

8.13 If picric acid is to be extracted with benzene from a water solution continuously in a well-mixed tank-type contactor, Equations 8.33 and 8.34 describe the steady state operation.

(a) Show that you can derive Equation 8.36 without a knowledge of the form of $\mathbf{r}_A$. Why is it necessary to show this for this particular system?

(b) If equilibrium can be achieved in a continuous contactor operating under the following conditions, find $c_A{}^{\mathrm{I}}$.

$$q^{\mathrm{I}} = 50 \text{ liters/min} \qquad q^{\mathrm{II}} = 50 \text{ liters/min}$$

$$c^{\mathrm{I}}_{Af} = 0.00 \text{ g-moles/liter} \qquad c^{\mathrm{II}}_{Af} = 0.02 \text{ g-moles/liter}$$

(c) Develop a graphical procedure for solving this problem and clearly indicate how construction should be carried out on the $c_A{}^{\mathrm{I}}$ versus c^{II}_A graph.

8.14 A second equilibrium stage is to be used in the picric acid extraction process described in Problem 8.13. Find the concentration of picric acid in both the water and benzene streams if the water stream leaving the first stage is fed to a second stage where it is contacted with a fresh pure benzene stream flowing at 50 liters/min.

Develop a graphical procedure that will allow one to find exit concentrations for any number of stages if flow rates and feed compositions to stage 1 are given and if fresh solvent is always used in each stage.

8.15 Trace amounts of phenol are to be removed continuously from a water stream by extraction with pure xylene. Denote the aqueous phase as I and the organic phase as II. Then from the value in Perry's *Chemical Engineers' Handbook* we have $M = 1.4$.

(a) If 90 percent of the phenol is to be removed in a single equilibrium stage what xylene/water flow rate ratio is required?

(b) If two equilibrium stages are to be used with the effluent from stage 1 as a feed to the second stage what xylene/water flow rate ratio is required for each stage?

8.16 In an experiment carried out by one of our undergraduate students a water stream and a chloroform stream were fed continuously to a 1000 cm³ cylindrical stirred tank of cross-sectional area 81 cm² with the exit at one-fourth the total height. Ammonia in amounts up to 1.7 g/liter was dissolved in the chloroform feed stream and extracted by the water stream. The fractional approach to equilibrium, $\mathcal{E}$, was measured in the water effluent as a function of mixer RPM. The chloroform stream is denoted as phase I, the water stream as phase II. Data obtained are as follows:

RPM	q^{I} cm³/sec	q^{II} cm³/sec	$\mathscr{E}$
0	3.7	13.0	0.71
0	3.7	8.5	0.98
0	2.4	7.6	0.82
0	2.5	6.5	0.51
0	2.4	8.4	0.89
0	2.4	8.4	0.95
0	2.4	8.4	0.87
0	2.4	8.4	0.92
0	3.9	18.2	0.93
182	3.9	18.2	0.97
235	3.9	18.2	0.97
305	3.9	18.2	0.98
370	3.9	18.2	0.98
500	3.9	18.2	0.99
505	3.9	18.2	0.98
580	3.9	18.2	0.97
630	3.9	18.2	0.99

It can be assumed that the densities of phases I and II are essentially independent of NH_3 concentration: $\rho_{\mathrm{I}} = 1.47$ g/cm³, $\rho_{\mathrm{II}} = 1.0$ g/cm³. The equilibrium distribution coefficient for ammonia, $M = c_{Ae}^{\mathrm{I}}/c_{Ae}^{\mathrm{II}}$, was found to have a nearly constant value of 0.044 at room temperature over the range studied.

(a) At 0 RPM the phase interface in the tank is nearly a flat horizontal plane. Estimate the mass transfer coefficient, K_m. Compare with the value obtained for the batch salt experiment in Section 8.3.4. What assumptions have you made? Comment on their validity, and determine whether your estimate of K_m provides some sort of bound on the actual value.

(b) Correlate $K_m a$ with RPM for this system, and in that way obtain a correlation for the efficiency of the mixer. Can you use your estimate of K_m from part (a) to obtain a meaningful correlation for surface area?

(c) Suppose that you wished to add another species to the aqueous phase which reacts with ammonia and is not soluble in chloroform. What terms in Equation 8.45, if any, can be neglected? Does this correspond to either the reaction limiting or mass transfer limiting case?

8.17   Read and discuss in the context of Chapter 8 the paper by R. L. Dedrick and K. B. Bischoff, "Pharmacokinetics in Applications of the Artificial Kidney," *Chem. Eng. Progress Symposium Series No.* 84, **64,** 32 (1968). Specifically, what assumptions are made about mass transfer? Discuss Appendix B particularly in this regard. Can you relate this physiological model to a limiting case in Section 8.5?

CHAPTER 9

Equilibrium Staged Processes

9.1 INTRODUCTION

The implementation of a great many processes depends ultimately on the ability to separate various species from one another. In this chapter we will build on the basic principles of interphase mass transfer developed in Chapter 8 to explore some of the ideas involved in the design of a separation process. The essential component of the analysis is the *equilibrium stage* defined in Section 8.4.1. The design problem can be roughly broken down into the actual mechanical implementation of an equilibrium stage and the computation of the number of equilibrium stages needed to effect a desired degree of separation. We shall deal only with the latter; the former remains very much an art and beyond the scope of this introductory text. For simplicity we will restrict this chapter to separation by liquid-liquid extraction. The overall approach and the solution techniques have much greater generality and are applicable to nearly all separation processes, phase-change and nonphase-change alike.

Liquid-liquid extraction is a process in which a solute is transferred between two solvents. For most of the chapter we will assume that the two solvents are absolutely insoluble in one another. This is a convenient approximation, although only sometimes realistic, and in Section 9.6 we will see how it may be relaxed. It is traditional to do separation process calculations using mass fractions instead of concentrations. We shall follow this practice, and Section 9.2 is devoted to reformulating the notion of an equilibrium stage in terms of mass fractions.

9.2 EQUILIBRIUM STAGE

A mass transfer stage for extraction is shown in Figure 9.1. The two phases with dissolved solute A are fed to a well stirred contactor, where transfer of

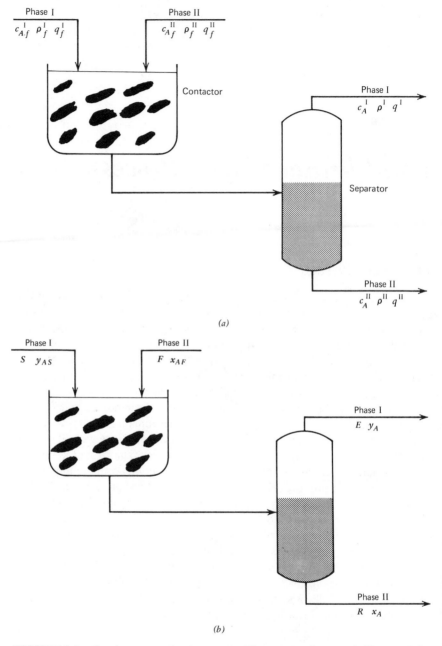

FIGURE 9.1 Continuous extraction stage. The nomenclature of Chapter 8 is shown in (a), traditional extraction nomenclature in (b).

A across the interface occurs. In the separator the phases separate because of density differences. An *equilibrium stage* has a sufficiently large residence time in the contactor to ensure that solute A has its equilibrium distribution between the phases. Figure 9.1(a) is identical to Figure 8.4 and shows the nomenclature used in Chapter 8. Figure 9.1(b) shows the nomenclature traditionally used in extraction, as follows:

S: solvent (phase I) mass flow rate $S = \rho_f^{\,I} q_f^{\,I}$

F: feed (phase II) mass flow rate $F = \rho_f^{\,II} q_f^{\,II}$

E: extract (phase I) mass flow rate $E = \rho^I q^I$

R: raffinate (phase II) mass flow rate $R = \rho^{II} q^{II}$

y_A and x_A denote the mass fractions of solute A in phases I and II, respectively. Since c_A denotes mass of A per volume and ρ denotes total mass per volume, the relations between concentrations and mass fractions must be as follows:

$$y_A = \frac{c_A^{\,I}}{\rho^I} \qquad x_A = \frac{c_A^{\,II}}{\rho^{II}}$$

$$y_{AS} = \frac{c_{Af}^{\,I}}{\rho_f^{\,I}} \qquad x_{AF} = \frac{c_{Af}^{\,II}}{\rho_f^{\,II}}$$

The equilibrium expression for solute concentrations in the two phases was written in Equation 8.12 as

$$c_A^{\,I} = M c_A^{\,II}$$

We will not use the subscript e to denote equilibrium since *all* concentrations in this chapter are equilibrium values. In terms of mass fractions the equilibrium expression becomes

$$y_A = K x_A$$

$$K = \frac{M \rho^{II}}{\rho^I} \tag{9.1}$$

In general, K may not be a constant. Density data for solutions of acetone in water and 1,1,2-trichloroethane are shown in Figure 9.2, and using these data the equilibrium data for this three component system shown in Figure 8.2 are replotted in Figure 9.3 as y_A versus x_A. (Actually, the data were originally reported in mass fractions and recalculated for Figure 8.2.) Notice that there is less curvature in this plot than in Figure 8.2 and a constant value $K = 1.5$ provides an excellent fit to the data over a fairly wide range of mass fractions.

We saw in Section 8.4.1 that the assumption that the phases are in equilibrium removes the need to consider each phase as a separate control

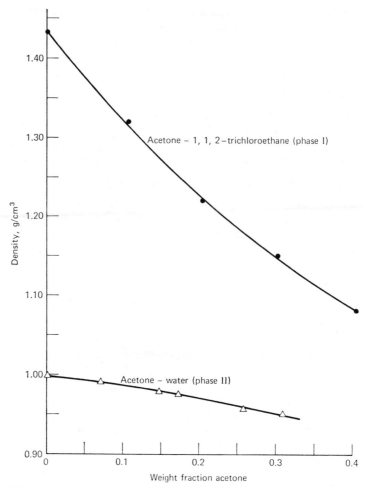

FIGURE 9.2   Density of acetone-1,1,2-trichloroethane and acetone-water solutions. [Data of Treybal et al., *Ind. Eng. Chem.*, **38**, 817 (1946), reproduced by permission.]

volume, and we can take the entire stage as the control volume. There are three species, the solute A, the feed solvent, and the extracting solvent. We will assume steady state, in which case the overall mass balance equation is

$$0 = F + S - E - R \qquad (9.2)$$

This is simply the sum of Equations 8.33I and 8.33II rewritten in the nomenclature of this chapter. Application of conservation of mass to species A gives

$$0 = Fx_{AF} + Sy_{AS} - Ey_A - Rx_A \qquad (9.3)$$

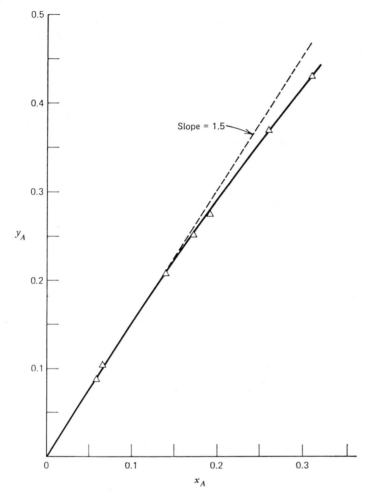

FIGURE 9.3   Equilibrium mass fraction of acetone in 1,1,2-trichloroethane (y_A) versus equilibrium mass fraction of acetone in water (x_A). [Data of Treybal et al., *Ind. Eng. Chem.*, **38,** 817 (1946), reproduced by permission.]

The assumption of immiscible solvents means that for each phase there are only two components and the mass fraction of solvent is simply one minus the mass fraction of A. Thus, for the feed solvent, conservation of mass requires

$$0 = F[1 - x_{AF}] - R[1 - x_A] \tag{9.4}$$

and for the extracting solvent

$$0 = S[1 - y_{AS}] - E[1 - y_A] \tag{9.5}$$

Only three of the four mass balance equations are independent, since Equation 9.2 is equal to the sum of Equations 9.3 to 9.5. Together with the equilibrium expression, Equation 9.1, these define the separation in an equilibrium stage.

9.2.1 Small Solute Transfer

The simplest situation to deal with is one in which the amount of solute transferred between phases is small. This approximation has wide applicability in practical situations since the error in using it is usually quite small, as we shall see in a subsequent section.

If the amount transferred is small then $1 - x_{AF}$ and $1 - x_A$ are nearly equal, so Equation 9.4 reduces to $F \simeq R$. Similarly, Equation 9.5 becomes $S \simeq E$. These are simply statements that, despite the transfer of solute, the mass flow rates of phases I and II are essentially unchanged. This approximation was made previously in deriving Equations 8.35. Equation 9.3 then simplifies to

$$0 = F[x_{AF} - x_A] + S[y_{AS} - y_A] \qquad (9.6)$$

This is identical to Equation 8.36 in the different nomenclature. By combining Equation 9.6 with the equilibrium expression, Equation 9.1, we can solve for x_A

$$x_A = \frac{\Lambda x_{AF} + y_{AS}}{\Lambda + K}$$

$$\Lambda = \frac{F}{S} \qquad (9.7)$$

Equation 9.7 is identical to Equation 8.38.

Example 9.1

An aqueous solution containing 200 g/liter of acetone is to be purified by continuous extraction with pure trichloroethane. How much acetone can be removed if the flow rate of both aqueous and organic streams is 10 liters/min?

This is identical to Example 8.1. Since we have pure solvent, $y_{AS} = 0$. From Figure 9.3 we take $K = 1.5$. Calculation of x_{AF} requires trial and error, since the data in Figure 9.2 show ρ^{II} versus x_A, while we are given only $c_{Af}^{II} = \rho_f^{II} x_{AF}$. We will use the method of direct substitution, Section 16.3. As a first approximation we take $\rho_f^{II} \simeq 1$ g/cm³, in which case $x_{AF} = c_{Af}^{II}/\rho_f^{II} = 0.2/\rho_f^{II} \simeq 0.2$. For $x_{AF} = 0.2$, $\rho_f^{II} \simeq 0.97$, giving

$$x_{AF} = c_{Af}^{II}/\rho_f^{II} = 0.200/0.97 = 0.206$$

Further calculation is unnecessary, since ρ_f^{II} is essentially unchanged. $\rho_f^{I} = 1.43$ for pure trichlorobenzene, so we can compute Λ as

$$\Lambda = \frac{F}{S} = \frac{\rho_f^{II} q_f^{II}}{\rho_f^{I} q_f^{I}} = \frac{0.97 \times 10^3 \times 10}{1.43 \times 10^3 \times 10} = 0.68$$

Then

$$x_A = \frac{\Lambda x_{AF}}{\Lambda + K} = \frac{0.68 \times 0.206}{0.68 + 1.50} = 0.064$$

From Figure 9.2, $\rho^{II} = 0.99$. Thus

$$c_A^{II} = \rho^{II} x_A = 0.99 \times 0.064 = 0.063 \text{ g/cm}^3 = 63 \text{ g/liter}$$

In Example 8.1 we computed $c_A^{II} = 66.7$ g/liter. This is good agreement, and the difference reflects inaccuracies in reading data for density, K, and M from graphs and round-off in retaining only two significant figures in the calculation. Two significant figures is all that is reasonable here since ρ^{II} cannot be determined with any greater accuracy.

9.2.2 Finite Solute Transfer

When the amount of solute transferred between phases is not small we need to use the full set of independent mass balance equations. Equations 9.4 and 9.5, together with the equilibrium Equation 9.1, can be written

$$R = F\left[\frac{1 - x_{AF}}{1 - x_A}\right]$$

$$E = S\left[\frac{1 - y_{AS}}{1 - Kx_A}\right]$$

Substitution into Equation 9.2 then gives

$$0 = F + S - S\left[\frac{1 - y_{AS}}{1 - Kx_A}\right] - F\left[\frac{1 - x_{AF}}{1 - x_A}\right]$$

With some slight rearrangement, using $\Lambda = F/S$, we obtain, finally

$$\frac{[Kx_A - y_{AS}][1 - x_A]}{[x_{AF} - x_A][1 - Kx_A]} = \Lambda \tag{9.8}$$

or, in standard quadratic form,

$$x_A^2 - \left\{\frac{\Lambda + K\Lambda x_{AF} + K + y_{AS}}{K[\Lambda + 1]}\right\} x_A + \frac{\Lambda x_{AF} + y_{AS}}{K[\Lambda + 1]} = 0 \tag{9.9}$$

Before solving Equation 9.9 for particular cases it is useful to examine the approximation involved in the small solute transfer assumption. We expand $[1 - x_A]/[1 - Kx_A]$ in a Taylor series about $x_A = 0$ (Section 15.8) to give

$$\frac{1 - x_A}{1 - Kx_A} = 1 + [K - 1]x_A + \cdots \tag{9.10}$$

In the expansion in Equation 9.10 we have neglected the $x_A{}^2$ terms compared to x_A. For x_A of order 0.1, say, and K of order two the error is less than one percent. (Compare the discussion at the end of Section 2.3.) Equation 9.8 can then be rearranged for x_A as

$$x_A = \frac{\Lambda x_{AF} + y_{AS}\{1 + [K - 1]x_A + \cdots\}}{\Lambda + K\{1 + [K - 1]x_A + \cdots\}} \tag{9.11}$$

For K of order two and x_A of order 0.1, Equation 9.11 is equivalent to Equation 9.7 with an uncertainty in y_{AS} and K of ten percent at most. y_{AS} will generally be small or zero, while the experimental error in determining K may be comparable to the error introduced by the factor

$$1 + [K - 1]x_A + \cdots$$

Thus, the small solute transfer equation can be expected to give reliable results even in many cases in which solute transfer does not really appear to be small.

Example 9.2

Using the data in Example 9.1 compute x_A without the small solute transfer approximation.

We were given $y_{AS} = 0$, $K = 1.5$, and computed $x_{AF} = 0.206$, $\Lambda = 0.68$ from available data. Equation 9.9 then becomes

$$x_A{}^2 - 0.95x_A + 0.056 = 0$$

The positive root between zero and one is $x_A = 0.063$. This is essentially the same as the solution using the small solute transfer approximation, though 13 percent of the mass of the aqueous phase was transferred to the organic phase.

9.3   TWO-STAGE EXTRACTION

The separation obtained in an equilibrium stage depends only on the separation factor K, the input mass fractions, and the relative mass flow rates. We showed in Examples 8.2 and 8.3 for a particular case that for a given system, fixed inputs, and a specified solvent flow rate a better separation could be obtained by splitting the solvent between two successive stages than

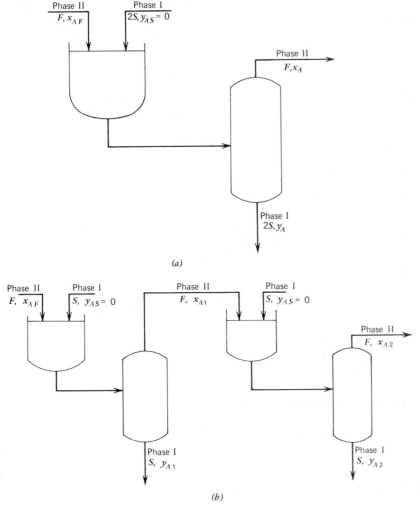

(a)

(b)

FIGURE 9.4   (a) One-stage extraction with solvent flow rate $2S$. (b) Two-stage extraction with solvent flow rate S to each stage. The total solvent used is the same in the two cases.

by using it all in one stage. We shall establish that result generally here and then extend it to see how the idea can be exploited with a substantial saving in solvent inventory. For algebraic simplicity it is assumed in everything that follows that solute transfer is small, phase mass flow rates are constant, and the extracting solvent is pure ($y_{AS} = 0$).

The two situations that we wish to compare are shown in Figures 9.4(a) and (b). The first configuration is the case that we have considered in the preceding section, but we have shown the solvent stream as having a mass flow rate $2S$ to simplify later comparisons. In the second configuration we take the raffinate stream leaving the first stage and use it as the feed to a second stage. Solvent mass flow is S in each stage, so that the same total amount of solvent is used. The subscript 1 is used for streams leaving the first stage and 2 for streams leaving the second. The configurations are shown schematically in Figures 9.5(a) and (b), and we shall use the schematic representation henceforth.

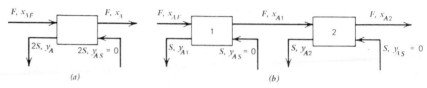

(a)   (b)

FIGURE 9.5   (a) Schematic diagram of one-stage extraction with solvent flow rate $2S$. (b) Schematic diagram of two-stage extraction with solvent flow rate S to each stage.

For the one-stage process the separation is determined by Equation 9.7. If we continue to define $\Lambda = F/S$ then, because we are using $2S$ solvent flow, we obtain for $y_{AS} = 0$

$$x_A = \frac{[1/2]\Lambda x_{AF}}{[1/2]\Lambda + K}$$

or, defining the separation ratio s_1,

$$s_1 = \frac{x_A}{x_{AF}} = \frac{\Lambda}{\Lambda + 2K} \tag{9.12}$$

For the two-stage process, Figure 9.5(b), each stage is described by Equation 9.7. Thus

$$x_{A1} = \frac{\Lambda x_{AF}}{\Lambda + K}$$

The feed to stage two has mass fraction x_{A1}, so Equation 9.7 gives

$$x_{A2} = \frac{\Lambda x_{A1}}{\Lambda + K}$$

Combining these two equations we obtain the two stage separation ratio, s_2

$$s_2 = \frac{x_{A2}}{x_{AF}} = \frac{\Lambda^2}{[\Lambda + K]^2} \tag{9.13}$$

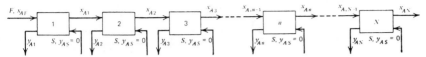

FIGURE 9.6   Schematic diagram of N-stage extraction with solvent flow rate S to each stage. Stage n denotes a typical stage.

It is easy to show that s_2 is always less than s_1, and hence that the two-stage process is more efficient, by taking the ratio:

$$\frac{s_2}{s_1} = \frac{\Lambda^2/[\Lambda + K]^2}{\Lambda/[\Lambda + 2K]} = \frac{\Lambda^2 + 2K\Lambda}{\Lambda^2 + 2K\Lambda + K^2} < 1 \tag{9.14}$$

If we were to use N consecutive stages, as shown in Figure 9.6, the separation ratio follows in a similar way as

$$s_N = \frac{\Lambda^N}{[\Lambda + K]^N} \tag{9.15}$$

Compared to one stage with solvent flow rate NS the ratio of separation ratios is

$$\frac{s_N}{s_1} = \frac{\Lambda^N + NK\Lambda^{N-1}}{[\Lambda + K]^N}$$

$$= \frac{\Lambda^N + NK\Lambda^{N-1}}{\Lambda^N + NK\Lambda^{N-1} + \dfrac{N[N-1]}{2!}K^2\Lambda^{N-2} + \cdots} < 1 \tag{9.16}$$

Multistage operation is an appealing alternative to using all of the solvent in one stage, since a better separation can be obtained with the same amount of solvent. There is still something wasteful about the process, however. In the first stage the mass fraction of solute is decreased to a fraction $\Lambda/[\Lambda + K]$ of that in the feed. In the next stage the further fractional reduction is the same. Thus, if 50 percent of the solute is removed in the first stage, only 25 percent more is removed in the second, 12-1/2 percent in the third, and so on, although the same amount of solvent is required in each stage. The solvent *leaving* the latter stages will be quite dilute, and it may be necessary to handle large volumes of solvent containing only small amounts of solute to achieve a desired separation.

A nice compromise that combines the advantages of multistage operation with the use of a small amount of solvent readily suggests itself. Consider Figure 9.5(*b*). The extract stream leaving stage two contains only a small amount of solute, y_{A2}, because the greatest amount of solute was extracted in stage one. Thus, there would be little penalty if the solvent stream feeding stage one were to contain mass fraction y_{A2} of solute instead of zero. This

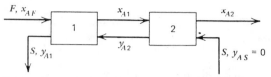

FIGURE 9.7   Schematic diagram of two-stage countercurrent extraction.

countercurrent operation is shown schematically in Figure 9.7. The raffinate stream from stage one is the feed stream to stage two and the extract stream from stage two is the solvent stream for stage one. Equation 9.7, with appropriate nomenclature, applies to each stage. In the first stage the solvent stream contains mass fraction y_{A2}, so we have

$$x_{A1} = \frac{\Lambda x_{AF} + y_{A2}}{\Lambda + K}$$

In the second stage the feed has mass fraction x_{A1}, and the solvent mass fraction y_{AS} is zero, so

$$x_{A2} = \frac{\Lambda x_{A1}}{\Lambda + K}$$

Together with the equilibrium relation $y_{A2} = K x_{A2}$ these can be combined to give

$$S_{c2} = \frac{x_{A2}}{x_{AF}} = \frac{\Lambda^2}{[\Lambda + K]^2 - \Lambda K} \tag{9.17}$$

S_{c2} denotes "separation ratio for two-stage countercurrent operation." Comparison of Equations 9.12 and 9.17 shows that for $K > \Lambda$ the separation in a two-stage countercurrent process is better than in a single stage which uses twice as much solvent. Comparison with Equation 9.13 gives the relation between the separation ratios in countercurrent and noncountercurrent two stage extraction:

$$\frac{S_{c2}}{S_2} = 1 + \frac{K\Lambda}{\Lambda^2 + K\Lambda + K^2} \tag{9.18}$$

Example 9.3

Using the data in Example 9.1, compute the separation for two-stage countercurrent and noncountercurrent operation.

We are given $\Lambda = 0.68$, $K = 1.5$. Using Equations 9.13 and 9.17

$$\text{noncountercurrent:} \quad S_2 = \frac{x_A}{x_{AF}} = \frac{\Lambda^2}{[\Lambda + K]^2} = 0.097$$

$$\text{countercurrent:} \quad S_{c2} = \frac{x_{A2}}{x_{AF}} = \frac{\Lambda^2}{[\Lambda + K]^2 - \Lambda K} = 0.123$$

In agreement with Equation 9.18, $s_{c2}/s_2 = 1.27$. The countercurrent operation uses 10 liters/min of trichloroethane, the noncountercurrent 20 liters/min.

Example 9.4

Using the data in Example 9.1, how much solvent will be needed to obtain a separation ratio of 0.097 in a two-stage countercurrent process?

According to Equation 9.17, we must solve for Λ from the equation

$$0.097 = \frac{\Lambda^2}{[\Lambda + K]^2 - \Lambda K}$$

For $K = 1.5$ the positive root is $\Lambda = 0.58$. Since $\Lambda = F/S$ and $S = \rho_f^I q_f^I$ and we are given $F = 9700$ g/min and $\rho_f^I = 1.43$ g/cm³, it then follows that $q_f^I = 11.8$ liters/min. That is, 12 liters/min in two-stage countercurrent operation gives a separation equal to that achieved using 20 liters/min in two-stage noncountercurrent operation. In a single stage 42 liters/min of solvent would be required.

9.4 MULTISTAGE COUNTERCURRENT EXTRACTION

Design equations for a multistage countercurrent extraction process follow directly from the considerations in the preceding section. The process is shown schematically in Figure 9.8 with N stages. It is assumed that solute

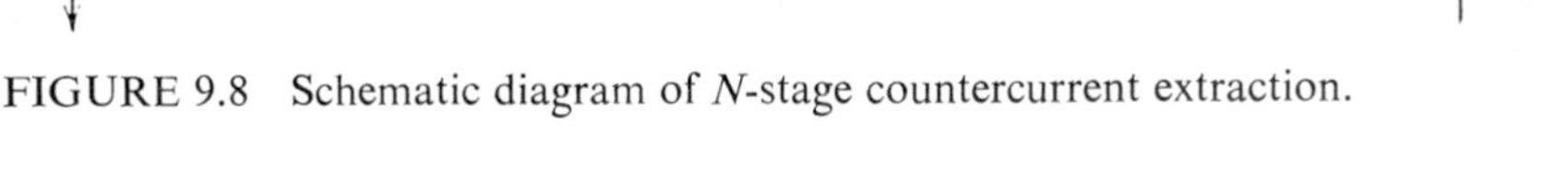

FIGURE 9.8 Schematic diagram of N-stage countercurrent extraction.

transfer is small, so that the mass flow rates F and S are constant throughout all the stages. Consider stage n, where n is any integer between 2 and $N - 1$. Conservation of mass applied to solute A leads to the equation

$$0 = Fx_{A,n-1} + Sy_{A,n+1} - Fx_{An} - Sy_{An} \tag{9.19}$$

The equilibrium relation is still taken as $y_A = Kx_A$, so

$$y_{An} = Kx_{An}, \qquad y_{A,n+1} = Kx_{A,n+1} \tag{9.20}$$

Combination of Equations 9.19 and 9.20 with some rearrangement leads to the analog of Equation 9.7

$$x_{An} = \frac{\Lambda x_{A,n-1} + Kx_{A,n+1}}{\Lambda + K} \tag{9.21}$$

It is convenient to introduce the ratio

$$\mu = \frac{\Lambda}{K}$$

in which case we can write Equation 9.21 in the form

$$2 \leq n \leq N - 1: \quad x_{A,n+1} - [1 + \mu]x_{An} + \mu x_{A,n-1} = 0 \quad (9.22)$$

For $n = 1$ and $n = N$ we have the input streams to take into account, and conservation of mass leads to

$$\text{stage 1:} \qquad x_{A2} - [1 + \mu]x_{A1} + \mu x_{AF} = 0 \qquad (9.23)$$

$$\text{stage } N: \qquad \frac{y_{AS}}{K} - [1 + \mu]x_{AN} + \mu x_{A,N-1} = 0 \qquad (9.24)$$

Equation 9.22 relates mass fractions x_A in successive stages $n - 1$, n, and $n + 1$, and is called a *finite difference equation*. It can be manipulated together with Equations 9.23 and 9.24 in a manner identical to that used for linear differential equations in Section 17.3 in order to obtain a solution for x_{An} explicitly in terms of N, μ, K, x_{AF}, and y_{AS}:

$$x_{An} = \frac{[x_{AF} - y_{AS}/K]\mu^n + y_{AS}/K - x_{AF}\mu^{N+1}}{1 - \mu^{N+1}} \qquad (9.25)$$

We shall not derive Equation 9.25, but we simply demonstrate that it is indeed equivalent to Equation 9.22. By replacing n with $n - 1$ and $n + 1$ we can rewrite Equation 9.25 as

$$x_{A,n-1} = \frac{[x_{AF} - y_{AS}/K]\mu^{n-1} + y_{AS}/K - x_{AF}\mu^{N+1}}{1 - \mu^{N+1}} \qquad (9.26)$$

$$x_{A,n+1} = \frac{[x_{AF} - y_{AS}/K]\mu^{n+1} + y_{AS}/K - x_{AF}\mu^{N+1}}{1 - \mu^{N+1}} \qquad (9.27)$$

When Equations 9.25 to 9.27 are substituted into Equation 9.22 the terms do sum to zero, as required. Equations 9.23 and 9.24 are verified in the same way.

For algebraic simplicity we will consider from here on only the case of pure solvent, $y_{AS} = 0$. By setting $n = N$ in Equation 9.25 we then obtain an expression for the separation ratio in an N-stage countercurrent process

$$s_{cN} = \frac{x_{AN}}{x_{AF}} = \frac{\mu^N - \mu^{N+1}}{1 - \mu^{N+1}} \qquad (9.28)$$

This can be rearranged to solve for N in terms of μ and s_{cN} for cases in which we wish to compute the number of equilibrium stages needed to achieve a given separation:

$$N = \frac{\log\left[\dfrac{s_{cN}}{1 - \mu + \mu s_{cN}}\right]}{\log \mu} \tag{9.29}$$

Equation 9.29 is a version of what is often called the *Kremser equation*.

Example 9.5

For $\Lambda = 0.68$ and $K = 1.5$ compute the separation in one-, two-, and three-stage countercurrent processes.

$$\mu = \frac{\Lambda}{K} = 0.45$$

$$N = 1: \quad s_{c1} = \frac{0.45 - 0.45^2}{1 - 0.45^2} = 0.31$$

$$N = 2: \quad s_{c2} = \frac{0.45^2 - 0.45^3}{1 - 0.45^3} = 0.12$$

$$N = 3: \quad s_{c3} = \frac{0.45^3 - 0.45^4}{1 - 0.45^4} = 0.05$$

The result for $N = 2$ agrees, of course, with the calculation in Example 9.3.

Example 9.6

For $\mu = 0.45$ compute the number of equilibrium stages needed to remove 90 and 99 percent of the dissolved solute in the feed stream.

The calculation is carried out using Equation 9.29. For 90 percent removal we have $s_{cN} = 0.10$ and

$$N = \frac{\log\left[\dfrac{0.10}{1 - 0.45 + 0.045}\right]}{\log[0.45]} = 2.2$$

We cannot build 2.2 stages, of course, so we need 3 perfect stages to achieve 90 percent removal. From the calculation in Example 9.5, we see that we will in fact remove 95 percent of the solute. We might wish to decrease μ by a small amount by increasing solvent flow rate in order to carry out the desired separation in two stages. We saw in Example 9.4 that a 20 percent increase in S would achieve the removal of 90 percent of solute in two stages.

For 99 percent removal we have $s_{cN} = 0.01$ and

$$N = \frac{\log\left[\dfrac{0.01}{1 - 0.45 + 0.005}\right]}{\log[0.45]} = 4.95 \simeq 5$$

Twice as many stages are required to pass from 90 to 99 percent removal. This reflects the logarithmic dependence of the number of stages on separation in Equation 9.29.

9.5   GRAPHICAL SOLUTION

The method leading to Equation 9.29 for computing the number of stages required to carry out a specified separation depends critically on the fact that all of the equations involved are linear in x_A and y_A. Nonlinearities can occur because the equilibrium relation is not linear or because changes in mass flow rates resulting from interphase transfer must be accounted for. Although the full set of nonlinear equations would normally be solved using a digital computer for precise results, a number of simple graphical procedures have been developed for rapid estimation of the type needed in preliminary design considerations. We shall discuss here the *McCabe-Thiele method*, a graphical method which is suitable when the assumption of constant mass flow rates is valid but the equilibrium is not linear.

The development of the graphical method of solving the extractor equations centers around the choice of a control volume. Instead of doing the mass

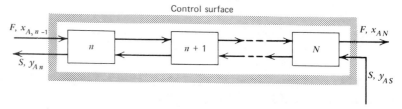

FIGURE 9.9   Control surface used for mass balances in developing the McCabe-Thiele method of graphical solution.

balance on stage n, the control volume is chosen as stages n to N. See Figure 9.9. Conservation of mass applied to component A then becomes

$$0 = Fx_{A,n-1} + Sy_{AS} - Sy_{An} - Fx_{AN}$$

or

$$y_{An} = \Lambda x_{A,n-1} - \Lambda x_{AN} + y_{AS} \tag{9.30}$$

In addition we have the equilibrium relation

$$y_A = f(x_A) \tag{9.31}$$

The construction is then carried out as shown on Figure 9.10. The equilibrium line, Equation 9.31, is drawn, as is the *operating line*,

$$y_A = \Lambda x_A - \Lambda x_{AN} + y_{AS} \tag{9.32}$$

If we set $n = 1$, then $x_{A,n-1}$ is x_{AF}. According to Equation 9.30, y_{A1} is the value of the operating line when $x_A = x_{AF}$ (point 1). y_{A1} is in equilibrium with x_{A1}, so the equilibrium line, Equation 9.31, must give x_{A1} for $y = y_{A1}$ (point 2). Given x_{A1} we find y_{A2} from the operating line (point 3), then x_{A2} from the equilibrium line (point 4), and so on. Ultimately we reach x_{AN} in

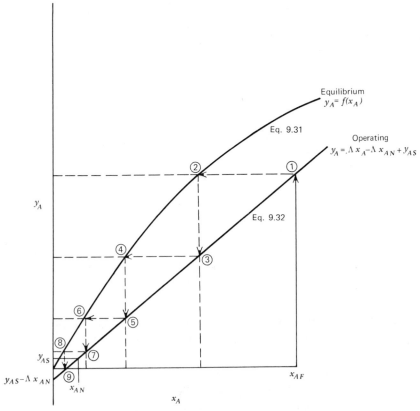

FIGURE 9.10   Calculation of the number of stages required for separation using the McCabe-Thiele method.

this manner. The number of steps on the diagram corresponds to the number of stages required. In Figure 9.10 four stages would be needed, since x_{AN} lies between the third and fourth step.

Example 9.7

For the equilibrium data shown in Figure 9.3 how many equilibrium stages are required to reduce the acetone mass fraction in the aqueous phase from 0.30 to 0.03 using $\Lambda = 0.68$ and pure solvent ($y_{AS} = 0$)?

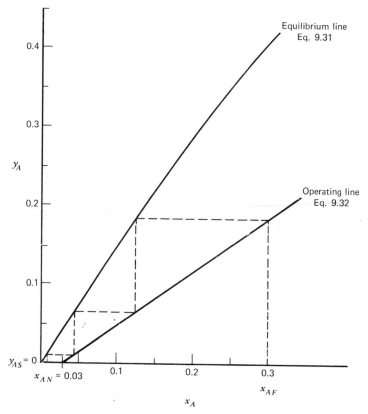

FIGURE 9.11   Calculation of the number of stages required to reduce x_A from 0.3 to 0.03 using $\Lambda = 0.68$ and the equilibrium data of Figure 9.3.

The construction is shown in Figure 9.11. Between two and three (and hence three) stages are needed. The equilibrium line is nearly linear in this range, so Equation 9.29 should provide a close approximation to the solution. In Example 9.6 a separation ratio of 0.10 was found to require 2.2 (hence three) stages for this system using the constant value $K = 1.5$.

Example 9.8

What is the absolute minimum pure solvent flow which can be used to reduce the acetone mass fraction of an aqueous phase from 0.30 to 0.03?

The solvent flow rate is contained in the ratio $\Lambda = F/S$, which is the slope of the operating line in the McCabe-Thiele method. The smaller S is, the larger the slope. Figure 9.12 shows a sequence of operating lines with increasing slope. As Λ increases, the number of stages increases. The number of stages goes to infinity as the operating line *intersects* the equilibrium line at

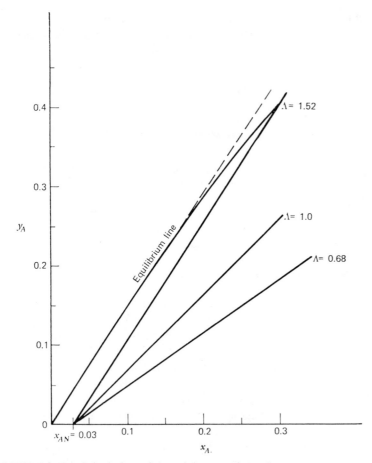

FIGURE 9.12   Calculation of the minimum solvent flow rate necessary to reduce x_A from 0.30 to 0.03.

x_{AF}, and for Λ greater than this value the construction cannot be carried out. Thus, the minimum solvent flow corresponds to the slope Λ such that the two lines intersect at x_{AF}. For the given data and three component system this is

$$\Lambda = \frac{F}{S_{min}} = 1.52$$

$$S_{min} = 0.66F$$

The desired separation can never be carried out unless S is greater than $0.66F$ Beyond that point an economic balance must be carried out between capital investment for the larger number of stages and the cost of handling larger volumes of solvent in the smaller number of stages.

If the equilibrium line is nearly straight, as in the acetone-water-trichloro-ethane system, the calculation can be carried out algebraically with ease. From Equations 9.31 and 9.32 the two lines intersect when

$$y_A = Kx_{AF} = x_{AF} - x_{AN} + y_{AS}$$

$$\Lambda = \frac{K}{1 - s_{cN}} - \frac{y_{AS}}{x_{AF} - x_{AN}}$$

For the case $y_{AS} = 0$ the same result is obtained from Equation 9.29 by setting $1 - \mu + \mu s_{cN}$ to zero so that N goes to infinity. For $K = 1.5$, $s_{cN} = 0.1$, $y_{AS} = 0$, and this gives $\Lambda = 1.67$, $S_{min} = 0.60F$. It is obvious from Figure 9.12 that taking the equilibrium slope as constant will always lead to too low an estimate of the minimum solvent flow rate.

9.6   TRIANGULAR DIAGRAMS

All of the calculations carried out in this chapter have assumed that the two solvents are completely immiscible, and most have assumed negligible solute transfer. All of these restrictions can be relaxed without difficulty, though the resulting equations are exceedingly complex and large amounts of data may need to be stored. In this section we shall open the discussion of rapid graphical methods for general three-component solvent extraction systems in which miscibility of all species in any amount must be accounted for. The purpose here is to be suggestive only, and more specialized texts should be consulted for a complete treatment.

Three-component systems in which the component species are soluble in one another are often dealt with on equilateral triangular coordinates as a convenient means of representing data and of doing rapid graphical cal-culation. (Other types of triangles may be used with only slight modification.) As shown in Figure 9.13, each apex represents a pure species, and the mass fraction of that species is measured along the altitude from the apex to the opposite side. The geometry of the triangle allows the scale to be marked along the side instead of the altitude in order to minimize confusion within the triangle. This is shown in the figure for species A, where it is obvious that the A scale along the altitude and its projection along the AB side are equiv-alent. The mass fraction of each component is then found by drawing the line parallel to the side opposite the apex to the appropriate scale along the triangle side. Point X in the figure has mass fractions $x_A = 0.2$, $x_B = 0.3$, and $x_C = 0.5$.

Clearly it is necessary that each point in the triangle interior have the property that the mass fractions sum to unity. That this is true can be seen by

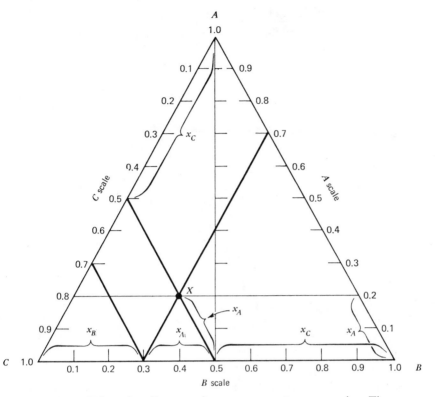

FIGURE 9.13 Triangular diagram. Apexes represent pure species. The mass fraction of each specie at any point is found by drawing a line parallel to the side opposite the apex.

again referring to point X in Figure 9.13. Along side BC are marked off distances x_A, x_B, and x_C, which together take up exactly the side of length unity. The length x_C is geometrically similar to the equivalent length on the C scale, since the line through X is parallel to the face. The length x_A is the base of an equilateral triangle whose right side is the side of a parallelogram opposite x_A on the A scale. The length x_B is on the B scale. A similar construction could be carried out on any side.

The property of the triangular diagram which makes graphical computation of staged separations easy is that if two streams of differing composition are mixed together the composition of the mixture must lie on the straight line between them, with the location on that line fixed by the relative amounts of each stream. Equivalently, if a mixture of given composition splits into two immiscible phases the same relation applies. Consider a single-stage extraction process as shown in Figure 9.1. The solute is A while B and C are solvents. B is soluble in an A–C mixture and C is the soluble in an A–B

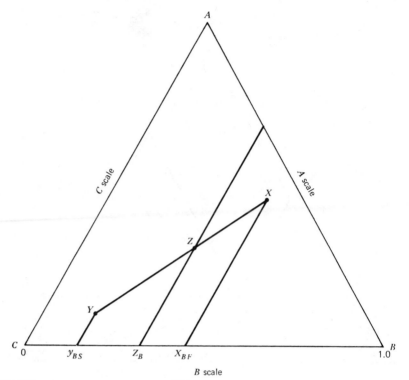

FIGURE 9.14   Construction showing that the composition of any mixture of X and Y lies along the straight line between X and Y.

mixture. The mass balance equations, one of which is redundant, can then be written

$$\text{overall:} \qquad F \quad + S \quad = E \quad + R \tag{9.33}$$

$$\text{component } A: \qquad Fx_{AF} + Sy_{AS} = Ey_A + Rx_A \tag{9.34a}$$

$$\text{component } B: \qquad Fx_{BF} + Sy_{BS} = Ey_B + Rx_B \tag{9.34b}$$

$$\text{component } C: \qquad Fx_{CF} + Sy_{CS} = Ey_C + Rx_C \tag{9.34c}$$

X and Y in Figure 9.14 can refer to either the input streams (x_{AF}, x_{BF}, x_{CF} and y_{AS}, y_{BS}, y_{CS}) or the effluent streams (x_A, x_B, x_C and y_A, y_B, y_C). Z is the composition that would result from the mixture of feed streams if they were miscible. We shall show that Z lies on the line $\overline{YX}$.

We shall carry out the proof for the feed streams. An identical proof applies to the effluent. Define mass fractions z_A, z_B, and z_C by

$$Fx_{AF} + Sy_{AS} = [F + S]z_A \tag{9.35a}$$

$$Fx_{BF} + Sy_{BS} = [F + S]z_B \tag{9.35b}$$

$$Fx_{CF} + Sy_{CS} = [F + S]z_C \tag{9.35c}$$

Also define $\xi = F/[F + S]$. Assume Z lies on $\overline{YX}$. From the properties of a trapezoid we have

$$\frac{\overline{YZ}}{z_B - y_{BS}} = \frac{\overline{ZX}}{x_{BF} - z_B}$$

Substituting Equation 9.35b this becomes

$$\frac{\overline{YZ}}{\overline{ZX}} = \frac{\xi x_{BF} + [1 - \xi]y_{BS} - y_{BS}}{x_{BF} - \xi x_{BF} - [1 - \xi]y_{BS}}$$

$$= \frac{\xi[x_{BF} - y_{BS}]}{[1 - \xi][x_{BF} - y_{BS}]} = \frac{\xi}{1 - \xi} \tag{9.36}$$

A similar construction shows that the line passing through z_A parallel to side $\overline{CB}$ intersects $\overline{YX}$ in the same place, as does the line through z_C parallel to $\overline{AB}$. Thus, point Z does lie on $\overline{YX}$, a fraction $F/[F + S]$ from Y to X. Similarly, Z lies on the line between y_A and x_A a fraction $R/[R + E]$ from y_A to x_A.

We can see more clearly both how the triangular diagram is developed and how to use it of we consider the experimental procedure which is followed to obtain the data shown in Figure 9.15 for the acetone, water, and trichloroethane system. If a mixture of 80 percent acetone, 10 percent water, and 10 percent trichloroethane is made up and thoroughly mixed for a period of time we find that a single homogeneous liquid phase results. This is shown on Figure 9.15 as point A. Any mixture that contains over 60 percent acetone will be single phase no matter how long the three materials remain in contact. Points B, a mixture of 60 percent acetone, 30 percent water, and 10 percent trichloroethane, and C, a mixture of 70 percent acetone, 10 percent water, and 20 percent trichloroethane are other examples of single phase mixtures. If we make up a mixture of 20 percent acetone, 50 percent water, and 30 percent trichloroethane, thoroughly agitate for a period of time and then stop the mixing, two phases will settle out, a water phase and a trichloroethane phase. Point Z in Figure 9.15 represents the initial mixture for this experiment. Each phase is then analyzed for each of the three components (if the water and trichloroethane are completely immiscible, it is only necessary to analyze for acetone). With the 20-50-30 initial mixture the water phase is found to contain both acetone and a small amount of trichloroethane. Likewise the trichloroethane phase contains both acetone and a small amount of water. The composition of the water phase is found by measurement to be about 82 percent water, 17 percent acetone, and 1 percent trichloroethane. This point is established on the triangular diagram as point X. Point Y

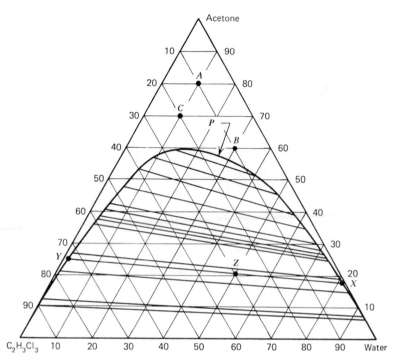

FIGURE 9.15 Equilibrium data for the system acetone-water-1,1,2-trichloro-ethane. Compositions in the aqueous and organic phases which are in equilibrium are connected by the lines. [Data of Treybal et al., *Ind. Eng. Chem.*, **38**, 817 (1946), reproduced by permission.]

designates the composition of the trichloroethane phase, a mixture of about 74 percent trichloroethane, 25 percent acetone, and 1 percent water.

A *tieline* is drawn to pass through points X, Z, and Y. It connects the equilibrium compositions on opposite sides of the miscibility line. The distance $\overline{YZ}$ divided by the total length of the tieline $\overline{YX}$ represents the fraction of total mixture with composition X. A similar set of data can be collected starting with mixtures of different initial compositions and the curved line and the tie lines shown on the triangular plot established. As the percentage of acetone in the mixture is increased, a situation can be reached where the composition of the two phases approach each other (a tieline of zero length). This point is designated on Figure 9.15 as P and is referred to as the *critical* or *plait point*. Once a number of tielines and the curved line establishing the two-phase region have been established by experiment for any system the triangular diagram can be used by others for design purposes. Application to a separation calculation is demonstrated in the following example.

Example 9.9

An aqueous stream of 9700 g/min containing 0.20 mass fraction acetone is mixed with 14,300 g/min of pure 1,1,2-trichloroethane. Determine the composition and amount of the effluent streams. (With slight rounding off these are the data for Examples 9.1 and 9.2.)

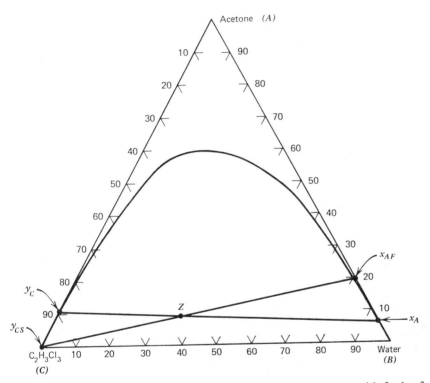

FIGURE 9.16   Computation of separation in a one-stage extractor with feeds of 0.20 acetone in water and pure 1,1,2-trichloroethane.

The system is shown in Figure 9.16 with only one tieline for clarity. The feed mixture lies along the line $\overline{y_{CS}x_{AF}}$ (pure trichloroethane − 0.20 acetone in water) with

$$\frac{\overline{y_{CS}z}}{\overline{y_{CS}x_{AF}}} = \frac{F}{F+S} = \frac{9700}{9700 + 14,300} = 0.40$$

The tieline passing through Z interacts the envelope at x_A (acetone) $\simeq 0.07$, x_B (water) $\simeq 0.92$, $x_C(C_2H_3Cl_3) \simeq 0.01$ for the water-rich phase and

$y_A \simeq 0.11, y_B \simeq 0, y_C \simeq 0.89$ in the organic phase. The ratio $\overline{y_C z}/\overline{y_C x_A}$ is 0.37. Thus

$$\frac{R}{R + E} = 0.37$$

$$R + E = F + S = 2400$$

$$R = 8900 \text{ g/min}$$

$$E = 15{,}100 \text{ g/min}$$

The separation is essentially that computed in the earlier examples.

This graphical construction can be extended to develop a method for multistage countercurrent calculations for miscible solvents and measurable transfer between phases similar to the McCabe-Thiele method. We shall go no further here, however.

9.7   THE DESIGN PROBLEM

The use of the equilibrium stage is not a serious restriction, since everything done in this chapter can be reformulated in nearly the same form to nonequilibrium situations by using the *efficiency* introduced in Section 8.4.2. The essence of the analytical design question is expressed in Example 9.8. For a given separation a certain minimum amount of solvent is required. Beyond the minimum there is a trade-off between costs associated with handling larger volumes and costs associated with adding additional stages. Given capital and operating costs, the computation of an optimum follows in an identical manner to that for reactors in Chapter 7. The same comments apply to other separation processes such as gas absorption, ion exchange, and distillation. Indeed, with appropriate changes of nomenclature to account for the different physical processes *the equations and techniques developed in the first five sections of this chapter carry over to these and other separation processes almost without change.*

The engineering art in separation begins with the choice of a process and, in the case of extraction, the choice of a solvent. We cannot deal with that crucial aspect of design here, for it is highly coupled with specific experience. For difficult separations in which a large number of stages is required, calculation uncertainty can be compensated for at minimal extra cost by adding extra stages, and this provides a useful safety factor if throughput must subsequently be increased because, say, of inaccurate market forecasts. The design of the configuration of a stage that will provide equilibrium, or at least high efficiency, is the significant engineering problem. This may involve the arrangement of baffling to induce adequate phase contact, the construction of efficient and inexpensive mechanical agitation for developing interfacial area, the design of phase separators which minimize the entrainment of one phase in the other, and so on. These considerations are beyond our analytical abilities here, and they still depend to a large extent on the designer's skill and appreciation of the particular system at hand.

9.8 CONCLUDING REMARKS

The concept of a multistage countercurrent process is the most important idea introduced in this chapter. You should be certain that you understand the reasoning underlying its use. The relationship of this chapter to Chapter 8 is essentially that of Chapter 7 to Chapter 5. The principles developed in one are applied in the other. Introduction of economics would have been easy here. Notice, however, the comments in Section 9.7 regarding the design problem for separation.

The algebraic manipulations in this chapter are sometimes tedious, and there are a number of new symbols to keep track of. Furthermore, the use of graphical methods to study the behavior of the system equations is probably new to you. In all cases avoid getting bogged down in the detail and note the essential simplicity of the basic equations. At this point you are capable of modeling rather complex situations and obtaining a model with a simple structure.

The triangular diagram is a useful device. Data are often reported in that way. A graphical procedure very like the one used for the triangular diagram is introduced in Chapter 11, so you should be sure that you fully understand the details of the construction.

A number of texts are available which go into details of multistage equilibrium system design. See, for example,

9.1 C. J. King, *Separation Processes*, McGraw-Hill, New York, 1971.

9.2 W. L. McCabe and J. C. Smith, *Unit Operations of Chemical Engineering*, McGraw-Hill, New York, 1965.

9.3 B. Smith, *Design of Equilibrium Stage Processes*, McGraw-Hill, New York, 1963.

9.4 R. E. Treybal, *Liquid Extraction*, McGraw-Hill, New York, 1951.

9.9 PROBLEMS

9.1 The following equilibrium data were obtained at 20°C for the system acetic acid-water-isopropylether by Othmer et al., *Industrial & Engineering Chemistry*, **33**, 1240 (1941). x_A is the mass fraction of acid in the water phase and y_A is the mass fraction of acid in the ether phase. (Notice that x_A and y_A are multiplied by 100 to obtain the values in the table.)

$100x_A$	0.7	1.4	2.9	6.4	13.3	25.5	36.7
$100y_A$	0.2	0.4	0.8	1.9	4.8	11.4	21.6

(a) Plot the data. Estimate K from the data at small mass fraction.

(b) *In a batch operation* 1000 lb of an aqueous solution of 15 percent by weight acetic acid is to be contacted with 1000 lb of pure isopropyl ether. If equilibrium is achieved in this batch operation, how much acetic acid can be removed? Use the assumption that only a small amount of solute is transferred. Is the assumption of a constant value of K justified?

(c) If the resulting water mixture is extracted two more times, each with 1000 lb of fresh isopropyl ether, how much acid would be removed?

9.2 (a) Repeat the calculations in Problem 9.1 (b), (c) for
  (i) methylcyclohexane as the solvent. $K = 1.0$.
  (ii) di-isobutyl ketone as the solvent. $K = 0.23$.

(b) Which of the three solvents extracts the most acid? What other factors would you consider in selection of the solvent?

9.3 It is necessary to process 1000 lb/min of water containing 15 percent acetic acid. How much acid can be extracted using (a) 1000 lb/min of isopropyl ether in a single stage? (b) 500 lb/min? Make the assumption of small solute transfer. (c) Repeat (a) and (b) for methylcyclohexane as the solvent. (Equilibrium data are in Problems 9.1 and 9.2.)

9.4 Repeat the calculation in Problem 9.3 without making the assumption of small solute transfer.

9.5 Repeat the calculations in Problem 9.3 for the case where the ether contains 1.5 percent acetic acid.

9.6 The acetic acid-water mixture described in Problem 9.3 is to be extracted in a two-stage device with methylcyclohexane. Compute the amount and composition of the extract and raffinate which can be produced by

(a) 500 lb/min of fresh solvent in each stage
(b) countercurrent operation with 1000 lb/min of solvent.

9.7 (a) What is the minimum amount of isopropyl ether that can be used in countercurrent operation to remove 75 percent of the acid from the feed stream described in Problem 9.3?

(b) Repeat for methylcyclohexane.

9.8 How many countercurrent stages are required to remove 75 percent of the acid from the feed stream described in Problem 9.3 using 50 percent more than the minimum solvent computed in Problem 9.7 for (a) isopropyl ether (b) methylcyclohexane?

9.9 (a) A separation is to be carried out in N consecutive stages with the same feed/solvent ratio Λ in each stage. The total feed/solvent ratio is then Λ/N. For a given separation ratio s_N compute the minimum solvent required. (That is, find the limit as $N \to \infty$. It helps to recall that $\beta^{1/N} = \exp\{[1/N]\ln\beta\} = 1 + [\ln\beta]/N + \cdots$.)

  (b) Repeat the calculations in Problem 9.7 for this case and compare with the minimum solvent required in countercurrent operation.

9.10 Refer to Problem 8.15. What is the minimum amount of xylene required to remove 90 and 99 percent of the phenol? Does extraction with xylene seem like a feasible approach for purification?

9.11 The following equilibrium data have been obtained for the recovery of methyl ethyl ketone from water by solvent extraction [Newman, Hayworth, and Treybal, *Industrial & Engineering Chemistry*, **41**, 2039 (1949)]:

Water-rich phase (weight percentage)			Solvent-rich phase (weight percentage)		
Ketone	Solvent	ρ (g/cm³)	Ketone	Solvent	ρ (g/cm³)
Trichloroethylene as solvent					
20.71	0.18	0.97	81.90	13.10	0.87
16.10	0.22	0.98	71.65	25.31	0.91
11.03	0.32	0.98	58.10	41.68	0.98
8.50	0.36	0.99	43.70	56.00	1.07
6.75	0.38	0.99	22.10	77.72	1.23
1,1,2 Trichloroethane as solvent					
18.15	0.11	0.97	75.00	19.92	0.89
12.78	0.16	0.98	58.62	38.65	0.97
9.23	0.23	0.99	44.38	54.14	1.06
6.00	0.30	0.99	31.20	67.80	1.14
2.83	0.37	0.99	16.90	82.58	1.26
1.02	0.41	1.00	5.58	94.42	1.36
Chlorobenzene as solvent					
18.10	0.05	0.97	75.52	20.60	0.86
13.10	0.08	0.98	58.58	39.28	0.90
9.90	0.12	0.98	43.68	55.15	0.95
7.65	0.16	0.99	29.65	69.95	0.99
5.52	0.21	0.99	17.40	82.15	1.03
3.64	0.28	0.99	8.58	91.18	1.06

One thousand pounds per hour of an aqueous 18 percent by weight ketone solution is to be processed, with 170 lb/hr of ketone removed.

  (a) Find the minimum amount of each solvent required to effect the separation. State any assumptions carefully.

  (b) Find the number of stages required for each solvent if (i) 1.2 times (ii) 1.5 times the minimum solvent is to be used.

9.12 Obtain the current cost of each of the solvents considered in Problem 9.11 and describe how you would incorporate economic considerations into your design.

9.13 The following equilibrium data for the system water-benzene-ethyl alcohol at 25°C have been reported by Chang and Moulton [*Industrial & Engineering Chemistry*, **45**, 2350 (1953)].

Benzene-rich phase (weight percentage)		Water-rich phase (weight percentage)	
Ethyl alcohol	Benzene	Ethyl alcohol	Benzene
1.86	98.00	15.61	0.19
3.85	95.82	30.01	0.65
6.21	93.32	38.50	1.71
7.91	91.25	44.00	2.88
11.00	87.81	49.75	8.95
14.68	83.50	52.28	15.21
18.21	79.15	51.72	22.73
22.30	74.00	49.95	29.11
23.58	72.41	48.85	31.85
30.85	62.01	43.42	42.49
37.65	52.10	37.65	52.10

(a) Plot the data on a triangular diagram.
(b) Locate the following points on the triangular diagram. Determine if the mixture is one phase or two. If it is two phase, determine the composition of each.

   (i) 60% alcohol, 20% benzene, 20% water
   (ii) 70% alcohol, 10% benzene, 20% water
   (iii) 20% alcohol, 40% benzene, 40% water
   (iv) 50% alcohol, 20% benzene, 30% water

9.14 Five hundred pounds per minute of a stream containing 10 percent ethanol and 90 percent benzene is to be purified in a single stage using 100 lb/min of water to remove the alcohol. Calculate the separation by each of the following methods and compare the results:
(a) Estimate a constant value for K and use the small solute transfer assumption.
(b) Construct a y–x curve and solve graphically using the small solute transfer assumption.
(c) Solve graphically on the triangular diagram.

9.15 Repeat all parts of Problem 9.14 if the feed stream contains 40 percent ethanol, 60 percent benzene, and 250 lb/min of water is used.

PART III

NonIsothermal Systems

CHAPTER 10

Law of Conservation of Energy

10.1 INTRODUCTION

In dealing with applications of the law of conservation of mass we were able to presume a familiarity with mass and how it is transported. When mass is the fundamental dependent variable of interest the characterizing variables are easily determined, since it is usually obvious how the total mass or the mass of a particular component is measured. In problems in which energy is our fundamental dependent variable, selection of the characterizing variables is not so readily done. There is no such device as an energy meter, and it is thus necessary as a first step to discuss the various kinds of energy and to consider what characterizing variables are employed to measure the energy of a system. This chapter is devoted to this task and provides an introduction to the field of thermodynamics. We will introduce the concept of internal energy, write a general energy balance in terms of the fundamental dependent variables for a tank-type device, and then discuss the experimental procedures that allow us to write the energy balance in terms of characterizing dependent variables.

10.2 INTERNAL ENERGY

It was not necessary in our discussion of problems where mass is the fundamental dependent variable to identify or define different "kinds" of mass. Mass is a quantity with which we are familiar from our everyday experience. Energy is not as familiar a concept and, furthermore, it is necessary to identify and define various kinds of energy. Some of these kinds of energy are familiar to us from studies of mechanics, the most common being the potential energy (PE). The potential energy per unit mass ($\underline{PE}$) of a substance can be readily expressed in terms of variables which can be measured, since it is defined as the work necessary to raise a certain mass to a given height:

$$\underline{PE} = \frac{hg}{g_c} \tag{10.1}$$

$\underline{PE}$ has units of ft lb_f/lb_m. h is the height of the material above some arbitrary datum, g the gravitational acceleration, and g_c the dimensional constant required when lb_m and lb_f are used. g_c is discussed in Chapter 2.

The kinetic energy (KE) is also a type of energy with which we are reasonably familiar. It is defined as the work required to accelerate a constant mass from rest to a velocity, v. The kinetic energy per unit mass ($\underline{KE}$) can be written as follows:

$$\underline{KE} = \frac{v^2}{2g_c} \tag{10.2}$$

Units of $\underline{KE}$ are also ft lb_f/lb_m. v is the velocity, in this case expressed in feet per second. Since velocities and heights are readily determined experimentally the kinetic and potential energies of a system are easily expressed in terms of characterizing variables. We can, of course, work equally well with the metric system of units, in which case the units of both Equations 10.1 and 10.2 would be dynes per gram or Newtons per kilogram.

The kinetic and potential energies are defined by the amount of work that must be done for a material to achieve a certain height or a certain velocity. Our experience with the physical world also tells us, however, that if we perform work on a system we can detect changes under conditions when the kinetic and potential energies remain constant. If, for example, we compress a gas in a cylinder we can measure a change in pressure and sometimes also a change in temperature. Since work has been done on the gas (the piston moving through some distance) and no change has occurred in either the kinetic or potential energy we must identify another type of energy change which is related in some way to the temperature and pressure of the system. We call such an energy the *internal energy* of the system. As with kinetic and potential energies, the change in value of the internal energy depends only on the initial and final states of the system and not on the means of getting there.

The physical motivation for defining the internal energy was provided by a series of experiments carried out by Joule during the nineteenth century. In these experiments water was maintained in a well-insulated tank and work was done on the system. In the most famous case a paddle was immersed in the water and made to turn by a series of pulleys attached to a falling weight. The work done by the falling weight was easily calculated. It was observed that the water temperature increased as a result of the work done, one pound of water increasing one degree Fahrenheit for each 773 ft lb_f of work done on the system. Other experiments with different means of doing work were as follows:

1. Mechanical work was performed to compress a gas in a cylinder immersed in a well-insulated vessel containing water.
2. Mechanical work was done on two pieces of iron rubbed together beneath the water surface.
3. An electric current was generated by mechanical work and a coil carrying the current was immersed in water.

In experiment (1) it was found that 795 ft lb_f were required to raise 1 lb_m of water 1°F. In experiment (2) 775 ft lb_f were required and in experiment (3) 838 ft lb_f. To within the accuracy that could be achieved in these early experiments it is clear that each of the four different methods gave essentially the same value. In each of the experiments the water was enclosed in a well-insulated vessel and little or no heat was lost to the surroundings. In thermodynamic terms we call such a system *adiabatic*, and we can make the following statement based on the early experiments and verified many times since:

The change of a body inside an adiabatic enclosure from a given initial state to a given final state involves the same amount of work by whatever means the process is carried out.

We can express this idea in symbols as follows:

$$w = -[U_B - U_A] \tag{10.3}$$

U is the internal energy, subscript A indicates the initial state and B the final state. By convention w is the work performed by the system. If the system does work then U_B is less than U_A and w is positive. Both sides of Equation 10.3 have the dimensions of energy. We usually work with the units ft lb_f for work and BTU (British Thermal Units) for energy; the presently accepted value for the conversion is that 1 BTU is equal to 778 ft lb_f. In the metric system 1 calorie is equal to 4.184 joules.

We know from the early experiments that work is a mode of energy transfer, but our experience tells us that we can also raise the temperature of a body if we bring the body into contact with another body at higher temperature. We must therefore postulate another mode of energy transfer different from work, called heat and represented by the symbol **q**. (The bold face is to distinguish the symbol for heat from that for volumetric flow rate.) If we perform a thought experiment and consider a batch process whereby a body both absorbs heat and performs work, then the change in the internal energy can be expressed as follows:

$$U_B - U_A = \mathbf{q} - w \tag{10.4}$$

This equation is the symbolic representation of the *first law of thermodynamics* and states that the change in internal energy, U, of a batch system is the algebraic sum of the heat and work effects. By convention we take **q** to be positive if heat is added to the system and w to be positive if work is done by the system. U is a *state function*, its value depending only on the initial and final states, and **q** and w are modes of transfer of energy between one system and another.

Equation 10.3 provides us with an experimental means of determining changes in internal energy provided we can measure the work done. The batch system must be well enough insulated so that there is no transfer of heat to or from the system. If the system is not well insulated and there is a transfer of heat, we can use Equation 10.4 to find **q** if we have determined $U_B - U_A$ by the adiabatic experiment and if we can measure the work done. Conversely, Equation 10.4 can be used to determine internal energy changes if **q** and w can be measured.

It is most important to note that the definitions of both internal energy and heat are *constructive*, in that they are defined only in terms of measurable quantities. The internal energy change is defined by Equation 10.3 through the measurable work done by an adiabatic system. The heat is then defined by Equation 10.4 in terms of work and internal energy change, the latter now measurable as a result of the definition 10.3. Notice also that, as defined, internal energy is like potential energy in that only relative values can be determined. Thus a description of a physical process can never depend on the absolute magnitude of the internal energy, but only on changes.

Finally, it is important to note that internal energy must depend not only on temperature but on volume and composition as well, while internal energy per unit mass must depend on density (reciprocal of volume per unit mass) and composition. We draw on experience to observe a fact which we shall use without a formal proof, that temperature, density, and composition in a single phase uniquely determine the pressure. Thus, to establish the internal energy it must be sufficient to establish the temperature, density or volume, and composition. This observation plays an important role in applications of the principle of conservation of energy and we shall return to it repeatedly.

10.3 A GENERAL ENERGY BALANCE

Our strategy in applying the law of conservation of energy will be somewhat different from that used for mass balances. We will develop an energy balance for a well-stirred flow system (one in which we allow mass to cross the control surface) and then simplify the general balance as we apply it to various problems. The system and control surface shown in Figure 10.1 is general enough to be of considerable use in many problems in chemical

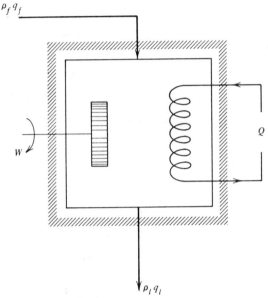

FIGURE 10.1 System for energy balance. Heat is added at a rate Q, work done at a rate W.

engineering without becoming confusingly complex. The figure is schematic only. It represents some general system in which work can be done either by or on the system (represented schematically by the paddle) and in which heat can be added or removed (represented schematically by the coil). Material flows into the system at some specified mass flow rate, $\rho_f q_f$ $\text{lb}_\text{m}/\text{time}$, out at a rate $\rho_l q_l$ $\text{lb}_\text{m}/\text{time}$. Notice that we are changing our previous convention by using a subscript "l" to denote the stream leaving the tank. Quantities in the tank remain unsubscripted. The reason will be apparent subsequently.

The energy contained within the control volume at time $t + \Delta t$ must be equal to the energy contained in the control volume at time t plus the total amount of energy which has appeared in the control volume in the interval Δt by all processes less the total amount of energy which has disappeared from the control volume in the time interval Δt by all processes.

We will consider that the total energy can be represented by the sum of the internal energy, the kinetic energy, and the potential energy. Other forms of energy, such as surface energy and electromagnetic energy, are not considered, although they may be of importance in problems not considered in this text. We let the total internal energy be represented by U and the internal energy per unit mass by $\underline{U}$. Similarly, KE is used to represent the kinetic energy, PE the potential energy, and $\underline{KE}$ and $\underline{PE}$ the respective energies per unit mass. We can now write the energy balance as follows:

$$[U + KE + PE]\big|_{t+\Delta t} = [U + KE + PE]\big|_t$$

+ total energy in with convective flow during Δt

− total energy out by convective flow during Δt

+ total energy added as heat (schematically, through the coil) during Δt

− total energy removed because the system does work (schematically, with the paddle) during Δt

The total energy which comes in with the feed stream can be written in terms of the value per unit mass

$$\rho_f q_f [\underline{U}_f + \underline{KE}_f + \underline{PE}_f] \, \Delta t$$

Similarly, the energy leaving by flow is written in terms of the value per unit mass

$$\rho_l q_l [\underline{U}_l + \underline{KE}_l + \underline{PE}_l] \, \Delta t$$

Let us designate the *rate* at which energy is added to our system as heat by Q, BTU/time. Then $Q = d\mathbf{q}/dt$. By convention, Q is positive if heat is added and negative if heat is removed. We have shown a coil in Figure 10.1 as the means of adding heat, but this is a schematic representation only. Q is understood to be the representation for any energy added or removed from the system as heat. Similarly, we use the symbol W for the rate at which energy disappears from the system in the form of work, $W = dw/dt$. If work is done on the system by some device, W is taken by convention to be negative. The energy balance can then be written entirely in symbols as

$$[U + KE + PE]\big|_{t+\Delta t} = [U + KE + PE]\big|_t$$
$$+ \rho_f q_f [\underline{U}_f + \underline{KE}_f + \underline{PE}_f] \, \Delta t$$
$$- \rho_l q_l [\underline{U}_l + \underline{KE}_l + \underline{PE}_l] \, \Delta t$$
$$+ Q \, \Delta t - W \, \Delta t \; .$$

or, dividing by Δt and taking the limit as $\Delta t \to 0$,

$$\frac{d[U + KE + PE]}{dt} = \rho_f q_f [\underline{U}_f + \underline{KE}_f + \underline{PE}_f]$$

$$- \rho_l q_l [\underline{U}_l + \underline{KE}_l + \underline{PE}_l] + Q - W \quad (10.5)$$

It is often convenient to separate the work term into two parts. Work must be done on the system to push material in in the feed stream, and work must be done by the system to push material out in the effluent. All other work except this *flow* work (e.g., useful work for turning a turbine) is called *shaft work*. The rate of doing shaft work is designated W_s. The flow work for the stream entering can be calculated as follows:

$$\text{work} = F\,\Delta l$$

where F is force and Δl denotes the small distance a mass is pushed over the system boundary. $F = p_f A_f$, where p_f is the pressure at the entrance and A_f the cross-sectional area. Thus

$$\text{work} = p_f A_f\,\Delta l = p_f\,\Delta V$$

where ΔV is the volume of the material pushed in. Then

$$\frac{\text{work}}{\text{mass}} = p_f\frac{\Delta V}{\text{mass}} = p_f/\rho_f$$

since mass per volume is simply density. Rate of doing work is work per mass times mass per time, or $\rho_f q_f p_f/\rho_f$. Finally, since work is being done on the system we need a negative sign. In a similar way, the rate of work done *by* the system to push mass out is $\rho_l q_l p_l/\rho_l$. Thus

$$W = W_s - \rho_f q_f p_f/\rho_f + \rho_l q_l p_l/\rho_l \tag{10.6}$$

and an alternate form for the energy equation, Equation 10.5, is

$$\frac{d[U + \text{KE} + \text{PE}]}{dt} = \rho_f q_f[\underline{U}_f + p_f/\rho_f + \underline{\text{KE}}_f + \underline{\text{PE}}_f]$$

$$- \rho_l q_l[\underline{U}_l + p_l/\rho_l + \underline{\text{KE}}_l + \underline{\text{PE}}_l]$$

$$+ Q - W_s \tag{10.7}$$

The combination $U + pV$ often appears in engineering problems and is given a special name, *enthalpy*, with a symbol H. Enthalpy per unit mass, $\underline{H}$, then equals $\underline{U} + p/\rho$:

$$H = U + pV \tag{10.8}$$

$$\underline{H} = \underline{U} + \frac{p}{\rho} \tag{10.9}$$

Equation 10.7 can then be written in the form commonly seen

$$\frac{d[U + \text{KE} + \text{PE}]}{dt} = \rho_f q_f[\underline{H}_f + \underline{\text{KE}}_f + \underline{\text{PE}}_f]$$

$$- \rho_l q_l[\underline{H}_l + \underline{\text{KE}}_l + \underline{\text{PE}}_l]$$

$$+ Q - W_s \tag{10.10}$$

A word about units is appropriate here. Internal energy, enthalpy, and heat are normally measured in BTU or calories, while kinetic and potential energy and work are customarily reported in ft lb_f. Thus, the numerical conversion must always be accounted for properly when dealing with numbers.

10.4 DRAINING TANK

We can illustrate the application of the energy equation nicely by returning to the tank-draining problem considered in Chapters 2 and 3. Recall that the problem was one in which a cylindrical tank open to the atmosphere empties through a small circular hole in the bottom. We found that in order to solve the equation of conservation of mass for the liquid level at any time it was necessary to obtain an independent relation between the flow rate through the orifice and the liquid head. By adding the energy equation to the analysis the correct form of this additional relation can be derived. This illustrates the step outlined on Figure 3.5 in which an additional fundamental dependent variable is added to obtain a sufficient number of equations for solution of the problem.

The system is shown in Figure 10.2. The tank cross-sectional area is A, the orifice area is A_0, the liquid height above the bottom at any time is h,

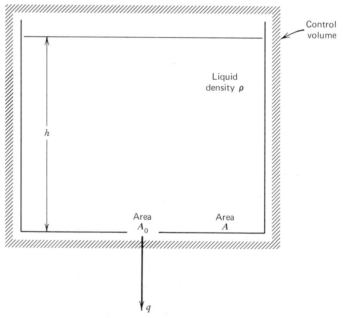

FIGURE 10.2   Control volume for application of the principle of conservation of energy to the emptying tank problem.

and the liquid density ρ. The control volume which we have used previously is drawn on the figure.

It is assumed that there is negligible heat gain or loss from the surroundings, so $Q = 0$, and that the only work being done is that required to move mass into and out of the control volume. Thus, $W_s = 0$ and the energy equation, in the form expressed in Equation 10.7, becomes

$$\frac{d[U + KE + PE]}{dt} = \rho_f q_f [\underline{U}_f + p_f/\rho_f + \underline{KE}_f + \underline{PE}_f]$$

$$- \rho_l q_l [\underline{U}_l + p_l/\rho_l + \underline{KE}_l + \underline{PE}_l] \quad (10.11)$$

This simplifies considerably by noting that air is entering the tank while liquid is leaving, so $\rho_f \ll \rho_l$. Thus, the only term in the first bracket on the right that will not be negligible is the pressure term $\rho_f q_f p_f/\rho_f = q_f p_f$. Similarly, only terms involving the liquid in the tank need be taken into account in computing the energies on the left-hand side of the equation because of the large density difference. Also, air flows in at the same rate as liquid leaves, so $q_f = q_l = q$. Thus the working form of the energy equation becomes

$$\frac{d[U + KE + PE]}{dt} = qp_f - \rho q \underline{U}_l - qp_l - \rho q \underline{KE}_l - \rho q \underline{PE}_l$$

$$(10.12)$$

Here we have dropped the subscript "l" on the liquid density since $\rho_l = \rho$, the liquid density in the tank.

The liquid jet at the exit is presumed to be at the same pressure as the surroundings, so $p_f = p_l$ for a tank that is open to the atmosphere and the pressure terms cancel one another. If we select the bottom of the tank as a reference level for potential energy then $\underline{PE}_l = 0$. Also, it is convenient to express the energies on the left on a per unit mass basis by $U = \rho A h \underline{U}$, etc., so we have

$$\frac{d\{\rho A h[\underline{U} + \underline{KE} + \underline{PE}]\}}{dt} = -\rho q \underline{U}_l - \rho q \underline{KE}_l \quad (10.13)$$

We must solve this together with the equation for conservation of mass which was first derived as Equation 2.1

$$\frac{d[\rho A h]}{dt} = -\rho q \quad (10.14)$$

The liquid density and the cross-sectional area are constant and can be removed from the derivative. Thus

$$\frac{d\{h[\underline{U} + \underline{KE} + \underline{PE}]\}}{dt} = [\underline{U} + \underline{KE} + \underline{PE}]\frac{dh}{dt}$$

$$+ h\left[\frac{d\underline{U}}{dt} + \frac{d\underline{KE}}{dt} + \frac{d\underline{PE}}{dt}\right]$$

$$= -\frac{q}{A}\underline{U}_l - \frac{q}{A}\underline{KE}_l \qquad (10.15)$$

$$\frac{dh}{dt} = -\frac{q}{A} \qquad (10.16)$$

It is convenient now to deal with the remaining terms in Equation 10.15 one at a time. The internal energy per unit mass, $\underline{U}$, is a constant, since the temperature and density of the liquid in the tank do not change. Thus

$$\frac{d\underline{U}}{dt} = 0$$

Furthermore, the exit stream will be at the same temperature and density as the liquid in the tank, so $\underline{U} = \underline{U}_l$. Together with Equation 10.16 this means

$$\underline{U}\frac{dh}{dt} = -\frac{q}{A}\underline{U}_l$$

All of the liquid in the tank is moving at a uniform velocity q/A (volumetric flow rate/area), so

$$\underline{KE} = \frac{1}{2g_c}\left[\frac{q}{A}\right]^2$$

It is necessary to recall from the study of mechanics in elementary physics that the potential energy of a finite mass is expressed in terms of the center of mass

$$\underline{PE} = \frac{g}{g_c}\text{ [height of center of mass above datum]}$$

The center of mass at any time is at a height $h/2$, so

$$\underline{PE} = \frac{gh}{2g_c}$$

Finally, the velocity of the exit stream is volumetric flow rate/area of exit jet. If the jet were to have the same area as the orifice this would be q/A_0. In fact, it is known experimentally (and requires very difficult application of the conservation of momentum to prove theoretically) that the jet diameter below a sharp hole is only about 80 percent of the orifice diameter. Thus the jet area is 60 to 65 percent of the orifice area and we write

$$\underline{KE}_t = \frac{1}{2g_c}\left[\frac{q}{C_0 A_0}\right]^2$$

where C_0 is a number approximately 0.6 to 0.65.

Putting all of the preceding relations into Equation 10.15 gives

$$\left\{\frac{1}{2g_c}\left[\frac{q}{A}\right]^2 + \frac{gh}{2g_c}\right\}\frac{dh}{dt} + h\left\{\frac{1}{2g_c}\frac{d}{dt}\left[\frac{q}{A}\right]^2 + \frac{q}{2g_c}\frac{dh}{dt}\right\}$$

$$= -\frac{q}{2g_c A}\left[\frac{q}{C_0 A_0}\right]^2 \quad (10.17)$$

Using Equation 10.16 for conservation of mass we write

$$\frac{d}{dt}\left[\frac{q}{A}\right]^2 = 2\frac{q}{A}\frac{d}{dt}\left[\frac{q}{A}\right] = 2\frac{q}{A}\frac{d}{dt}\left[-\frac{dh}{dt}\right] = -2\frac{q}{A}\frac{d^2h}{dt^2}$$

A term $[dh/dt]/2g_c$ factors out of Equation 10.17, giving

$$\left[\frac{q}{A}\right]^2 + gh + 2h\frac{d^2h}{dt^2} + gh\frac{dh}{dt} = \left[\frac{q}{C_0 A_0}\right]^2$$

or

$$\left[\frac{q}{C_0 A_0}\right]^2\left\{1 - C_0^2\left[\frac{A_0}{A}\right]^2\right\} = 2gh\left[1 + \frac{1}{g}\frac{d^2h}{dt^2}\right] \quad (10.18)$$

The term $C_0^2 A_0^2/A^2$ can be neglected compared to one with negligible error. If $A_0 = 0.25A$ then $A_0^2/A^2 = 0.063$; if $A_0 = 0.1A$ then $A_0^2/A^2 = 0.01$; and so on. On the other hand, the term d^2h/dt^2 appears to present some problems, for the presence of a second derivative means two integrations, hence two constants of integration and the necessity of knowing *two* things about the system at some time. We probably know the height, h_0, at $t = 0$, but it would be surprising if the solution depended on the flow rate, or dh/dt, at $t = 0$. This intuitive observation is equivalent to stating that we expect the term $[1/g][d^2h/dt^2]$ to be unimportant in most cases, or

$$\left|\frac{d^2h}{dt^2}\right| \ll g \quad (10.19)$$

This assumption needs to be verified subsequently, though it would indeed be surprising to find accelerations in an emptying tank which are of the order of 32 ft/sec/sec. Then, Equation 10.18 finally simplifies to

$$\left[\frac{q}{C_0A_0}\right]^2 = 2gh$$

$$q = C_0A_0\sqrt{2gh} \tag{10.20}$$

This is the relation deduced in Example 3.4 using dimensional analysis. This proportionality of q with the square root of h was also developed experimentally in Section 2.3. The value of C_0 computed there from the experimental data was 0.65.

The critical assumption that needs to be verified is that expressed by Equation 10.19. Substituting Equation 10.20 into Equation 10.16 for conservation of mass yields

$$\frac{dh}{dt} = -\frac{C_0A_0}{A}\sqrt{2gh}$$

$$h^{-1/2}\,dh = -\frac{C_0A_0}{A}[2g]^{1/2}\,dt$$

This leads, on integration, to

$$h = h_0\left[1 - \frac{C_0A_0\sqrt{2g}t}{2A\sqrt{h_0}}\right]^2$$

which is, of course, the form found to fit the experimental data in Figure 2.6. Then

$$\frac{1}{g}\frac{d^2h}{dt^2} = C_0^2\left[\frac{A_0}{A}\right]^2 \ll 1$$

That is, the assumption that d^2h/dt^2 can be neglected follows from being able to neglect $C_0^2A_0^2/A^2$ compared to 1.

The result derived here would be of only passing interest were it not for the fact that, on reflection, we recognize that the "driving force" for flow through the orifice is the pressure difference. Inside the tank just below the orifice the pressure is $p_A + \rho gh/g_c$, where p_A is the ambient pressure outside the tank and $\rho gh/g_c$ is the extra pressure from the weight of the column of liquid. Outside the pressure is simply p_A. Thus, the pressure difference, denoted by Δp, is

$$\Delta p = \rho gh/g_c$$

and Equation 10.20 can be rewritten

$$q = C_0A_0\sqrt{2g_c\,\Delta p/\rho} \tag{10.21}$$

(This was also obtained by dimensional analysis in Example 3.4.) Thus, by considering the special problem of flow from a tank we are led to a relation that must always hold between the flow rate and pressure difference for the flow of a liquid through an obstruction, where the single parameter C_0 must be obtained experimentally. This is the basis for the design of *orifice flowmeters*, where the flow rate can be obtained by measuring the pressure drop across an orifice in a flowing liquid.

10.5   HEAT CAPACITY

The purpose of this section is to show how the enthalpy and internal energy may be related to quantities that can be measured experimentally. We first consider an experimental means of determining a quantity called the *heat capacity* which is related to H and U as shown below.

If a material absorbs a certain quantity of heat, Δq, we know from experience that the temperature increases by a definite amount, ΔT. A quantity called the average heat capacity can be defined for the material in question as follows:

$$\Delta q = C_{AV} \Delta T$$

or

$$C_{AV} = \frac{\Delta q}{\Delta T}$$

An instantaneous heat capacity may be defined as Δq and ΔT become very small:

$$C \equiv \lim_{\Delta T \to 0} \frac{\Delta q}{\Delta T} = \frac{dq}{dT} \tag{10.22}$$

A batch experiment is the most convenient way to determine C. For a batch system, $q_f = q_i = 0$, and our general energy balance, Equation 10.5, simplifies to

$$\frac{d[U + KE + PE]}{dt} = Q - W \tag{10.23}$$

Since a batch system is a tank of some sort in a fixed position in the laboratory, the kinetic and potential energy changes will be zero and the mathematical description becomes

$$\frac{dU}{dt} = Q - W$$

or

$$Q = \frac{dU}{dt} + W \tag{10.24}$$

Q is the rate at which heat is put into the system, dq/dt, and W is the rate of doing work, dw/dt. Thus

$$\frac{dq}{dt} = \frac{dU}{dt} + \frac{dw}{dt}$$

Each term is a derivative, so we can integrate to obtain

$$\Delta q = \Delta U + \Delta w \tag{10.25}$$

where Δq, ΔU, and Δw refer, respectively, to the changes in q, U, and w over the length of the time interval.

Experiments can be performed in this batch system in two ways. If we have a totally enclosed system, the volume will remain constant and no work can be performed. Equation 10.25 becomes

$$\Delta q = \Delta U \tag{10.26}$$

When the volume of the system remains constant (e.g., a filled, sealed tank) and the total mass and composition is unchanged, we define the *heat capacity at constant volume*, C_V, from Equation 10.22 as

$$C_V = \lim_{\Delta T \to 0} \frac{\Delta q}{\Delta T}\bigg)_{V,n_i} = \lim_{\Delta T \to 0} \frac{\Delta U}{\Delta T}\bigg)_{V,n_i} = \frac{\partial U}{\partial T}\bigg)_{V,n_i} \tag{10.27}$$

The symbol $(\partial U/\partial T)_{V,n_i}$ means the partial derivative of U with respect to T holding volume (V) and number of moles of each species (n_i) constant. The definition of the partial derivative is contained in Section 15.11. Units of C_V in Equation 10.27 are BTU/degree F or calories/degree C. It is more common to work with the *heat capacity per unit mass*, $\underline{c}_V$, or *heat capacity per unit mole*, $\underset{\sim}{c}_V$, obtained by dividing each side of Equation 10.27 by the total mass or number of moles, respectively:

$$\underline{c}_V = \frac{\partial \underline{U}}{\partial T}\bigg)_{\rho,c_i} \tag{10.28}$$

$$\underset{\sim}{c}_V = \frac{\partial \underset{\sim}{U}}{\partial T}\bigg)_{\rho,c} \tag{10.29}$$

Here $\underset{\sim}{U}$ refers to internal energy per unit mole and, since volume is fixed, the partial derivatives in Equations 10.28 and 10.29 are taken at constant density and concentrations. Units of $\underline{c}_V$ are BTU/lb$_m$ °F or cal/g °C and of $\underset{\sim}{c}_V$ BTU/lb-mole °F or cal/g-mole °C.

A second controlled manner in which the batch experiment can be carried out is at constant pressure. This can be done as shown in Figure 10.3 by closing the cylinder with a movable piston, so that the volume can change if necessary to allow the pressure to remain equal to a specified value. In that

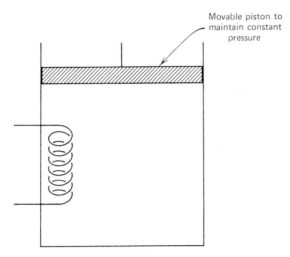

FIGURE 10.3   Constant pressure system.

case W_s is not zero, since the system will do work on the surroundings if the material expands with addition of heat. We then define the *heat capacity at constant pressure*, C_p, from Equation 10.22 as

$$C_p = \lim_{\Delta T \to 0} \frac{\Delta q}{\Delta T}\bigg)_{p,n_i} = \lim_{\Delta T \to 0} \frac{\Delta U + \Delta w}{\Delta T}\bigg)_{p,n_i}$$

The relation $\Delta q = \Delta U + \Delta w$ is Equation 10.25. Δw can be expressed in terms of the pressure against which expansion or contraction takes place and the resulting volume change:

$$\Delta w = p\,\Delta V$$

Since p is constant, $p\,\Delta V = \Delta pV$. Thus

$$\Delta U + \Delta w = \Delta U + \Delta pV = \Delta\,[U + pV] = \Delta H$$

where the enthalpy, $H = U + pV$, was first introduced in Equation 10.8. Thus

$$C_p = \lim_{\Delta T \to 0} \frac{\Delta H}{\Delta T}\bigg)_{p,n_i} = \frac{\partial H}{\partial T}\bigg)_{p,n_i} \tag{10.30}$$

Heat capacities per unit mass, $\underline{c}_p$, and per unit mole, $\underline{c}_p$, then follow as

$$\underline{c}_p = \frac{\partial \underline{H}}{\partial T}\bigg)_{p,c_i} \tag{10.31}$$

$$\underline{c}_p = \frac{\partial \underline{H}}{\partial T}\bigg)_{p,c_i} \tag{10.32}$$

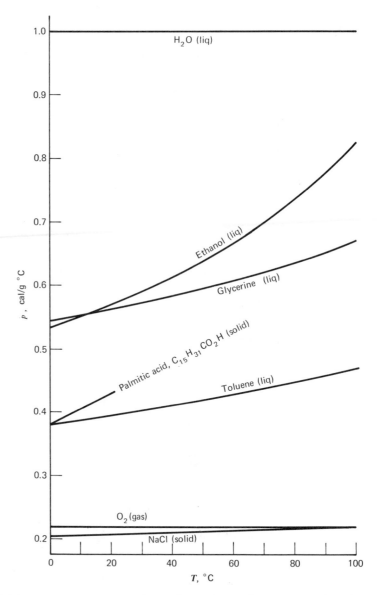

FIGURE 10.4 Heat capacities $\underline{c}_p$ for various substances as a function of temperature.

Here $\underline{H}$ refers to the enthalpy per mole. Typical heat capacity data are shown in Figure 10.4 and Table 10.1. Heat capacities have been tabulated for many materials of interest, and extensive compilations are available in the *International Critical Tables*, *Handbook of Chemistry and Physics*, and *Chemical Engineers' Handbook*.

TABLE 10.1 Heat capacities per unit mass at constant pressure and volume for typical materials

Material	State	T, °C	c_p cal/g °C	c_V cal/g °C
Silver	solid	20	0.032	
Steel	solid	20	0.11	
Charcoal	solid	10	0.16	
Sodium chloride	solid	0	0.20	
Urea	solid	20	0.32	
Water (ice)	solid	−10	0.53	
Paraffin	solid	20	0.69	
Mercury	liquid	20	0.033	
Sulfur dioxide	liquid	20	0.33	
Sulfuric acid	liquid	20	0.34	
Toluene	liquid	50	0.42	
Turpentine	liquid	18	0.42	
Glycerine	liquid	15	0.56	
Olive oil	liquid	15	0.57	
Ethylene glycol	liquid	15	0.57	
Ethanol	liquid	23	0.58	
Water	liquid	18	1.00	
Sulfur dioxide	gas, 1 atm	15	0.15	0.12
Oxygen	gas, 1 atm	15	0.22	0.16
Nitrogen	gas, 1 atm	15	0.25	0.18
Ethylene	gas, 1 atm	15	0.36	0.29
Water (steam)	gas, 1 atm	100	0.48	0.36
Hydrogen	gas, 1 atm	15	3.39	2.42

In Table 10.1 only c_p data are shown for solids and liquids. Clearly the change in volume will usually be negligible for these phases in the measurement of c_p, so Δq should not differ measurably in constant volume and constant pressure experiments. Thus, for solids and liquids $c_p = c_V$, $\underline{c}_p = \underline{c}_V$. For ideal monatomic gases it is shown in physical chemistry that $\underline{c}_p = \underline{c}_V + R$, where R is the gas constant, and for many real gases the ratio $\underline{c}_p/\underline{c}_V$ is approximately 1.4. It should also be noted that the definitions of BTU and calorie are such that

$$1\,\frac{\text{BTU}}{\text{lb}_m\,°\text{F}} = 1\,\frac{\text{cal}}{\text{g}\,°\text{C}}$$

Occasionally heat capacities are reported in units of $PCU/lb_m \, ^\circ C$. This is a mixed metric-engineering unit, where $1 \ PCU = 1.8 \ BTU$ and, as a consequence,

$$\frac{1 \ PCU}{lb_m \, ^\circ C} = \frac{1 \ BTU}{lb_m \, ^\circ F}$$

Heat capacities are often reported without units as *specific heat*, which is the ratio of heat capacity to that of liquid water at 1 atmosphere pressure and 20°C. The value of the latter is 0.99947 cal/g °C, so the specific heat, with proper attention to units, can be used directly for heat capacity.

Knowledge of the heat capacity enables us to calculate changes in enthalpy or internal energy as a result of temperature changes. For example, Equation 10.31 can be rewritten

$$\text{constant } p, c_i: \qquad d\underline{H} = \underline{c}_p \, dT$$

and integrating between temperatures T_1 and T_2,

$$\text{constant } p, c_i: \quad \underline{H}(T_2) = \underline{H}(T_1) + \int_{T_1}^{T_2} \underline{c}_p \, dT \tag{10.33}$$

If $\underline{c}_p$ is a constant over the given temperature range, as is often the case for liquids,

$$\underline{c}_p = \text{constant}: \qquad \underline{H}(T_2) = \underline{H}(T_1) + \underline{c}_p[T_2 - T_1] \tag{10.34}$$

Data are often fit with the empirical equation

$$\underline{c}_p = a + bT + cT^2 \tag{10.35}$$

where T is absolute temperature, °R or °K. Then Equation 10.33 becomes

$$\underline{H}(T_2) = \underline{H}(T_1) + \int_{T_1}^{T_2} [a + bT + cT^2] \, dT$$

$$= \underline{H}(T_1) + a[T_2 - T_1] + \frac{b}{2}[T_2^2 - T_1^2] + \frac{c}{3}[T_2^3 - T_1^3]$$

$$\tag{10.36}$$

In a similar manner we obtain from Equations 10.28, 10.29, and 10.32 the equations for $\underline{U}$, $\underset{\sim}{U}$, and $\underset{\sim}{H}$ that are analogous to Equation 10.33:

$$\text{constant } \rho, c_i: \quad \underline{U}(T_2) = \underline{U}(T_1) + \int_{T_1}^{T_2} \underline{c}_V \, dT \tag{10.37}$$

$$\text{constant } \rho, c_i: \quad \underset{\sim}{U}(T_2) = \underset{\sim}{U}(T_1) + \int_{T_1}^{T_2} \underset{\sim}{c}_V \, dT \tag{10.38}$$

$$\text{constant } p, c_i: \quad \underset{\sim}{H}(T_2) = \underset{\sim}{H}(T_1) + \int_{T_1}^{T_2} \underset{\sim}{c}_p \, dT \tag{10.39}$$

Example 10.1

The heat capacity of liquid water at 1 atm. is nearly constant at 1 BTU/lb$_m$ °F. What is the change in enthalpy of 3 lb$_m$ when the temperature is changed from 77°F to 40°F?

From Equation 10.34 for constant $\underline{c}_p$,

$$H(32°F) = H(77°F) + 1 \frac{BTU}{lb_m \, °F} [32 - 77]$$

$$H(32°F) - H(77°F) = -45 \text{ BTU/lb}_m$$

$$H(32°F) - H(77°F) = [3 \text{ lb}_m][-45 \text{ BTU/lb}_m] = -135 \text{ BTU}$$

Example 10.2

The heat capacity of liquid toluene at atmospheric pressure in the range 280°K $\leq T \leq$ 360°K is well represented by the equation

$$\underline{c}_p = -0.28 + 2.3 \times 10^{-3} \, T, \text{ cal/g °C}$$

where T is in degrees Kelvin. What is the enthalpy change per gram in heating the liquid from 10°C to 60°C?

$$10°C = 283°K, \qquad 60°C = 333°K$$

From Equation 10.33

$$H(333°K) = H(283°K) + \int_{283}^{333} [-0.28 + 2.3 \times 10^{-3}T] \, dT$$

$$= H(283°K) - 0.28[333 - 283]$$

$$+ \frac{2.3 \times 10^{-3}}{2} [333^2 - 283^2]$$

(Note that $333^2 - 283^2 = [333 - 283][333 + 283] = 50 \times 616$.)

$$H(333°K) - H(283°K) = 21.4 \text{ cal/g}$$

Values of $\underline{c}_p$ and $\underline{c}_V$ are easily obtained from the literature for pure materials and certain common mixtures. For most mixtures which might be of interest, however, heat capacity data are not tabulated, and it is necessary to make experimental measurements over the entire range or to develop rational procedures for calculating heat capacities of mixtures from knowledge of pure component heat capacities. We can deal with the latter problem following a discussion of the concept of the partial molar enthalpy.

10.6 PARTIAL MOLAR QUANTITIES

When we defined the internal energy we noted that, since at a fixed composition the temperature and volume uniquely fix the pressure, it suffices to consider the internal energy as a function of temperature, volume, and composition. We choose temperature and volume rather than temperature and pressure because the heat capacity at constant volume, C_V, enables us to compute the change in internal energy when volume is fixed. In a similar way, since the enthalpy change with temperature can be computed from knowledge of C_p, the heat capacity at constant pressure, it is natural to take the enthalpy as a function of temperature, pressure, and composition. In functional notation this is written

$$H = H(T, p, n_1, n_2, \ldots, n_S) \tag{10.40}$$

$$U = U(T, V, n_1, n_2, \ldots, n_S) \tag{10.41}$$

Here $n_1, n_2, \ldots, n_S$ refer to the number of moles of each of S species which make up the system.

If we introduce small changes $\Delta T, \Delta p, \Delta n_1, \Delta n_2, \ldots, \Delta n_S$ in $T, p, n_1, n_2, \ldots, n_S$, respectively, then the small change in H, denoted by ΔH, follows from the definition of the partial derivative, Section 15.11:

$$\Delta H = \frac{\partial H}{\partial T}\bigg)_{p, n_i} \Delta T + \frac{\partial H}{\partial p}\bigg)_{T, n_i} \Delta p + \frac{\partial H}{\partial n_1}\bigg)_{\substack{T, p, n_i \\ i \neq 1}} \Delta n_1$$

$$+ \frac{\partial H}{\partial n_2}\bigg)_{\substack{T, p, n_i \\ i \neq 2}} \Delta n_2 + \cdots + \frac{\partial H}{\partial n_S}\bigg)_{\substack{T, p, n_i \\ i \neq S}} \Delta n_S \tag{10.42}$$

The notation T, p, n_i (with $i \neq j$) means a partial derivative holding T, p, and the number of moles of each species except species j constant. The coefficient of ΔT is, of course, C_p. The coefficient of Δp will not be derived here in terms of experimentally measurable quantities, but it is shown in textbooks on thermodynamics to equal

$$\frac{\partial H}{\partial p}\bigg)_{T, n_i} = V - T\frac{\partial V}{\partial T}\bigg)_{p, n_i} \tag{10.43}$$

We will show at the end of the chapter that this term is small for liquids and solids, where the density is essentially constant, and that is the reason that the enthalpy in a liquid does not depend strongly on pressure. The remaining coefficients in Equation 10.42 can be measured experimentally and are given the name *partial molar enthalpy*, $\tilde{H}_j$:

$$\tilde{H}_j = \frac{\partial H}{\partial n_j}\bigg)_{\substack{T, p, n_i \\ i \neq j}} \tag{10.44}$$

The partial molar enthalpy has units of BTU/lb-mole or cal/g-mole. Equation 10.42 can then be written

$$\Delta H = C_p \Delta T + \left[V - T \frac{\partial V}{\partial T} \right)_{p,n_i} \right] \Delta p$$
$$+ \tilde{H}_1 \Delta n_1 + \tilde{H}_2 \Delta n_2 + \cdots + \tilde{H}_S \Delta n_S \quad (10.45)$$

If the small changes $\Delta T, \Delta p, \Delta n_1, \Delta n_2, \ldots, \Delta n_S$ take place over a time interval of Δt we write

$$\frac{\Delta H}{\Delta t} = C_p \frac{\Delta T}{\Delta t} + \left[V - T \frac{\partial V}{\partial T} \right)_{p,n_i} \right] \frac{\Delta p}{\Delta t}$$
$$+ \tilde{H}_1 \frac{\Delta n_1}{\Delta t} + \tilde{H}_2 \frac{\Delta n_2}{\Delta t} + \cdots + \tilde{H}_S \frac{\Delta n_S}{\Delta t}$$

and in the limit as $\Delta t \to 0$

$$\frac{dH}{dt} = C_p \frac{dT}{dt} + \left[V - T \frac{\partial V}{\partial T} \right)_{p,n_i} \right] \frac{dp}{dt}$$
$$+ \tilde{H}_1 \frac{dn_1}{dt} + \tilde{H}_2 \frac{dn_2}{dt} + \cdots + \tilde{H}_S \frac{dn_S}{dt} \quad (10.46)$$

Equation 10.46 is an application of the chain rule, Equation 15.29.

In a similar manner we note

$$\Delta U = \frac{\partial U}{\partial T} \right)_{V,n_i} \Delta T + \frac{\partial U}{\partial V} \right)_{T,n_i} \Delta V + \frac{\partial U}{\partial n_1} \right)_{\substack{T,V,n_i \\ i \neq 1}} \Delta n_1$$
$$+ \frac{\partial U}{\partial n_2} \right)_{\substack{T,V,n_i \\ i \neq 2}} \Delta n_2 + \cdots + \frac{\partial U}{\partial n_S} \right)_{\substack{T,V,n_i \\ i \neq S}} \Delta n_S \quad (10.47)$$

The first coefficient is C_V. We note without proof the thermodynamic equation

$$\frac{\partial U}{\partial V} \right)_{T,n_i} = T \frac{\partial p}{\partial T} \right)_{V,n_i} - p \quad (10.48)$$

and define the *partial molar internal energy*, $\tilde{U}_j$:

$$\tilde{U}_j = \frac{\partial U}{\partial n_j} \right)_{\substack{T,V,n_i \\ i \neq j}} \quad (10.49)$$

Then Equation 10.47 can be written

$$\Delta U = C_V \Delta T + \left[T \frac{\partial p}{\partial T}\bigg)_{V, n_i} - p \right] \Delta V$$

$$+ \tilde{U}_1 \Delta n_1 + \tilde{U}_2 \Delta n_2 + \cdots + \tilde{U}_S \Delta n_S$$

or, in terms of changes over a time interval Δt which tends to zero,

$$\frac{dU}{dt} = C_V \frac{dT}{dt} + \left[T \frac{\partial p}{\partial T}\bigg)_{V, n_i} - p \right] \frac{dV}{dt}$$

$$+ \tilde{U}_1 \frac{dn_1}{dt} + \tilde{U}_2 \frac{dn_2}{dt} + \cdots + \tilde{U}_S \frac{dn_S}{dt} \quad (10.50)$$

The coefficient of dV/dt is negligible for most liquids and solids, so U may be considered to depend on temperature and composition for these phases.

The defining equations for the partial molar quantities, Equations 10.44 and 10.49, contain within them the description of the experiment required to measure $\tilde{H}_i$ and $\tilde{U}_i$. At a given composition the temperature and pressure or temperature and volume are fixed and then a small amount of species i (Δn_i) is added and the change in H or U is measured. This would have to be done for every composition in every mixture of interest and that is clearly not practical. It turns out, however, that certain groupings of partial molar quantities arise frequently and can often be measured and recorded without great difficulty. We shall see shortly how this happens.

For subsequent calculations it is useful to derive some further relations between the enthalpy of a mixture and the partial molar enthalpies. We will carry out a thought experiment at constant temperature and pressure. In Equation 10.45 this means $\Delta T = \Delta p = 0$. In an empty tank we mix together Δn_A moles of A and Δn_B moles of B. The total enthalpy is then

$$\Delta H = \tilde{H}_A \Delta n_A + \tilde{H}_B \Delta n_B$$

To this we add an equal amount of a mixture of identical concentration. Since everything exactly doubles, the enthalpy doubles to an amount $2 \Delta H$, while the number of moles of A and B are now $2 \Delta n_A$ and $2 \Delta n_B$. Thus

$$2 \Delta H = \tilde{H}_A [2 \Delta n_A] + \tilde{H}_B [2 \Delta n_B]$$

$\tilde{H}_A$ and $\tilde{H}_B$ clearly remain the same, so they depend only on concentrations, not total number of moles. The process is repeated N times to give finally

$$N \Delta H = \tilde{H}_A [N \Delta n_A] + \tilde{H}_B [N \Delta H_B]$$

Since we started with an empty tank the total number of moles is $n_A = N \Delta n_A$, $n_B = N \Delta n_B$, while the total enthalpy is $N \Delta H$. Thus

$$H = n_A \tilde{H}_A + n_B \tilde{H}_B$$

More generally, for a mixture of S components,

$$H = n_1\tilde{H}_1 + n_2\tilde{H}_2 + \cdots + n_S\tilde{H}_S \tag{10.51}$$

An equivalent relation for U is,

$$U = n_1\tilde{U}_1 + n_2\tilde{U}_2 + \cdots + n_S\tilde{U}_S \tag{10.52}$$

By noting that $\underline{H} = H/\rho V$, $c_i = n_i/V$, these equations can be rewritten

$$\underline{H} = \frac{1}{\rho}[c_1\tilde{H}_1 + c_2\tilde{H}_2 + \cdots + c_S\tilde{H}_S] \tag{10.53}$$

$$\underline{U} = \frac{1}{\rho}[c_1\tilde{U}_1 + c_2\tilde{U}_2 + \cdots + c_S\tilde{U}_S] \tag{10.54}$$

ρ is the density of the mixture at the specified temperature and concentrations.

10.7 HEAT OF SOLUTION

The *heat of solution* or *heat of mixing* is a quantity related to the partial molar enthalpy which is frequently required in situations where materials are to be mixed together. Consider an experiment in which n_A moles of pure A and n_B moles of pure B are to be mixed together in a system in which we can accurately measure the heat input or heat removal required to maintain the temperature at a constant value. The physical situation is shown in Figure 10.5. Data are usually collected at 25°C.

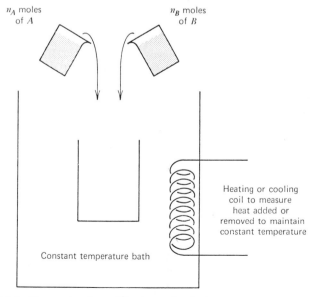

FIGURE 10.5 Determination of the heat of solution.

It is convenient to work in molar units. Application of the law of conservation of mass to each species yields

$$M_{wA}\frac{dn_A}{dt} = \rho_A q_{Af} \tag{10.55A}$$

$$M_{wB}\frac{dn_B}{dt} = \rho_B q_{Bf} \tag{10.55B}$$

M_{wA} and M_{wB} are molecular weights and q_{Af} and q_{Bf} are the volumetric flow rates of A and B into the system, respectively. The energy balance is obtained by modifying Equation 10.10 to account for the fact that there are two inlet streams. Notice that $q_l = 0$, $W_S = 0$, so neglecting kinetic and potential energy terms,

$$\frac{dU}{dt} = \rho_A q_{Af} H_A + \rho_B q_{Bf} H_B + Q \tag{10.56}$$

Now, $\underline{H}_A = H_A/M_{wA}$, $\underline{H}_B = H_B/M_{wB}$, and on substitution of Equations 10.55 the energy balance can be written

$$\frac{dU}{dt} = \underline{H}_A\frac{dn_A}{dt} + \underline{H}_B\frac{dn_B}{dt} + Q \tag{10.57}$$

We will presume that dpV/dt is negligible compared to dH/dt, an estimate which we will justify later in this chapter for liquid systems. Thus

$$\frac{dU}{dt} = \frac{dH}{dt} - \frac{dpV}{dt} \simeq \frac{dH}{dt} \tag{10.58}$$

The enthalpies per mole of pure material, $\underline{H}_A$ and $\underline{H}_B$, are constant as long as the temperature is fixed, while $Q = d\mathbf{q}/dt$. Thus, Equation 10.57 can be written

$$\frac{d}{dt}[H - \underline{H}_A n_A - \underline{H}_B n_B - \mathbf{q}] = 0$$

$$H - \underline{H}_A n_A - \underline{H}_B n_B - \mathbf{q} = \text{constant}$$

where $\mathbf{q}$ is the total amount of heat added to the system. When the tank is empty $H = 0$, so the constant is zero, and we have

$$H = \underline{H}_A n_A + \underline{H}_B n_B + \mathbf{q} \tag{10.59}$$

From Equation 10.51 we can express H in terms of the partial molar enthalpies

$$H = \tilde{H}_A n_A + \tilde{H}_B n_B$$

Thus

$$\tilde{H}_A n_A + \tilde{H}_B n_B = \underline{H}_A n_A + \underline{H}_B n_B + \mathbf{q} \tag{10.60}$$

Remember, $\underline{H}_A$ is the enthalpy per mole of pure A, $\tilde{H}_A$ the partial molar enthalpy of A in the mixture of A and B.

Sometimes when two substances are mixed together no heat need be added or removed in order to maintain the temperature constant. Then $\mathbf{q} = 0$ and Equation 10.60 requires $\tilde{H}_A = \underline{H}_A$, $\tilde{H}_B = \underline{H}_B$. That is, the partial molar enthalpies are equal to the enthalpies per mole of pure component. Such solutions are called *ideal*. When a solution is ideal the analysis is greatly simplified, since pure component enthalpies are substituted for partial molar enthalpies and the former can be related directly to pure component heat capacities, which are extensively tabulated.

Many mixtures of pragmatic interest do show heat effects upon mixing. The measured amount of heat required to maintain the temperature is called the *heat of solution* and denoted ΔH_s. Equation 10.60 is rewritten

$$\Delta H_s = \mathbf{q} = n_A[\tilde{H}_A - \underline{H}_A] + n_B[\tilde{H}_B - \underline{H}_B] \tag{10.61}$$

Data are usually recorded on a "per mole of solute" basis

$$\Delta \underline{H}_s = \frac{\mathbf{q}}{n_A} = \tilde{H}_A - \underline{H}_A + \frac{n_B}{n_A}[\tilde{H}_B - \underline{H}_B] \tag{10.62}$$

$\Delta \underline{H}_s$ is called the *integral heat of solution*. Typical data for a number of aqueous solutions at 25°C are shown in Figure 10.6. Recall that 1 kcal = 1000 cal. If $\Delta \underline{H}_s$ is positive then, in accordance with the convention on $\mathbf{q}$, it means that heat must be added during mixing to maintain the temperature. This is called *endothermic* mixing. When $\Delta \underline{H}_s < 0$ heat must be removed to maintain the temperature. This is called *exothermic* mixing.

Heat of solution data for the sulfuric acid-water system are useful for more detailed examination. In our nomenclature A is sulfuric acid, B is water. The heat of solution for this system is strongly concentration dependent up until a mole ratio of 30 moles of water to 1 mole of acid. After this point $\Delta \underline{H}_s$ becomes nearly constant at -17.5 Kcal/mole acid (1 Kcal = 1000 cal). This constant value is often referred to as the *heat of solution at infinite dilution* and is represented by the term $\tilde{H}_A - \underline{H}_A$ in Equation 10.62. At infinite dilution the second term in the equation, representing the difference between the partial molar enthalpy of water and the enthalpy per mole of water, approaches zero since the solution is almost all water. At the other end of the concentration scale $\Delta \underline{H}_s$ is equal to zero. There is no water in the system ($n_B = 0$) and the partial molar enthalpy of sulfuric acid $\tilde{H}_A$ in a pure sulfuric acid mixture becomes equal to $\underline{H}_A$, the enthalpy per mole of sulfuric acid.

The heat of solution at 25°C is tabulated for a number of solutions at infinite dilution and can be found in any of the standard tables of physical constants. Some representative data are shown in Table 10.2. In the absence of data at the needed concentrations, heat of solution at infinite dilution can often be employed to yield limiting values for design purposes.

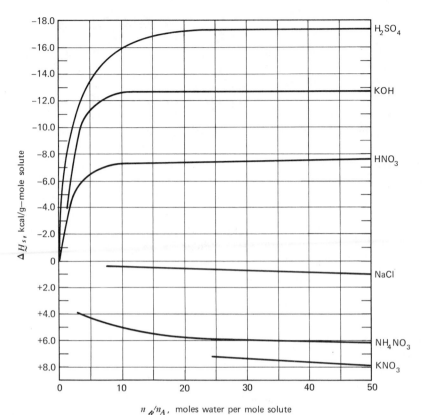

FIGURE 10.6 Heats of solution in water at 25°C. From selected values of Chemical Thermodynamic Properties, Nat. Bur. Standards Circ. #500 and supplements, 1952.

TABLE 10.2 Heat of solution in water at infinite dilution for selected compounds at 25°C

Compound	ΔH_s at infinite dilution (kcal/g-mole)
Acetic acid	+2.3
Ammonium nitrate	+6.5
Cuprous sulfate	−11.6
Magnesium iodide	−50.2
o-nitrophenol	+6.3
p-nitrophenol	+4.5
Potassium hydroxide	−12.9
Potassium iodide	+5.2
Sodium chloride	+1.2
Sodium citrate	−5.3
Sucrose	+1.3

Example 10.3

How much heat has to be removed to maintain the temperature at 25°C when 700 g of water and 100 g of sulfuric acid, each at 25°C, are mixed together?

$$M_{wA} = 98 \qquad\qquad M_{wB} = 18$$

$$n_A = \tfrac{100}{98} = 1.02 \qquad n_B = \tfrac{700}{18} = 38.9$$

$$\frac{n_B}{n_A} = \frac{38.9}{1.02} = 38.1$$

From Figure 10.6, at a mole ratio of 38.1,

$$\Delta H_s = q/n_A = -17.5 \text{ Kcal/g-mole}$$

$$q = -17.5 \times 1.02 = -17.8 \text{ Kcal}$$

17,800 calories must be removed to maintain the temperature at 25°C.

Example 10.4

How much heat has to be removed to maintain the temperature at 25°C when 600 g of water is mixed with 200 g of a 50 percent by weight sulfuric acid-water solution?

In the 50 percent by weight mixture

$$n_A = \tfrac{100}{98} = 1.02 \qquad n_B = \tfrac{100}{18} = 5.55$$

$$\frac{n_B}{n_A} = \frac{5.55}{1.02} = 5.44$$

At a mole ratio of n_B/n_A the integral heat of mixing is

$$\Delta H_s = q/n_A = -14 \text{ Kcal/g-mole}$$

The heat removed in making 200 g of the 50 percent by weight mixture was then

$$q = -14 \times 1.02 = -14.3 \text{ Kcal}$$

In the previous example the heat removed to make a mixture of 700 g water, 100 g acid was computed as −17.8 Kcal. The additional heat to be removed by adding the remaining 600 g of water is then $-17.8 - [-14.3] = -3.5$ Kcal.

10.8 HEAT CAPACITIES OF MIXTURES

The heat capacity of a binary mixture can be expressed in terms of the heat capacities of the pure materials and the heat of mixing. From Equation 10.30

$$C_p = \frac{\partial H}{\partial T}\bigg)_{p,n_i} \tag{10.30}$$

Equation 10.60 can be written

$$H = \Delta H_s + n_A \underline{H}_A + n_B \underline{H}_B \tag{10.63}$$

Thus, substituting for H in Equation 10.30 we have

$$C_p = \frac{\partial \Delta H_s}{\partial T}\bigg)_{p,n_i} + n_A \frac{\partial \underline{H}_A}{\partial T}\bigg)_{p,n_i} + n_B \frac{\partial \underline{H}_B}{\partial T}\bigg)_{p,n_i} \tag{10.64}$$

The temperature derivatives of $\underline{H}_A$ and $\underline{H}_B$ are simply the molar heat capacities of pure A and B, respectively, as shown in Equation 10.32. Thus we can write

$$C_p = \frac{\partial \Delta H_s}{\partial T}\bigg)_{p,n_i} + n_A \underline{c}_{pA} + n_B \underline{c}_{pB} \tag{10.65}$$

We obtain a somewhat more usable form by deriving an expression for the heat capacity per unit mass, $\underline{c}_p$. Note that

$$\rho V \underline{c}_p = C_p$$

$$\Delta H_s = n_A \Delta \underline{H}_s$$

Thus

$$\rho V \underline{c}_p = n_A \frac{\partial \Delta \underline{H}_s}{\partial T}\bigg)_{p,n_i} + n_A \underline{c}_{pA} + n_B \underline{c}_{pB}$$

$$\underline{c}_p = \frac{1}{\rho}\left[c_A \frac{\partial \Delta \underline{H}_s}{\partial T} + c_A \underline{c}_{pA} + c_B \underline{c}_{pB}\right] \tag{10.66}$$

For an ideal solution, in which $\Delta \underline{H}_s$ is zero, the heat capacity of a mixture is simply a weighted sum of pure component heat capacities. This is also true in a nonideal mixture when $\Delta \underline{H}_s$ is a weak function of temperature. Heat capacities of mixtures are much more frequently tabulated than heats of solution as a function of temperature, so Equation 10.66 is not a particularly useful equation in fact. Figure 10.7 shows the heat capacity for the sulfuric acid-water system. Equation 10.66 with $\partial \Delta H_s/\partial T$ set equal to zero often provides an adequate estimate of $\underline{c}_p$, although experimentally measured heat capacities should be used whenever possible. When the heat of solution can

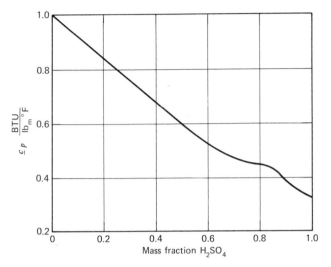

FIGURE 10.7   Heat capacity of sulfuric acid-water mixtures at 20°C.

be neglected, the heat capacity for a mixture of S species can be written

$$\underline{c}_p = \frac{1}{\rho} [c_1 \underline{c}_{p1} + c_2 \underline{c}_{p2} + \cdots + c_S \underline{c}_{pS}] \tag{10.67}$$

It should be noted that the results of this section can be used to obtain a relation for the variation of heat of solution with temperature. Equation 10.66 can be written as

$$c_A \frac{\partial \Delta \underline{H}_s}{\partial T} = \rho \underline{c}_p - c_A \underline{c}_{pA} - c_B \underline{c}_{pB}$$

This can be integrated from the temperature T° at which data are reported to any other temperature to yield

$$c_A \Delta \underline{H}_s(T) = c_A \Delta \underline{H}_s(T^\circ)$$

$$+ \int_{T^\circ}^{T} [\rho \underline{c}_p(T) - c_A \underline{c}_{pA}(T) - c_B \underline{c}_{pB}(T)] \, dT \tag{10.68}$$

For constant heat capacities this reduces to

$$c_A \Delta \underline{H}_s(T) = c_A \Delta \underline{H}_s(T^\circ) + [\rho \underline{c}_p - c_A \underline{c}_{pA} - c_B \underline{c}_{pB}][T - T^\circ] \tag{10.69}$$

Example 10.5

Compute the heat capacity of a 50 percent by weight sulfuric acid-water solution at 77°F (20°C). Data for heat of solution *per pound of mixture* versus temperature are shown in Figure 10.8.

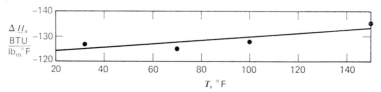

FIGURE 10.8   Heat of mixing per pound of a 50 percent by weight sulfuric acid-water mixture as a function of temperature.

Because of the per mass basis of the data it is necessary to reexpress Equation 10.66. Note the following:

$$\Delta H_s = \rho V \Delta \underline{H}_s \qquad \rho_A \underline{c}_{pA} = c_A \underline{c}_{pA} \qquad \rho_B \underline{c}_{pB} = c_B \underline{c}_{pB}$$

Here we have used ρ_A and ρ_B to denote the concentration of A and B in mass units in order to avoid confusion. Then Equation 10.65 can be written

$$\rho V \underline{c}_p = \rho V \frac{\partial \Delta \underline{H}_s}{\partial T} + \rho_A V \underline{c}_{pA} + \rho_B V \underline{c}_{pB}$$

$$\underline{c}_p = \frac{\partial \Delta \underline{H}_s}{\partial T} + x_A \underline{c}_{pA} + x_B \underline{c}_{pB} \qquad (10.70)$$

x_A and $x_B = 1 - x_A$ are mass fractions of A and B, respectively.

The data in Figure 10.8 are few and scattered but they do seem to be adequately represented by a straight line. The best straight line computed from Equation 6.9 has slope -0.081 BTU/lb$_m$ °F. Thus, $\partial \Delta \underline{H}_s / \partial T = -0.081$, $\underline{c}_{pA} = 0.335$ BTU/lb$_m$ °F, $\underline{c}_{pB} = 0.999$, $x_A = x_B = 0.5$, and

$$\underline{c}_p = -0.081 + 0.5 \times 0.335 + 0.5 \times 0.999 = 0.586 \, \text{BTU/lb}_m \, ^\circ\text{F}$$

This is in reasonable agreement with the measured value of 0.596. Neglect of the heat of mixing term would lead to a value of 0.667, a 14 percent difference.

10.9   PRESSURE DEPENDENCE OF ENTHALPY—I

There is a subtle point concerning the calculation of the enthalpy of a liquid in a tank which should be dealt with for completeness. Consider the definitions, Equations 10.8 and 10.9

$$H = U + pV \qquad (10.8)$$

$$\underline{H} = \underline{U} + p/\rho \qquad (10.9)$$

In a well-stirred system the internal energy is defined unambiguously, since temperature and volume have obvious meanings. The pressure varies with liquid height, however, so it is not clear what pressure is meant in Equation 10.8 when referring to the liquid in the tank. The problem is treated in the following way:

$$p = p_A + \frac{\rho g h}{g_c}$$

where p_A is the pressure at the exit of the tank. Then

$$\underline{H} = \underline{U} + \frac{p_A}{\rho} + \frac{g h}{g_c}$$

$$H = \int_0^h \rho A \underline{H} \, dh = \int_0^h \left[\rho A \underline{U} + p_A A + \frac{\rho g h A}{g_c} \right] dh \qquad (10.71)$$

$$H = \rho V \underline{U} + p_A V + \frac{\rho g A h^2}{2 g_c} = U + \left[p_A + \frac{\rho g h}{2 g_c} \right] V$$

Comparison of Equation 10.71 with Equation 10.8 shows that the pressure at the liquid midpoint should be used in computing the total enthalpy.

10.10 PRESSURE DEPENDENCE OF ENTHALPY—II

We have claimed on several occasions that pressure dependence of the enthalpy of a liquid is generally unimportant. We shall now justify this. Equation 10.45 can be written

$$\Delta H = \rho V \underline{c_p} \Delta T + \left[V - T \frac{\partial V}{\partial T} \bigg)_{p, n_i} \right] \Delta p + \cdots$$

The second term can be shown to be negligible compared to the first as follows:

For most liquids $T[\partial V / \partial T] \sim 10^{-3} VT$, where T is measured in °C. Thus, $V - T[\partial V / \partial T]$ is less than V, though perhaps of the same order of magnitude as V. We therefore need to show that $V \, \Delta p$ is negligible compared to $\rho V \underline{c_p} \Delta T$. Consider the ratio

$$\frac{V \, \Delta p}{\rho V \underline{c_p} \Delta T} = \frac{\Delta p}{\rho \underline{c_p} \Delta T}$$

Typically, $\rho \simeq 60$ lb_m/ft^3, $\underline{c}_p \simeq 0.8$ BTU/lb_m °F. If we take Δp as approximately atmospheric pressure, 14.7 $lb_f/in.^2 = [14.7 \times 144/778]$ BTU/ft^3, and ΔT only 1°F then

$$\frac{V \Delta p}{\rho V \underline{c}_p \Delta T} = \frac{14.7 \times 144}{778 \times 60 \times 0.8 \times 1} = 5.7 \times 10^{-2}$$

Thus, unless the pressure changes are quite large the pressure term is negligible in a liquid. This is not true, of course, in gases, where the pressure terms are of great importance.

In Equation 10.58 we claimed that for a liquid system

$$\frac{dU}{dt} = \frac{dH}{dt} - \frac{dpV}{dt} \simeq \frac{dH}{dt} \tag{10.58}$$

that is, that the rate of change of internal energy is nearly equal to the rate of change of enthalpy. If the liquid level is not changing, then pV is a constant, $dpV/dt = 0$, and the relation is exact! Otherwise, we need to compare dpV/dt with dH/dt and show that the latter dominates.

Consider a change over some interval Δt and compare ΔpV with ΔH. From Equation 10.45 the important term in ΔH is $\rho V \underline{c}_p \Delta T$. Thus, we need to estimate the ratio $\Delta pV/\rho V \underline{c}_p \Delta T$. If the total volume changes, then $\Delta pV \simeq pV$ and the ratio $\Delta pV/\rho V \underline{c}_p \Delta T \simeq p/ \rho \underline{c}_p \Delta T$. We have just shown that when p is of the order of atmospheric pressure this ratio is negligible for temperature changes of even 1°F. Thus, for a liquid system we can always interchange dU/dt and dH/dt without error. This is often not true for a gas system.

10.11   CONCLUDING REMARKS

A large number of quantities are defined in this chapter. In most cases the definitions are motivated by the experiment used to measure the quantity. These definitions are all essential to being able to write an energy equation in terms of measurable variables and you should commit them to memory. The heat capacity, for example, is an expression of the relation between energy and temperature. You should carefully review the discussion of internal energy, the derivation of the p/ρ term in the energy balance, the way in which enthalpy arises, and the discussion of heat capacity. The partial molar quantities are of great importance whenever mixtures are considered. Study the section on heat of solution carefully. In the next two chapters the way in which partial molar enthalpies combine, as in the heat of solution, is an essential part of the application of the principle of conservation of energy to practical situations.

In working with energy balances for various systems both mass and molar units are used. It is useful to have the following relations gathered together in one place for reference.

$$C_p = \rho_V \underline{c}_p \qquad C_V = \rho V \underline{c}_V$$

$$\underline{c}_{pi} = \frac{\underline{c}_{pi}}{M_{wi}} \qquad \underline{c}_{Vi} = \frac{\underline{c}_{Vi}}{M_{wi}}$$

$$\rho_i \underline{c}_{pi} = c_i \underline{c}_{pi} \qquad \rho_i \underline{c}_{Vi} = c_i \underline{c}_{Vi}$$

ideal mixture: $\underline{c}_p = \frac{1}{\rho}[c_1 \underline{c}_{p1} + c_2 \underline{c}_{p2} + \cdots + c_S \underline{c}_{pS}]$

$$\underline{H} = \frac{1}{\rho}[c_1 \tilde{H}_1 + c_2 \tilde{H}_2 + \cdots + c_S \tilde{H}_S]$$

$$\underline{H}_i = \frac{\underline{H}_i}{M_{wi}} \qquad U_i = \frac{U_{\sim i}}{M_{wi}}$$

This chapter provides an introduction to the subject of *thermodynamics*. The physical chemistry texts cited at the end of Chapter 5 all have one or more chapters on thermodynamics. The aspects of thermodynamics which we use in this text are expanded on in a particularly well written text by Denbigh:

10.1 K. G. Denbigh, *Principles of Chemical Equilibrium*, 3rd ed., Cambridge Univ. Press, Cambridge, 1971.

Other chemical engineering-oriented texts include:

10.2 R. Balzhiser, M. Samuels, and J. D. Eliassen, *Chemical Engineering Thermodynamics*, Prentice-Hall, Englewood Cliffs, 1971.

10.3 O. A. Hougen, K. M. Watson, and R. A. Ragatz, *Chemical Process Principles, Part II: Thermodynamics*, 2nd ed., Wiley, New York, 1959.

10.4 J. H. Smith and H. C. Van Ness, *Introduction to Chemical Engineering Thermodynamics*, 2nd ed., McGraw-Hill, New York, 1959.

10.12 PROBLEMS

10.1 Energy is designated by a variety of units with which one must become familiar. Some terms commonly appearing in energy balances are listed below. Find the numerical factor which must multiply each term if the equation is to have consistent units of BTU/lb$_\mathrm{m}$.

(a) p/ρ p [=] atm, ρ [=] lb$_\mathrm{m}$/ft^3

(b) p/ρ p [=] dynes/cm^2, ρ [=] g/cm^3

(c) hg/g_c $h [=] $ ft, $g [=] $ ft/sec^2
(d) $\underline{KE}$ $KE [=] $ ft lb$_f$/lb$_m$
(e) p/ρ $p [=] $ lb$_f$/in.^{2}, $\rho [=] $ lb$_m$/ft^3

10.2 If 10 lb of liquid with a heat capacity of 1 BTU/lb°F has its temperature raised by 10°F, the amount of energy required is 100 BTU. How high would we have to move this liquid to change its potential energy by 100 BTUs? To what velocity must we accelerate the liquid to change its kinetic energy by 100 BTUs?

10.3 Using chemical abstracts or the heat capacity tables in references 4.6, 4.7, or 4.8 find an original journal article describing how the heat capacity of some liquid was measured. Carefully read the paper in the journal in which it was published and describe the experimental procedure in your own words. Show clearly by means of a sample calculation how the heat capacity was calculated from the experimental data.

10.4 In carrying out engineering calculations it is important to have some idea of the range of values of the variables used. Make up a table showing the value of the heat capacity $\underline{c}_p$ in BTU/lb$_m$ °F for the following liquids, solids, and gases at 1 atm and 25°C:
(a) liquids: water, methanol, mercury, a light oil, benzene
(b) solids: wood, iron, aluminum, copper, asbestos
(c) gases: oxygen, ethylene, carbon dioxide, nitric oxide, nitrogen, ammonia, hydrogen

10.5 What numerical factor should a heat capacity in BTU/lb$_m$ °F be multiplied by to yield its value in the following units:
(a) cal/gram °C
(b) PCU/lb$_m$ °F
(c) BTU/lb-mole °F (assume molecular weight $= 18$)
(d) Kcal/gram-mole °C (assume molecular weight $= 18$)

10.6 Repeat Example 10.2 assuming that the heat capacity of toluene is constant, having a value equal to that computed at
(a) 10°C
(b) 60°C
(c) 35°C
Compute the difference between $\Delta \underline{H}$ obtained in each case with the value reported in Example 10.2.

10.7 Find the number of BTUs required to heat 1 lb of each of the following materials from 20°C to 90°C:
(a) graphite (solid carbon)
(b) malonic acid
(c) aluminum
(d) carbon dioxide at 1 atm

10.8 If the volume, V, of a system is a function of T, p, and the mole numbers n_1, n_2, and n_3 derive an expression for dV/dt in a manner similar to the development in Section 10.6.

Describe a simple experiment which could be performed to obtain the partial molar volumes, $\tilde{V}_j$.

10.9 By using Equations 10.46 and 10.51 prove that the following equation must always be true for batch binary systems at a constant temperature and pressure

$$n_1 \frac{d\tilde{H}_1}{dt} + n_2 \frac{d\tilde{H}_2}{dt} = 0$$

10.10 Compute the amount of heat in BTU which must be added or removed if the following operations are to be carried out under isothermal conditions at 25°C.

(a) 5 lb-moles of water added to 1 lb-mole of NH_4NO_3.

(b) 10 lb-moles of water added to 1 lb-mole of KOH.

(c) 30 lb-moles of water added to 1 lb-mole of HNO_3.

10.11 Compute the amount of heat in BTU which must be added or removed if the following operations are to be carried out under isothermal conditions at 25°C.

(a) 5 lb-moles of water added to a mixture of 5 lb-moles of water and 1 lb-mole of NH_4NO_3

(b) 1 lb mole of KOH added to a mixture of 10 lb moles of water and one lb mole of KOH.

10.12 (a) Compute the heat capacity at constant pressure, $\underline{c}_p$ for the following aqueous solutions using Equation 10.67.

(i) 20 mole % HCl at 20°C

(ii) 20 weight % HNO_3 at 20°C
40 weight % HNO_3 at 20°C
60 weight % HNO_3 at 20°C

(iii) 37 mole % ethanol at 20°C

(b) Find the heat capacity that has been measured experimentally for the above solutions and compute the percentage error between calculated and measured values.

10.13 Using Equation 10.66 and the results obtained in Problem 10.12 find $\partial\,\Delta H_s/\partial T$ for each system in Problem 10.12.

CHAPTER 11

Energy Balances for Nonreacting Liquid Systems

11.1 INTRODUCTION

The mixing of liquid streams of different composition and temperature is a common occurrence in processing operations and serves as an excellent illustration of the principles developed in the preceding chapter. We shall consider here the problem of determining the temperature variation in batch and continuous mixing in tank-type systems, including the development of shortcut procedures for frequently occurring calculations.

11.2 BATCH MIXING OF PURE STREAMS

Consider first the situation shown in Figure 11.1. Streams of pure A and B are mixed together in an insulated tank. We wish to determine the temperature in the tank as a function of time. (We refer to this, somewhat loosely, as batch mixing. It is, of course, a semibatch operation.)

Assuming perfect mixing, the mass balance equations are

$$\frac{dn_A}{dt} = \frac{dc_A V}{dt} = q_1 c_{A1} \tag{11.1a}$$

$$\frac{dn_B}{dt} = \frac{dc_B V}{dt} = q_2 c_{B2} \tag{11.1b}$$

Here, concentrations are measured in moles, and $c_{A1} = \rho_A / M_{wA}$, and so on, where ρ_A is the density of pure A and M_{wA} is the molecular weight. The energy equation follows directly from Equation 10.10 when kinetic and potential energy terms are neglected and we allow for the fact that there are two feed streams:

$$\frac{dU}{dt} = \rho_1 q_1 H_1 + \rho_2 q_2 H_2 \tag{11.2}$$

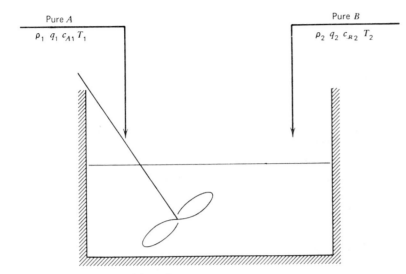

FIGURE 11.1   Batch mixing of two pure streams.

Since $\underline{H}_1 = \underline{H}_A = \underline{H}_A/M_{wA}$, $\underline{H}_2 = \underline{H}_B = \underline{H}_B/M_{wB}$, $\rho_1 = \rho_A = M_{wA}c_{A1}$, $\rho_2 = \rho_B = M_{wB}c_{B2}$, we can conveniently rewrite Equation 11.2 in molar units as

$$\frac{dU}{dt} = c_{A1}q_1\underline{H}_{A1} + c_{B2}q_2\underline{H}_{B2} \tag{11.3}$$

We have retained the extra subscript $\underline{H}_{A1}$ on $\underline{H}_A$ to emphasize that we are dealing with a property of the feed stream. The reason will become apparent subsequently. For this liquid system the rate of change of internal energy and enthalpy do not differ significantly, as shown in Section 10.10, so we can replace dU/dt by dH/dt. Thus, we can write

$$\frac{dH}{dt} = c_{A1}q_1\underline{H}_{A1} + c_{B2}q_2\underline{H}_{B2} \tag{11.4}$$

If the inlet temperatures T_1 and T_2 are constant, then the enthalpies per mole, $\underline{H}_{A1}$ and $\underline{H}_{B2}$ are constant, and we can obtain a solution to this problem in a particularly simple manner. Substituting Equations 11.1 for the mass balance into Equation 11.4 we have

$$\frac{dH}{dt} = \underline{H}_{A1}\frac{dn_A}{dt} + \underline{H}_{B2}\frac{dn_B}{dt}$$

$$= \frac{d}{dt}[n_A\underline{H}_{A1} + n_B\underline{H}_{B2}]$$

Since both sides are derivatives we can integrate at once. Taking the initial condition as an empty tank ($n_A = n_B = 0$, $H = 0$)

$$H = n_A \underset{\sim}{H}_{A1} + n_B \underset{\sim}{H}_{B2} \tag{11.5}$$

$\underset{\sim}{H}_{A1}$ and $\underset{\sim}{H}_{B2}$ are evaluated at inlet conditions, T_1 and T_2. It is convenient to have all enthalpies evaluated at the tank temperature, T, so we use Equation 10.39,

$$\underset{\sim}{H}_A(T) = \underset{\sim}{H}_{A1} + \int_{T_1}^{T} \underset{\sim}{c}_{pA} \, dT$$

$$\underset{\sim}{H}_B(T) = \underset{\sim}{H}_{B2} + \int_{T_2}^{T} \underset{\sim}{c}_{pB} \, dT$$

Equation 11.5 can then be written

$$H = n_A \underset{\sim}{H}_A(T) + n_B \underset{\sim}{H}_B(T) - n_A \int_{T_1}^{T} \underset{\sim}{c}_{pA} \, dT - n_B \int_{T_2}^{T} \underset{\sim}{c}_{pB} \, dT \tag{11.6}$$

The enthalpy of the mixture can always be expressed in terms of the partial molar enthalpies, $\tilde{H}_A$ and $\tilde{H}_B$. From Equation 10.51

$$H = n_A \tilde{H}_A(T) + n_B \tilde{H}_B(T)$$

Combined with Equation 11.6 we have, with slight rearrangement,

$$-n_A \left\{ \tilde{H}_A(T) - \underset{\sim}{H}_A(T) + \frac{n_B}{n_A} \left[\tilde{H}_B(T) - \underset{\sim}{H}_B(T) \right] \right\}$$

$$= n_A \int_{T_1}^{T} \underset{\sim}{c}_{pA} \, dT + n_B \int_{T_2}^{T} \underset{\sim}{c}_{pB} \, dT \tag{11.7}$$

The bracketed quantity on the left is the *integral heat of solution* at temperature T, discussed in Section 10.7, so Equation 11.7 can be written

$$n_A[-\Delta \underset{\sim}{H}_s(T)] = n_A \int_{T_1}^{T} \underset{\sim}{c}_{pA} \, dT + n_B \int_{T_2}^{T} \underset{\sim}{c}_{pB} \, dT \tag{11.8}$$

If the pure component heat capacities are essentially constant, then the integrations in Equation 11.8 can be easily carried out to give

$$n_A[-\Delta \underset{\sim}{H}_s(T)] = n_A \underset{\sim}{c}_{pA}[T - T_1] + n_B \underset{\sim}{c}_{pB}[T - T_2] \tag{11.9}$$

The heat of solution is generally temperature dependent, so Equation 11.9 cannot be solved directly for the tank temperature. Data are tabulated at a standard temperature T^0, and for constant heat capacity the relation between $\Delta \underset{\sim}{H}_s(T)$ and the standard heat of solution, $\Delta \underset{\sim}{H}_s{}^0$, is given by Equation 10.69

$$n_A \, \Delta \underset{\sim}{H}_s(T) = n_A \, \Delta \underset{\sim}{H}_s{}^0 + \{\rho V \underset{\sim}{c}_p - n_A \underset{\sim}{c}_{pA} - n_B \underset{\sim}{c}_{pB}\}[T - T^0]$$

Here $\underset{\sim}{c}_p$ is the heat capacity per unit mass of the mixture. After substitution into Equation 11.9 we can solve for the tank temperature at any time in terms of the number of moles of A and B which have been added:

$$T = T^0 + \frac{n_A[-\Delta \underset{\sim}{H}_s{}^0] + n_A \underset{\sim}{c}_{pA}[T_1 - T^0] + n_B \underset{\sim}{c}_{pB}[T_2 - T^0]}{\rho V \underset{\sim}{c}_p}$$

(11.10)

Two limiting cases are often useful. If $\Delta \underset{\sim}{H}_s$ is nearly constant, then it follows directly from Equation 11.9 that

$$\Delta \underset{\sim}{H}_s = \text{constant}: \quad T = \frac{n_A[-\Delta \underset{\sim}{H}_s] + n_A \underset{\sim}{c}_{pA} T_1 + n_B \underset{\sim}{c}_{pB} T_2}{\rho V \underset{\sim}{c}_p}$$

Furthermore, if $\Delta \underset{\sim}{H}_s$ is negligible,

$$\Delta \underset{\sim}{H}_s = 0: \quad T = \frac{n_A \underset{\sim}{c}_{pA} T_1 + n_B \underset{\sim}{c}_{pB} T_2}{\rho V \underset{\sim}{c}_p}$$

Data are often recorded on a mass fraction basis. In that case Equation 11.10 can be rewritten in an alternate form which may be more convenient for some calculations. Let m denote the total mass, m_A the mass of A, and m_B the mass of B in the tank. Then

$$n_A \underset{\sim}{c}_{pA} = m_A \underset{\sim}{c}_{pA}$$

$$n_B \underset{\sim}{c}_{pB} = m_B \underset{\sim}{c}_{pB}$$

Equation 11.10 becomes

$$T = T^0 + \frac{\lambda[-\Delta \underset{\sim}{H}_s{}^0]}{M_{wA} \underset{\sim}{c}_p} + \lambda \frac{\underset{\sim}{c}_{pA}}{\underset{\sim}{c}_p}[T_1 - T^0] + [1 - \lambda]\frac{\underset{\sim}{c}_{pB}}{\underset{\sim}{c}_p}[T_2 - T^0]$$

(11.11)

Here, $\lambda = m_A/m$. For computation it helps to note that the mole ratio can be written

$$\frac{n_B}{n_A} = \frac{M_{wA}}{M_{wB}}\left[\frac{1 - \lambda}{\lambda}\right]$$

(11.12)

Example 11.1

Compute the temperature that results when 50 lb_m each of water at $32°F$ and H_2SO_4 at $25°F$ are mixed together in an insulated tank.

Denote H_2SO_4 by A and H_2O by B. The integral heat of solution at $T° = 25°C = 77°F$ is shown in Figure 10.6 and the mixture heat capacity, which we will assume independent of temperature, in Figure 10.7. We have the following data:

$$T^0 = 77°F \qquad\qquad T_1 = 25 \quad T_2 = 32$$

$$m_A = 50 \text{ lb}_m \qquad\qquad m_B = 50 \quad m = 100 \quad \lambda = 0.5$$

$$\underline{c}_{pA} = 0.34 \text{ BTU/lb}_m \text{ °F} \quad \underline{c}_{pB} = 1.0 \quad \underline{c}_p = 0.60$$

$$M_{wA} = 98 \qquad\qquad M_{wB} = 18$$

$$\frac{n_B}{n_A} = \frac{98}{18} = 5.4 \quad \text{(Equation 11.2)}$$

Then substituting into Equation 11.11

$$T = 77 + \frac{0.5 \times 25{,}200}{98 \times 0.60} + 0.5 \times \frac{0.34}{0.60} [25 - 77]$$

$$+ 0.5 \times \frac{1.0}{0.60} [32 - 77]$$

$$= 77 + 214 - 15 - 38 = 238°F$$

It is generally unrealistic to use constant heat capacities over so wide a temperature range, but for this system the error is not particularly serious.

Equation 11.11 is valid, of course, only as long as the system remains a liquid. The boiling point of a 50 percent by weight sulfuric acid solution is about 253°F. Had we taken T_1 and T_2 as, say, 77°F, then the computed temperature after mixing would be 291°F. This is well above the boiling point and hence in error, since the equation is not applicable in this case.

It is important to emphasize that the equation which we have derived for temperature is a solution of Equations 11.1 and 11.4, which describe behavior at all times during the mixing process. Thus, we can follow the temperature change during the mixing. The result of such a calculation can be quite instructive. Consider the following.

Example 11.2

Compute the temperature as a function of time for the mixing described in Example 11.1 when the water and acid are added as follows:

I. 50 lb$_m$ acid in tank at $t = 0$, 50 lb$_m$ water added steadily over 10 min at a rate of 5 lb$_m$/min. Then $m_A = 50$, $m_B = 5t$, and

$$\lambda_I = \frac{m_A}{m_A + m_B} = \frac{1}{1 + 0.1t}$$

II. 50 lb_m water in tank at $t = 0$, 50 lb_m acid added steadily over 10 min at a rate of 5 lb_m/min. Then $m_A = 5t$, $m_B = 50$, and

$$\lambda_{II} = \frac{m_A}{m_A + m_B} = \frac{t}{t + 10}$$

Figure 11.2 shows λ as a function of t for the two cases. The other data are unchanged. ΔH_s^0 and $\underline{c}_p$ are determined from Figures 10.6 and 10.7 for various

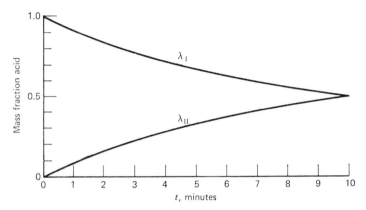

FIGURE 11.2   Mass fraction acid versus time. I: 50 lb_m water added steadily to 50 lb_m sulfuric acid over 10 min. II: 50 lb_m sulfuric acid added steadily to 50 lb_m water over 10 min.

values of λ as t varies from 0 to 10. T is then computed from Equation 11.11 at enough values of t to establish the temperature versus time curves shown in Figure 11.3. Notice that in case II, acid to water, the final temperature is approached gradually. In case I, water to acid, the temperature rises rapidly and reaches a maximum after 3 min which exceeds the final temperature by 85°, declining thereafter. The mixture does not boil in this case, since at all times the boiling point for a mixture of mass fraction λ_I is greater than T_I. This calculation dramatically demonstrates the reason for the chemistry laboratory safety rule of *add acid to water*.

11.3 BATCH MIXING OF MIXTURES

The analysis of the preceding section can be readily generalized to the case where streams 1 and 2 are not pure components, but are themselves mixtures. As shown in Figure 11.4, stream 1 is a mixture of A and B with concentrations c_{A1}, c_{B1}, while stream 2 has concentrations c_{A2}, c_{B2}. The

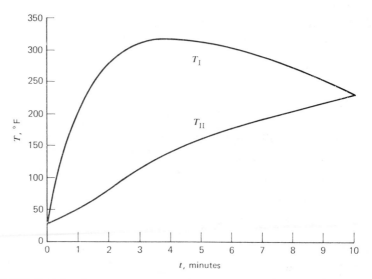

FIGURE 11.3 Temperature versus time. I: 50 lb$_m$ water at 32°F added steadily to 50 lb$_m$ sulfuric acid initially at 25°F. II: 50 lb$_m$ sulfuric acid at 25°F added steadily to 50 lb$_m$ water initially at 32°F.

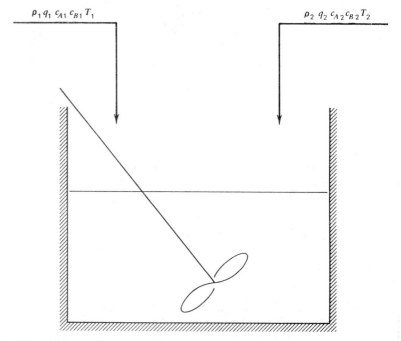

FIGURE 11.4 Batch mixing of two mixtures, each containing A and B.

equations for conservation of mass are then

$$\frac{dn_A}{dt} = \frac{dc_A V}{dt} = q_1 c_{A1} + q_2 c_{A2} \tag{11.13a}$$

$$\frac{dn_B}{dt} = \frac{dc_B V}{dt} = q_1 c_{B1} + q_2 c_{B2} \tag{11.13b}$$

If the tank is initially empty, then these integrate directly to

$$n_A = n_{A1} + n_{A2} \tag{11.14a}$$

$$n_B = n_{B1} + n_{B2} \tag{11.14b}$$

where the subscripts "1" and "2" denote the number of moles of A or B which entered in streams 1 and 2, respectively, during the time interval considered.

Neglecting kinetic and potential energy and the rate of change of pV the energy equation is unchanged from that used previously

$$\frac{dH}{dt} = \rho_1 q_1 H_1 + \rho_2 q_2 H_2 \tag{11.15}$$

For the special case in which the temperature and composition of each feed stream are constant, and therefore ρ_1, H_1, ρ_2, and H_2 are constant, this integrates immediately to

$$H(T) = m_1 H_1(T_1) + m_2 H_2(T_2) \tag{11.16}$$

where m_1 and m_2 denote the total mass which entered in streams 1 and 2, respectively. Notice that

$$\rho V = m = m_1 + m_2$$

For comparison, the enthalpies should all be evaluated at the tank temperature, T. Thus, assuming constant heat capacities,

$$H_1(T_1) = H_1(T) + c_{p1}[T_1 - T]$$

$$H_2(T_2) = H_2(T) + c_{p2}[T_2 - T]$$

The mixture enthalpies can be expressed in terms of partial molar enthalpies using the results of Section 10.6, as follows:

$$H = n_A \tilde{H}_A + n_B \tilde{H}_B$$

$$m_1 H_1 = n_{A1} \tilde{H}_{A1} + n_{B1} \tilde{H}_{B1}$$

$$m_2 H_2 = n_{A2} \tilde{H}_{A2} + n_{B2} \tilde{H}_{B2}$$

Thus, Equation 11.16 becomes

$$n_A \tilde{H}_A + n_B \tilde{H}_B = n_{A1} \tilde{H}_{A1} + n_{B1} \tilde{H}_{B1} + n_{A2} \tilde{H}_{A2} + n_{B2} \tilde{H}_{B2}$$
$$+ m_1 \underline{c}_{p1}[T_1 - T] + m_2 \underline{c}_{p2}[T_2 - T]$$

$$(11.17)$$

All partial molar enthalpies are evaluated at temperature T. The partial molar enthalpies in Equation 11.17 are conveniently related to the pure component enthalpies through the heats of solution of each mixture, Equation 10.62:

$$n_A \tilde{H}_A + n_B \tilde{H}_B = n_A \Delta \underline{H}_s + n_A \underline{H}_A + n_B \underline{H}_B$$
$$n_{A1} \tilde{H}_{A1} + n_{B1} \tilde{H}_{B1} = n_{A1} \Delta \underline{H}_{s1} + n_{A1} \underline{H}_A + n_{B1} \underline{H}_B$$
$$n_{A2} \tilde{H}_{A2} + n_{B2} \tilde{H}_{B2} = n_{A2} \Delta \underline{H}_{s2} + n_{A2} \underline{H}_A + n_{B2} \underline{H}_B$$

Equation 11.17 then becomes

$$n_A \Delta \underline{H}_s + n_A \underline{H}_A + n_B \underline{H}_B = n_{A1} \Delta \underline{H}_{s1} + n_{A1} \underline{H}_A + n_{B1} \underline{H}_B$$
$$+ n_{A2} \Delta \underline{H}_{s2} + n_{A2} \underline{H}_A + n_{B2} \underline{H}_B$$
$$+ m_1 \underline{c}_{p1}[T_1 - T] + m_2 \underline{c}_{p2}[T_2 - T]$$

$$(11.18)$$

or, since $n_A = n_{A1} + n_{A2}$, $n_B = n_{B1} + n_{B2}$, the terms $n_{A1} \underline{H}_A + n_{A2} \underline{H}_A$ on the right equal $n_A \underline{H}_A$ on the left, and similarly for $n_B \underline{H}_B$, giving

$$n_A \Delta \underline{H}_s - n_{A1} \Delta \underline{H}_{s1} - n_{A2} \Delta \underline{H}_{s2} = m_1 \underline{c}_{p1}[T_1 - T]$$
$$+ m_2 \underline{c}_{p2}[T_2 - T] \qquad (11.19)$$

Finally, we need to refer the enthalpy of mixing to the standard value, ΔH_s^0 at T°. For constant heat capacities Equation 10.69 is

$$n_A \Delta \underline{H}_s = n_A \Delta \underline{H}_s^0 + [m\underline{c}_p - n_A \underline{c}_{pA} - n_B \underline{c}_{pB}][T - T^0]$$
$$n_{A1} \Delta \underline{H}_{s1} = n_{A1} \Delta \underline{H}_{s1}^0 + [m_1 \underline{c}_{p1} - n_{A1} \underline{c}_{pA} - n_{B1} \underline{c}_{pB}][T - T^0]$$
$$n_{A2} \Delta \underline{H}_{s2} = n_{A2} \Delta \underline{H}_{s2}^0 + [m_2 \underline{c}_{p2} - n_{A2} \underline{c}_{pA} - n_{B2} \underline{c}_{pB}][T - T^0]$$

and substituting into Equation 11.19

$$n_A \Delta \underline{H}_s^0 - n_{A1} \Delta \underline{H}_{s1}^0 - n_{A2} \Delta \underline{H}_{s2}^0$$
$$+ [m\underline{c}_p - m_1 \underline{c}_{p1} - m_2 \underline{c}_{p2}][T - T^0]$$
$$= m_1 \underline{c}_{p1}[T_1 - T] + m_2 \underline{c}_{p2}[T_2 - T] \qquad (11.20)$$

It is convenient to define $\lambda = m_1/m$, $\nu = n_{A1}/n_A$, and $x_A = M_{wA} n_A / m$. x_A is the mass fraction of A in the tank. In the mixing of pure streams $\nu = 1$.

Equation 11.20 can then be solved for T to give

$$T = T^0 + \frac{x_A\{-\Delta \underline{H}_s^0 + \nu \Delta \underline{H}_{s1}^0 + [1 - \nu] \Delta \underline{H}_{s2}^0\}}{M_{wA}\underline{c}_p}$$

$$+ \lambda \frac{\underline{c}_{p1}}{\underline{c}_p}[T_1 - T^0] + [1 - \lambda]\frac{\underline{c}_{p2}}{\underline{c}_p}[T_2 - T^0] \quad (11.21)$$

When the streams are pure $\Delta \underline{H}_{s1}^0$ and $\Delta \underline{H}_{s2}^0$ are zero and Equation 11.21 reduces to Equation 11.11. In general it is necessary to subtract the heat of solution associated with each feed from the heat of solution of the final mixture. Compare Example 10.4.

Example 11.3

Equal amounts of a 10 percent by weight sulfuric acid solution at 32°F and a 90 percent by weight solution at 70°F are mixed together. Compute the final temperature.

Take the 90 percent stream as stream 1. Then

$$T^0 = 77°F \quad T_1 = 70° \quad T_2 = 32° \quad \lambda = 0.5 \quad x_A = 0.5$$

$$M_{wA} = 98 \quad M_{wB} = 18$$

From Figure 10.7

$$\underline{c}_p = 0.60 \text{ BTU/lb}_m \text{ °F} \quad \underline{c}_{p1} = 0.36 \quad \underline{c}_{p2} = 0.90$$

To compute ν note the following two relations:

50 percent acid: $n_A M_{wA} = 0.5m$

90 percent acid: $n_{A1} M_{wA} = 0.9m_1 = 0.9\lambda m = 0.9 \times 0.5m$

Thus

$$\nu = \frac{n_{A1}}{n_A} = 0.9$$

Mole ratios can be computed from Equation 11.12 for each stream and the corresponding heats of solution obtained from Figure 10.6:

$$\frac{n_{B1}}{n_{A1}} = \frac{98}{18}\left[\frac{1 - 0.9}{0.9}\right] = 0.6 \quad \Delta \underline{H}_{s1} = -8000 \text{ BTU/lb-mole acid}$$

$$\frac{n_{B2}}{n_{A2}} = \frac{98}{18}\left[\frac{1 - 0.1}{0.1}\right] = 49 \quad \Delta \underline{H}_{s2} = -35{,}000 \text{ BTU/lb-mole acid}$$

$$\frac{n_B}{n_A} = \frac{98}{18}\left[\frac{1 - 0.5}{0.5}\right] = 5.4 \quad \Delta \underline{H}_s = -25{,}200 \text{ BTU/lb-mole acid}$$

Then Equation 11.21 becomes

$$T = 77 + \frac{0.5\{+25{,}200 + 0.9[-8000] + 0.1[-35{,}000]\}}{98 \times 0.6}$$

$$+ 0.5 \times \frac{0.36}{0.60} [70 - 77] + 0.5 \times \frac{0.90}{0.60} [32 - 77]$$

$$T = 165°F$$

11.4 ENTHALPY-CONCENTRATION DIAGRAM

The computations carried out thus far on adiabatic mixing using Equation 11.21 can be done much more rapidly by a graphical procedure that utilizes the heat of solution and heat capacity. The chart, known as an *enthalpy-concentration diagram*, is a plot of the enthalpy change per pound of mixture, ΔH, versus weight fraction of pure acid (or other solute). We will first construct the chart and then demonstrate its use.

Consider first the enthalpy of mixing at $32°F$, which we have been measuring in BTU/lb-mole acid. The enthalpy change per pound of mixture, ΔH, is related to ΔH_s as follows:

$$\Delta H = \frac{n_A \Delta H_s}{m} = \frac{x_A \Delta H_s}{M_{wA}}$$

x_A is the mass fraction of A, m_A/m. For example, a 50 percent mixture of acid and water has a heat of mixing of $-25{,}000$ BTU/mole acid. Thus the enthalpy change per pound, ΔH, is

$$\Delta H = \frac{x_A \Delta H_s}{M_{wA}} = \frac{0.5[-25{,}000]}{98} = -127 \text{ BTU/lb}_m$$

Similarly, for a 90 percent acid solution we found the experimental value of ΔH_s in Example 11.3 to be -8000 BTU/mole acid. Thus

$$\Delta H = \frac{0.9[-8000]}{98} = -74 \text{ BTU lb}_m$$

These two points, together with the entire curve for $T = 32°F$, are shown in Figure 11.5.

Next, consider the enthalpy change per pound in taking a mixture of fixed composition at one temperature (say, $32°F$) and going to a new temperature (say, $70°F$). If we consider the heat capacity to be constant, then this enthalpy change is simply $c_p[T_2 - T_1]$, as noted in Equation 10.34. Thus, we construct a curve of ΔH at $70°F$ by adding 38 c_p to the value on the $32°F$ curve. For example, a 50 percent acid solution has a c_p of 0.60 BTU/lb°F, so

$$\Delta H \text{ @ } 70°F = \Delta H \text{ @ } 32°F + 0.6 \times 38 = -104 \text{ BTU/lb}$$

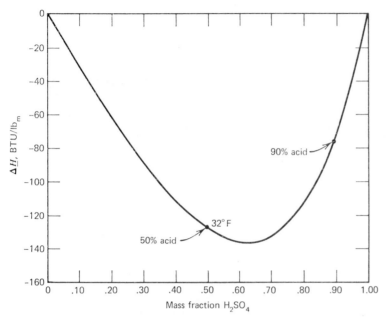

FIGURE 11.5 Enthalpy-concentration diagram for water-sulfuric acid at 32°F. The line is computed from the heat of mixing in Figure 10.6.

A 90 percent acid has $\underline{c}_p = 0.36$, so $\Delta \underline{H}$ @ 70°F $= -61$ BTU/lb. In a similar way the entire curve for 70°F is constructed, as shown in Figure 11.6. This is repeated for all temperatures, and the complete enthalpy concentration diagram is shown for liquid H_2SO_4—H_2O mixtures in Figure 11.7.

The 32° line was constructed by computing the difference between the mixture enthalpy and the pure component enthalpies at 32°. Thus

$$m \, \Delta \underline{H}(32) = m \underline{H}(32) - m_A \underline{H}_A(32) - m_B \underline{H}_B(32)$$

At some other temperature T, we have

$$m \, \Delta \underline{H}(T) = m \, \Delta \underline{H}(32) + m \underline{c}_p[T - 32]$$

The following relation, Equation 10.34, is always true.

$$m \underline{H}(T) = m \underline{H}(32) + m \underline{c}_p[T - 32]$$

These combine to give

$$m \, \Delta \underline{H}(T) = m \underline{H}(T) - m_A \underline{H}_A(32) - m_B \underline{H}_B(32) \qquad (11.22)$$

Thus, the chart is constructed in such a way that the value read on the ordinate represents the enthalpy change between the mixture of interest at temperature T and the enthalpy of pure A and pure B mixed at a temperature of 32°F

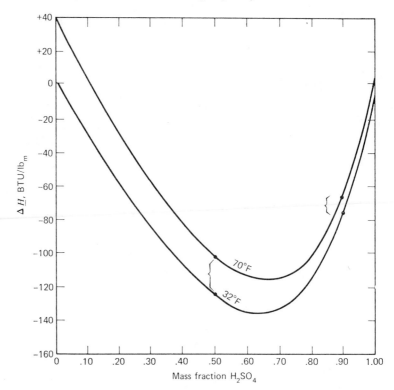

FIGURE 11.6   Enthalpy-concentration diagram for water-sulfuric acid at 32°F and 70°F. The 70° line is computed from the 32°F line and the heat capacity data in Figure 10.7.

If solutions of m_{A1}, m_{B1} and m_{A2}, m_{B2} are mixed together we have

$$m_1 \, \Delta \underline{H}_1(T_1) = m_1 \underline{H}_1(T_1) - m_{A1} \underline{H}_A(32) - m_{B1} \underline{H}_B(32) \quad (11.23a)$$

$$m_2 \, \Delta \underline{H}_2(T_2) = m_2 \underline{H}_2(T_2) - m_{A2} \underline{H}_A(32) - m_{B2} \underline{H}_B(32) \quad (11.23b)$$

From Equation 11.16 for batch mixing of mixtures

$$m \underline{H}(T) = m_1 \underline{H}_1(T_1) + m_2 \underline{H}_2(T_2) \quad (11.16)$$

Thus, substituting Equations 11.22 and 23 into Equation 11.16

$$m \, \Delta \underline{H}(T) = m_1 \, \Delta \underline{H}_1(T_1) + m_2 \, \Delta \underline{H}_2(T_2) + \underline{H}_A(32)[m_{A1} + m_{A2} - m_A]$$
$$+ \underline{H}_B(32)[m_{B1} + m_{B2} - m_B] \quad (11.24)$$

Since $m_A = m_{A1} + m_{A2}$, $m_B = m_{B1} + m_{B2}$, we get, dividing by m and letting $\lambda = m_1/m$

$$\Delta \underline{H}(T) = \lambda \, \Delta \underline{H}_1(T_1) + (1 - \lambda) \, \Delta \underline{H}_2(T_2) \quad (11.25)$$

The graphical solution of the mixing problem now follows immediately from Equation 11.25. The composition of the resulting mixture is fixed by

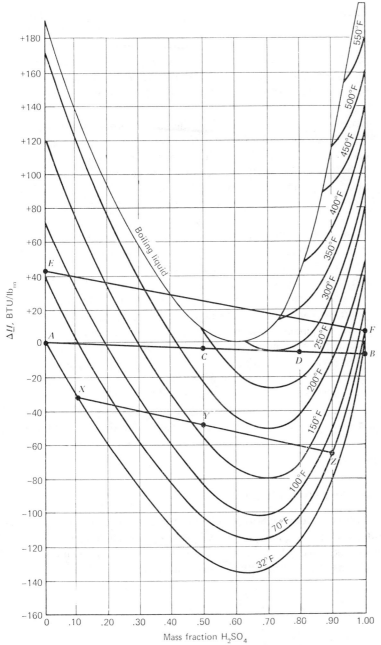

FIGURE 11.7 Enthalpy-concentration diagram for water-sulfuric acid in the liquid phase. (Reproduced from Hougen, Watson, and Ragatz, *Chemical Process Principles*, *part I*, John Wiley, New York, 1954, by permission.)

the law of conservation of mass when the compositions of the two mixtures making it up are specified. Once ΔH_1 and ΔH_2 are fixed on the enthalpy-concentration diagram, *the value of ΔH corresponding to any mixture of the two lies on the straight line formed as λ varies between 0 and 1.* In fact, the exact point is located a fraction λ of the distance from point 2 to point 1. Notice the similarity to the graphical procedure for two-phase mixture compositions developed in Section 9.6.

For instance, in Example 11.1 we considered the problem of mixing pure water at 32°F and pure acid at 25°F to form a 50 percent mixture. Line *AB* in Figure 11.7 connects these two points, and all values of the mixture must lie along that line. The midpoint, point *C*, corresponds to a 50 percent mixture and is at about 235°F, close to the value computed in Example 11.1. Point *D*, corresponding to $\lambda = 0.8$, has a temperature of slightly under 300°F. This is seen to agree with the corresponding value of T_1 in Figure 11.3 at $t = 2.5$ min. Notice that mixtures of acid and water each of 77°F (line *EF*) will boil in the range 38–75 percent acid. This checks with the computation made following Example 11.1. In order to avoid boiling, a very weak acid should always be concentrated by adding concentrated acid to dilute.

The computation carried out in Example 11.3 is shown by the line *XYZ*. *X* corresponds to 10 percent acid at 32°F, *Z* to 90 percent acid at 70°F. *Y* is the midpoint, 50 percent acid at about 165°F. The maximum temperature that can be achieved by adiabatic mixing of 10 and 90 percent acids at the given temperatures is about 190°F. The maximum temperature is found by determining the constant temperature line that is tangent to the mixing line.

Enthalpy-concentration diagrams are available for a number of common systems. They are clearly of value if a large number of computations are to be carried out on a system, for then the ultimate saving in the graphical solution of Equation 11.21 is worth the extensive calculation required in construction of the diagram.

11.5 CONTINUOUS MIXING

We can now consider the flow problem shown in Figure 11.8. Streams 1 and 2, mixtures of *A* and *B* at temperatures T_1 and T_2, respectively, are mixed in a well-stirred tank. We seek the relevant design equations for such a system.

The equations for conservation of mass are simply

$$\frac{dm}{dt} = \frac{d\rho V}{dt} = \rho_1 q_1 + \rho_2 q_2 - \rho q \tag{11.26}$$

$$\frac{dn_A}{dt} = \frac{dc_A V}{dt} = q_1 c_{A1} + q_2 c_{A2} - q c_A \tag{11.27a}$$

$$\frac{dn_B}{dt} = \frac{dc_B V}{dt} = q_1 c_{B1} + q_2 c_{B2} - q c_B \tag{11.27b}$$

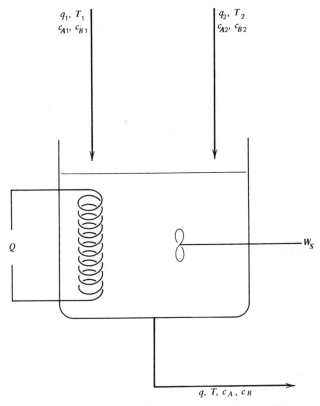

FIGURE 11.8 Continuous mixing of two streams, each containing A and B.

The energy equation, Equation 10.10, adjusted to account for two feed streams and neglecting kinetic and potential energy and the difference between rates of change of U and H is

$$\frac{dH}{dt} = \rho_1 q_1 \underline{H}_1 + \rho_2 q_2 \underline{H}_2 - \rho q \underline{H} + Q - W_s \tag{11.28}$$

$\underline{H}_1$ and $\underline{H}_2$ are evaluated at feed temperatures T_1 and T_2, respectively, while $\underline{H}$ is evaluated at the tank temperature, T. As in all continuous flow systems, the steady state equations provide the primary information needed for design. We will study them first and then return to the transient equations in a subsequent section.

At steady state the mass and energy equations are

$$\rho q = \rho_1 q_1 + \rho_2 q_2 \tag{11.29}$$

$$q c_A = q_1 c_{A1} + q_2 c_{A2} \tag{11.30a}$$

$$q c_B = q_1 c_{B1} + q_2 c_{B2} \tag{11.30b}$$

$$\rho q \underline{H} = \rho_1 q_1 \underline{H}_1 + \rho_2 q_2 \underline{H}_2 + Q - W_s \tag{11.31}$$

The enthalpies, $\underline{H}$, $\underline{H}_1$, and $\underline{H}_2$, need to be referred to the same temperature, and we have already seen that the reference temperature for heat of mixing is a convenient one. Assuming for simplicity that heat capacities are constant we get, using Equation 10.34,

$$\rho q \underline{H}(T) = \rho q \underline{H}(T^0) + \rho q \underline{c}_p[T - T^0]$$

$$\rho_1 q_1 \underline{H}_1(T_1) = \rho_1 q_1 \underline{H}_1(T^0) + \rho_1 q_1 \underline{c}_{p1}[T_1 - T^0]$$

$$\rho_2 q_2 \underline{H}_2(T_2) = \rho_2 q_2 \underline{H}_2(T^0) + \rho_2 q_2 \underline{c}_{p2}[T_2 - T^0]$$

Thus Equation 11.31 becomes

$$\rho q \underline{H}(T^0) + \rho q \underline{c}_p[T - T^0] = \rho_1 q_1 \underline{H}_1(T^0) + \rho_1 q_1 \underline{c}_{p1}[T_1 - T^0]$$

$$+ \rho_2 q_2 \underline{H}_2(T^0) + \rho_2 q_2 \underline{c}_{p2}[T_2 - T^0]$$

$$+ Q - W_s \qquad (11.32)$$

The enthalpies are related to pure component enthalpies and heat of solution by Equations 10.53 and 10.62

$$\rho \underline{H} = c_A \tilde{H}_A + c_B \tilde{H}_B = c_A \underline{H}_A + c_B \underline{H}_B + c_A \Delta \underline{H}_s^0$$

$$\rho_1 \underline{H}_1 = c_{A1} \tilde{H}_{A1} + c_{B1} \tilde{H}_{B1} = c_{A1} \underline{H}_A + c_{B1} \underline{H}_B + c_{A1} \Delta \underline{H}_{s1}^0$$

$$\rho_2 \underline{H}_2 = c_{A2} \tilde{H}_{A2} + c_{B2} \tilde{H}_{B2} = c_{A2} \underline{H}_A + c_{B2} \underline{H}_B + c_{A2} \Delta \underline{H}_{s2}^0$$

Substituting into Equation 11.32 and using the mass balance equations gives, after some manipulation,

$$T = T^0 + \frac{x_A}{M_{wA} \underline{c}_p} \{-\Delta \underline{H}_s^0 + \nu \Delta \underline{H}_{s1}^0 + [1 - \nu] \Delta \underline{H}_{s2}^0\}$$

$$+ \lambda \frac{\underline{c}_{p1}}{\underline{c}_p}[T_1 - T^0] + [1 - \lambda] \frac{\underline{c}_{p2}}{\underline{c}_p}[T_2 - T^0]$$

$$+ \frac{Q}{\rho q \underline{c}_p} - \frac{W_s}{\rho q \underline{c}_p} \qquad (11.33)$$

where

$$\lambda = \frac{\rho_1 q_1}{\rho q} \qquad \nu = \frac{q_1 c_{A1}}{q c_A}$$

and x_A is the mass fraction of A in the tank. With the addition of the Q and W_s terms (which could have been included previously) Equation 11.33 is identical to Equation 11.21 for batch mixing, where λ and ν have differing, but equivalent meanings.

Example 11.4

500 lb$_m$/hour of a 70 percent by weight sulfuric acid is to be produced at 100°F by diluting 80 percent acid at 200°F with water at 32°F. The system is agitated

with a two-horsepower motor. If we assume that mechanical linkages are about 50 percent efficient, this means W_s is about minus one horsepower. How much heat must be removed?

Call the 80 percent acid stream 1. Then we are given

$$T^0 = 77°F \qquad T_1 = 200 \qquad T_2 = 32 \qquad T = 100$$

$$x_A = 0.7 \qquad pq = 500 \text{ lb}_m/\text{hr}$$

$$W_s = -1 \text{ HP} = -2545 \text{ BTU/hr}$$

$$M_{wA} = 98 \qquad M_{wB} = 18$$

From Figure 10.7,

$$\underline{c}_{p1} = 0.45 \text{ BTU/lb}_m \text{ °F} \qquad \underline{c}_{p2} = 1.0 \qquad \underline{c}_p = 0.48$$

To solve Equation 11.33 for Q we require λ, ν, and the heats of solution. $c_{A2} = 0$ (pure water feed) so Equation 11.30a becomes

$$qc_A = q_1 c_{A1}$$

$$\nu = q_1 c_{A1}/qc_A = 1$$

Also

$$M_{wA} qc_A = 0.7 \, pq \qquad \text{(70 percent acid)}$$

$$M_{wA} q_1 c_{A1} = 0.8 \, p_1 q_1 \qquad \text{(80 percent acid)}$$

Thus

$$\lambda = \frac{p_1 q_1}{pq} = \frac{0.7}{0.8} \frac{q_1 c_{A1}}{qc_A} = 0.875$$

From Equation 11.12, the relation between mole ratio and mass fraction, and Figure 10.6 for heat of solution we obtain

$$\frac{n_{B1}}{n_{A1}} = \frac{98}{18} \left[\frac{1 - 0.8}{0.8} \right] = 1.36 \qquad \Delta \underline{H}_{s1} = -14{,}000 \text{ BTU/lb}_m$$

$$\frac{n_B}{n_A} = \frac{98}{18} \left[\frac{1 - 0.7}{0.7} \right] = 2.33 \qquad \Delta \underline{H}_s = -19{,}000 \text{ BTU/lb}_m$$

Since $\nu = 1$ we do not need $\Delta \underline{H}_{s2}$. Substituting all these numbers into Equation 11.33 and solving for Q we obtain $Q = -36{,}000$ BTU/hr. Heat must be removed at rate of 36,000 BTU/hr to obtain the desired exit temperature of 100°F.

11.6 CONTINUOUS MIXING VIA ENTHALPY-CONCENTRATION DIAGRAM

For systems for which an enthalpy-concentration diagram is available, computations of the type just described are easily carried out. Notice that

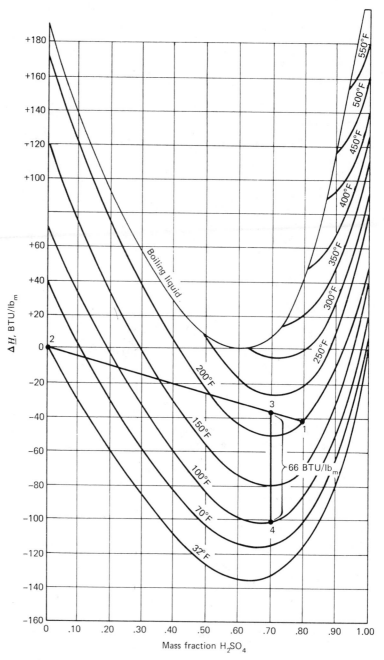

FIGURE 11.9 Continuous mixing calculation using the enthalpy-concentration diagram. Point 4, 70 percent acid at 100°F, lies approximately 66 BTU/lb$_m$ below point 3, 70 percent acid at about 225°F.

Equation 11.31 can be written

$$H = \lambda H_1 + [1 - \lambda] H_2 + \frac{Q}{\rho q} - \frac{W_s}{\rho q}$$

where $\lambda = \rho_1 q_1 / \rho q$. When Q and W_s are zero this is identical to the relation used for the batch system in Section 11.4, and the diagram is used in an identical fashion. Nonzero Q and W_s simply represent a different intercept, and hence a straight line parallel to the line between points 1 and 2.

Consider the problem just solved in Example 11.4. Points 1 and 2 in Figure 11.9 are 80 percent acid at 200°F and pure water at 32°F, respectively. Point 3 is the intersection of the line between them at 70 percent acid. In an adiabatic system the temperature would be about 225°F. The 100°F line intersects 70 percent acid at about 66 BTU/lb$_m$ less. Thus

$$\frac{1}{\rho q} [Q - W_s] = -66 \text{ BTU/lb}_m$$

But $\rho q = 500$ lb$_m$/hr, $W_s = -2545$ BTU/hr, so heat must be removed at a rate of about 35,500 BTU/hr, as computed previously. The slight difference in the two results reflects inaccuracies in making graphical measurements in both solutions.

11.7 TRANSIENT MIXING OF PURE MATERIALS

We have restricted our attention thus far to two special mixing situations, steady state continuous flow mixing and semibatch mixing at constant conditions. In semibatch mixing the mass and energy equations can be combined into a form that can be directly integrated, so that we obtain an explicit relation for the temperature in terms of the composition in the tank. The algebraic equations for steady continuous flow mixing also allow us to solve explicitly for temperature in terms of composition. The similarity in the mathematical structure of these two special cases is the reason that graphical calculations on an enthalpy-concentration diagram can be carried out in an identical fashion for both. Simplifications of the type encountered in these special cases are not to be expected in general, and transient operation of a flowing system will generally lead to a differential equation for temperature. In fact, Equation 11.21 is a solution of that equation for the special case of batch mixing.

The principles involved in the derivation are adequately demonstrated with a minimum of algebraic manipulation by considering continuous mixing of streams of pure A and pure B. The physical situation is then as shown in Figure 11.8 with $c_{B1} = c_{A2} = 0$. The component mass balances and the energy balance are then as obtained in Section 11.5, Equations 11.27 and 11.28:

$$\frac{dn_A}{dt} = \frac{dc_A V}{dt} = q_1 c_{A1} - q c_A \tag{11.34a}$$

$$\frac{dn_B}{dt} = \frac{dc_B V}{dt} = q_2 c_{B2} - q c_B \tag{11.34b}$$

$$\frac{dH}{dt} = q_1 c_{A1} \tilde{H}_A(T_1) + q_2 c_{B2} \tilde{H}_B(T_2) - \rho q \tilde{H}(T) + Q - W_s \tag{11.35}$$

Here we have made use of the equalities for pure systems

$$\rho_1 H_1(T_1) = c_{A1} \tilde{H}_A(T_1), \qquad \rho_2 H_2(T_2) = c_{B2} \tilde{H}_B(T_2)$$

We also have the following equations:

$$\text{Equation 10.46} \qquad \frac{dH}{dt} = \rho V \underline{c}_p \frac{dT}{dt} + \tilde{H}_A \frac{dn_A}{dt} + \tilde{H}_B \frac{dn_B}{dt}$$

$$= \rho V \underline{c}_p \frac{dT}{dt} + \tilde{H}_A [q_1 c_{A1} - q c_A]$$

$$+ \tilde{H}_B [q_2 c_{B2} - q c_B]$$

$$\text{Equation 10.53} \qquad \rho \tilde{H} = c_A \tilde{H}_A + c_B \tilde{H}_B$$

$$\text{Equation 10.39} \qquad \tilde{H}_A(T_1) = \tilde{H}_A(T) + \underline{c}_{pA}[T_1 - T]$$

$$\tilde{H}_B(T_2) = \tilde{H}_B(T) + \underline{c}_{pB}[T_2 - T]$$

Equation 10.46, which neglects the pressure term as written here, expresses the enthalpy in terms of characterizing variables. Equation 10.39 refers all of the enthalpies to the tank temperature. Equation 11.35 then becomes

$$\rho V \underline{c}_p \frac{dT}{dt} + \tilde{H}_A(T)[q_1 c_{A1} - q c_A] + \tilde{H}_B(T)[q_2 c_{B2} - q c_B]$$

$$= q_1 c_{A1} \tilde{H}_A(T) + q_1 c_{A1} \underline{c}_{pA}[T_1 - T] + q_2 c_{B2} \tilde{H}_B(T)$$

$$+ q_2 c_{B2} \underline{c}_{pB}[T_2 - T] - q[c_A \tilde{H}_A(T) + c_B \tilde{H}_B(T)]$$

$$+ Q - W_s$$

The terms multiplied by q appear on both sides of the equation and cancel. Thus we obtain

$$\rho V \underline{c}_p \frac{dT}{dt} = q_1 c_{A1} \underline{c}_{pA}[T_1 - T] + q_2 c_{B2} \underline{c}_{pB}[T_2 - T]$$

$$- q_1 c_{A1}[\tilde{H}_A - \tilde{H}_A] - q_2 c_{B2}[\tilde{H}_B - \tilde{H}_B] + Q - W_s \tag{11.36}$$

The terms involving differences between partial molar enthalpy and pure component enthalpy are very much like those that appear in the definition of heat of solution. Noting that $n_B/n_A = c_B/c_A$ we can write Equation 10.62 for the integral heat of solution as

$$\Delta \underline{H}_s = \tilde{H}_A - \underline{H}_A + \frac{c_B}{c_A}[\tilde{H}_B - \underline{H}_B]$$

Equation 11.36 for temperature is then

$$\rho V \underline{c}_p \frac{dT}{dt} = q_1 c_{A1} \underline{c}_{pA}[T_1 - T] + q_2 c_{B2} \underline{c}_{pB}[T_2 - T] - q_1 c_{A1} \Delta \underline{H}_s$$

$$+ \left[q_1 c_{A1} \frac{c_B}{c_A} - q_2 c_{B2} \right][\tilde{H}_B - \underline{H}_B] + Q - W_s$$

$$\tag{11.37}$$

All of these terms are familiar ones except the one involving $\tilde{H}_B - \underline{H}_B$. In fact, this term equals zero in a number of important practical situations:

  steady feed conditions on composition

$$\frac{dn_A}{dt} = \frac{dn_B}{dt} = 0 \rightarrow q_1 c_{A1} \frac{c_B}{c_A} = q_2 c_{B2}$$

  semibatch operation with $q_1 c_{A1}$ and $q_2 c_{B2}$ constant $\rightarrow$

$$q_1 c_{A1} \frac{c_B}{c_A} = q_2 c_{B2}$$

  "infinite dilution" $\tilde{H}_B = \underline{H}_B$

In the general case we can derive the following relationship by differentiating Equation 10.62 with respect to n_B/n_A

$$\tilde{H}_B - \underline{H}_B = \frac{\partial \Delta \underline{H}_s}{\partial [n_B/n_A]}$$

For small n_B/n_A this can be a very large quantity. The general form of the energy equation is then written

$$\rho V \underline{c}_p \frac{dT}{dt} = q_1 c_{A1} \underline{c}_{pA}[T_1 - T] + q_2 c_{B2} \underline{c}_{pB}[T_2 - T] - q_1 c_{A1} \Delta \underline{H}_s$$

$$+ \left[q_1 c_{A1} \frac{c_B}{c_A} - q_2 c_{B2} \right] \frac{\partial \Delta \underline{H}_s}{\partial [n_B/n_A]} + Q - W_s \quad (11.38)$$

Example 11.5

A continuous flow process mixing A and B normally operates in the steady state with feed temperatures T_1^* and T_2^*. If the temperature of stream 1 suddenly changes to $T_1 = T_1^* + \Delta$, how does the tank temperature change with time? Neglect shaft work.

We will assume that heat capacities and the heat of solution are independent of temperature. With steady feed conditions on composition the term $q_1 c_{A1} c_B / c_A - q_2 c_{B2}$ vanishes. The steady state temperature is obtained from Equation 11.38 by setting dT/dt to zero. This is equivalent to using Equation 11.33. The steady state temperature, T^*, is given by

$$T^* = \frac{q_1 c_{A1} \underline{c}_{pA} T_1^* + q_2 c_{B2} \underline{c}_{pB} T_2^* + Q - q \underline{c}_{A1} \Delta \underline{H}_s}{q_1 c_{A1} \underline{c}_{pA} + q_2 c_{B2} \underline{c}_{pB}}.$$

At some time the feed temperature T_1 changes. Call this time $t = 0$. Then $T(0) = T^*$. It is convenient to write $T = T^* + \delta(t)$, $\delta(0) = 0$. Equation 11.38 becomes

$$\rho V \underline{c}_p \left[\frac{dT^*}{dt} + \frac{d\delta}{dt} \right]$$
$$= q_1 c_{A1} \underline{c}_{pA} [T_1^* + \Delta - T^* - \delta]$$
$$+ q_2 c_{B2} \underline{c}_{pB} [T_2^* - T^* - \delta] - q_1 c_{A1} \Delta \underline{H}_s + Q$$

Substituting for T^* and noting that $dT^*/dt = 0$, since T^* is a constant, the equation simplifies to

$$\rho V \underline{c}_p \frac{d\delta}{dt} = -[q_1 c_{A1} \underline{c}_{pA} + q_2 c_{B2} \underline{c}_{pB}] \delta + q_1 c_{A1} \underline{c}_{pA} \Delta, \qquad \delta(0) = 0$$

This linear first-order equation with constant coefficients has a solution (Section 15.10)

$$\delta(t) = \Delta \frac{q_1 c_{A1} \underline{c}_{pA}}{q_1 c_{A1} \underline{c}_{pA} + q_2 c_{B2} \underline{c}_{pB}} [1 - e^{-[q_1 c_{A1} \underline{c}_{pA} + q_2 c_{B2} \underline{c}_{pB}/\rho V c_p]t}]$$

Unless some control action is taken the system exponentially approaches the new steady state $T = T^* + \Delta q_1 c_{A1} \underline{c}_{pA} / [q_1 c_{A1} \underline{c}_{pA} + q_2 c_{B2} \underline{c}_{pB}]$. Notice the similarity between this solution and the concentration dependence on time found for isothermal mixing of salt streams in Section 4.5.1.

11.8 THE RATE OF HEAT TRANSFER

To complete the discussion of conservation of energy in a nonreacting liquid system, we need to consider the problem of heat exchange. In Section 11.6, for example, we computed that it would be necessary to remove 36,000 BTU/hour from a tank in order to maintain the temperature at a desired level. How do we accomplish this task?

For simplicity, let us first examine the transfer of heat in the system shown in Figure 11.10. Two liquids at different temperatures are in well-stirred

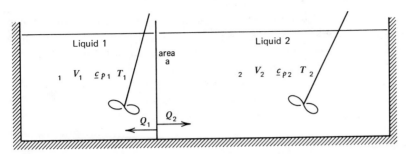

FIGURE 11.10   Heat transfer between liquids in adjacent well-stirred chambers.

adjacent chambers. There is no flow. The tanks are completely insulated except for the surface separating the two liquids. Taking each chamber as a separate control volume, the energy equation in each reduces to

$$\frac{dH_1}{dt} = Q_1 \qquad \frac{dH_2}{dt} = Q_2 \tag{11.39}$$

(In this section subscripts 1 and 2 refer to the chamber, not to flowing streams.) Because of the insulation, heat is transferred only between the two tanks. Thus the heat entering one must equal the heat leaving the other

$$Q_1 = -Q_2 \tag{11.40}$$

The concentration in each chamber remains constant, so Equation 10.46 relating enthalpy rate of change to characterizing variables reduces to

$$\frac{dH}{dt} = \rho V \underline{c}_p \frac{dT}{dt}$$

Then Equation 11.39 for each of the two liquids becomes

$$\rho_1 V_1 \underline{c}_{p1} \frac{dT_1}{dt} = Q_1 \tag{11.41a}$$

$$\rho_2 V_2 \underline{c}_{p2} \frac{dT_2}{dt} = Q_2 \tag{11.41b}$$

and, on substitution into Equation 11.40,

$$\rho_1 V_1 \underline{c}_{p1} \frac{dT_1}{dt} = -\rho_2 V_2 \underline{c}_{p2} \frac{dT_2}{dt} \tag{11.42}$$

Integrating this last equation, and assuming heat capacities independent of temperature for convenience, we obtain an equation relating T_1 and T_2 at any time,

$$\rho_1 V_1 \underline{c}_{p1}[T_1 - T_{10}] = \rho_2 V_2 \underline{c}_{p2}[T_{20} - T_2] \tag{11.43}$$

Here T_{10} and T_{20} are the temperatures at time zero. Notice the similarity between the steps leading to Equation 11.43 and the development of relations between concentrations in a single-phase batch reactor and between phase concentrations in a two-phase mass transfer system.

We know physically that after a long time the temperatures T_1 and T_2 will become equal. We can calculate this value, denoted by T_∞, by setting T_1 and T_2 in Equation 11.43 each to T_∞ and solving, yielding

$$T_\infty = \frac{\rho_1 V_1 c_{p1} T_{10} + \rho_2 V_2 c_{p2} T_{20}}{\rho_1 V_1 c_{p1} + \rho_2 V_2 c_{p2}} \tag{11.44}$$

Here, too, the development closely parallels equilibrium in a chemical reaction and phase equilibrium in mass transfer between immiscible liquid phases.

To determine the transient behavior it is necessary to solve Equation 11.41a for T_1 as a function of time, and hence, we need to know just what the rate of heat transfer, Q_1, is. Specifically, we need a constitutive relation between Q_1 and temperature. Now we find from experience that

1. All other things being constant, Q_1 is directly proportional to the area, a, through which heat is transferred.
2. All other things being equal, Q_1 is *roughly* proportional to $T_2 - T_1$.

Thus we are led to the relation

$$Q_1 = ha[T_2 - T_1] \tag{11.45}$$

where h, known as the *overall heat transfer coefficient*, has units BTU/hr ft² °F. (The symbol U is often used for the overall heat transfer coefficient. We use h to avoid confusion with the symbol for internal energy.) In a given experiment Q, a, and $T_2 - T_1$ can all be measured. Thus, Equation 11.45 *defines* h. The overall coefficient will depend on the liquids, material of construction, and degree of agitation, and relevant relationships will be derived in more advanced courses. For our purposes we can consider it to be an experimentally measured quantity. Experimental values for a variety of conditions are shown in Table 11.1.

The constitutive equation for the rate of heat transfer, Equation 11.45, has a form identical to the relationship developed in Chapter 8 for the rate of mass transfer between phases. In both relations the rate of transfer, **r** or Q, is proportional to the area available for transfer, a; a transfer coefficient, K_m or h; and a driving force, $c^I - Mc^{II}$ or $T_1 - T_2$. This analogy between heat transfer and mass transfer is far-reaching, and the subjects are often studied together. We showed in Chapter 8 how these simple relationships for the rate of transfer can be manipulated to yield useful results. Equation 11.41,

TABLE 11.1   Heat transfer coefficient under various conditions (selected from data in R. H. Perry, Ed., *Chemical Engineers' Handbook*, 3rd ed., McGraw-Hill, 1950)

Chamber 1	Chamber 2	Wall Material	Agitation in Chamber 2	h BTU/hr ft²°F
Steam	Water	Enameled cast iron	0–400 RPM	96–200
Steam	Milk	Enameled cast iron	None	200
Steam	Milk	Enameled cast iron	Stirring	300
Hot water	Warm water	Enameled cast iron	None	70
Ice water	Cold water	Stoneware	Agitated	7
Steam	Water	Copper	None	148
Steam	Water	Copper	Simple stirring	250

which describes this simple heat exchange experiment, can now be written as follows:

$$\rho_1 V_1 c_{p1} \frac{dT_1}{dt} = ha[T_2 - T_1] \tag{11.46}$$

Notice that the definition of h is such that it is always a positive number, for if T_2 is greater than T_1, heat is transferred into liquid 1 and the temperature will rise with time. Solving for T_2 from Equation 11.43 we can write Equation 11.46 as

$$\frac{dT_1}{dt} = -ha\left[\frac{1}{\rho_1 V_1 c_{p1}} + \frac{1}{\rho_2 V_2 c_{p2}}\right][T_1 - T_\infty] \tag{11.47}$$

where T_∞ is defined in Equation 11.44. Equation 11.47 can be formally separated

$$\frac{dT_1}{T_1 - T_\infty} = -ha\left[\frac{1}{\rho_1 V_1 c_{p1}} + \frac{1}{\rho_2 V_2 c_{p2}}\right]dt$$

and integrated to yield the exponential approach to equilibrium

$$T_1 = T_\infty + [T_{10} - T_\infty]e^{-ha\{[1/\rho_1 V_1 c_{p1}]+[1/\rho_2 V_2 c_{p2}]\}t} \tag{11.48}$$

T_{10} is the value of T_1 at $t = 0$.

Example 11.6

Liquids are contained in concentric cylinders of radii R_1 and λR_1 and height H (Figure 11.11). If the temperatures T_1 and T_2 are initially different, how long will it take to reach equilibrium?

$$a = 2\pi R_1 H$$

$$V_1 = \pi R_1^2 H = \tfrac{1}{2}R_1 a$$

$$V_2 = \pi[R_2^2 - R_1^2]H = \tfrac{1}{2}R_1 a[\lambda^2 - 1]$$

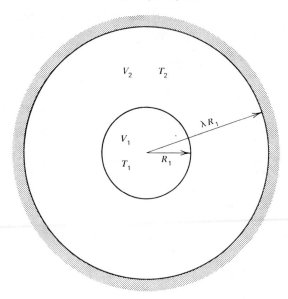

FIGURE 11.11   Top view, heat transfer between liquids in concentric cylindrical chambers.

For simplicity, take $\rho_1 = \rho_2 = \rho$, $\underline{c}_{p1} = \underline{c}_{p2} = \underline{c}_p$. Then Equation 11.48 becomes

$$T_1 = T_\infty + [T_{10} - T_\infty]e^{-[2h/\rho\underline{c}_p R_1]\lambda^2 t/[\lambda^2-1]}$$

The steady state is essentially reached when the exponent of e is -3. Thus, calling the time to reach equilibrium t_∞,

$$-\frac{2h}{\rho\underline{c}_p R_1}\left[\frac{\lambda^2}{\lambda^2 - 1}\right]t_\infty \simeq -3$$

$$t_\infty \simeq \frac{3\rho\underline{c}_p R_1}{2h}\left[\frac{\lambda^2 - 1}{\lambda^2}\right]$$

For water and a cast iron tank h will be about 75 BTU/hr ft^2°F, $\rho = 62.4$ lb$_m$/ft^3, $\underline{c}_p = 1$ BTU/lb$_m$ °F. Thus

$$t_\infty \text{ in minutes} \simeq \frac{4}{3}\left[1 - \frac{1}{\lambda^2}\right]R_1, \qquad R_1 \text{ in feet}$$

If the radius of the inner cylinder were one foot and the outer cylinder two feet, it would take about one hour to reach equilibrium. Notice that for λ greater than about 3 the approach to equilibrium is nearly independent of the size of the outer cylinder.

It is instructive to explore the last point in the example somewhat more generally. Suppose $V_2 \gg V_1$. Then Equations 11.44 and 11.48 can be written

$$T_\infty = \frac{T_{20}}{1 + [\rho_1 c_{p1}/\rho_2 c_{p2}][V_1/V_2]}$$

$$+ \frac{\rho_1 c_{p1}}{\rho_2 c_{p2}} \frac{V_1}{V_2} T_{10} \left\{ \frac{1}{1 + [\rho_1 c_{p1}/\rho_2 c_{p2}][V_1/V_2]} \right\} \xrightarrow[V_1/V_2 \to 0]{} T_{20}$$

$$T_1 = T_\infty + [T_{10} - T_\infty]e^{-[ha/\rho_1 V_1 c_{p1}]\{1+[\rho_1 c_{p1}/\rho_2 c_{p2}][V_1/V_2]\}t}$$

$$\xrightarrow[V_1/V_2 \to 0]{} T_{20} + [T_{10} - T_{20}]e^{-hat/\rho_1 V_1 c_{p1}}$$

That is, if V_2 is much larger than V_1 the temperature of liquid 2 is essentially unchanged and the approach to equilibrium depends only on properties of chamber 1. For comparison purposes later it is helpful to compute the heat transfer in this case when $T_2 = $ constant. From Equation 11.45

$$Q = ha[T_2 - T_1] = ha[T_2 - T_{10}]e^{-hat/\rho_1 V_1 c_{p1}}$$

The *average* rate of heat transfer from time zero to t, $\bar{Q}$, is

$$\bar{Q} = \frac{1}{t}\int_0^t Q(\tau)\,d\tau = \frac{\rho_1 V_1 c_{p1}}{t}[T_2 - T_{10}][1 - e^{-hat/\rho_1 V_1 c_{p1}}] \qquad (11.49)$$

One final comment on the similarity between heat and mass transfer is appropriate. Suppose the wall between the two liquids were removed and the components were mixed together. If the liquids are immiscible, then the physical situation is identical to the one considered in Chapter 8, where a now refers to the interfacial area between phases. For sufficiently large a, equilibrium will be reached rapidly. Heating of one liquid by contacting it with an immiscible liquid is called *direct contact heat transfer* and is carried out in either tank-type or tubular systems.

11.9   HEAT TRANSFER TO A JACKET

Heat transfer in a continuous flow-tank system can be achieved by putting a jacket around the vessel and passing the cooling (or heating) fluid through the jacket, as shown in Figure 11.12. The coolant in the jacket is assumed to be well mixed and hence at uniform temperature. Using the subscript c to denote properties of the coolant and the jacket we can take the jacket volume as a control volume and write the equation for energy conservation as

$$\rho_c V_c c_{pc} \frac{dT_c}{dt} = \rho_c q_c c_{pc}[T_{cf} - T_c] + Q_c \qquad (11.50)$$

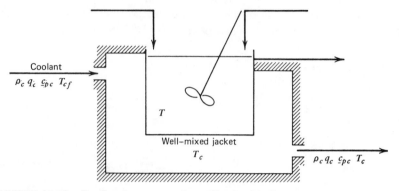

FIGURE 11.12 Cooling by means of a well-mixed jacket.

The temperature of the liquid in the tank is described by Equation 11.38. At steady state $dT/dt = dT_c/dt = 0$ and we can express Q_c in terms of an overall heat transfer coefficient as

$$Q_c = ha[T - T_c] = -Q$$

T is the temperature inside the tank. Thus, at steady state, Equation 11.50 becomes

$$0 = \rho_c q_c \mathcal{c}_{pc}[T_{cf} - T_c] + ha[T - T_c]$$

or, solving for T_c,

$$T_c = \frac{T_{cf}}{1 + K} + \frac{KT}{1 + K} \tag{11.51}$$

where $K = ha/\rho_c q_c \mathcal{c}_{pc}$. Then

$$Q = ha[T_c - T] = \frac{ha}{1 + K}[T_{cf} - T] \tag{11.52}$$

Equation 11.52 is the result needed for jacket design. Notice that two limiting estimates follow directly from Equations 11.51 and 11.52. For a given heat load, Q, the smallest surface area will occur when $|T - T_c|$ is as large as possible. This occurs when q_c gets large, for as $q_c \to \infty$, $K \to 0$, and, from Equation 11.51, $T_c \to T_{cf}$. Thus, the minimum surface area is

$$a_{\min} = \frac{Q}{h[T_{cf} - T]} \tag{11.53}$$

Similarly, the minimum coolant flow will correspond to $a \to \infty$, in which case $K \to \infty$ and, from Equation 11.51, $T_c \to T$. Also, $1 + K \to K$, $ha/[1 + K] \to ha/K = \rho_c q_c \mathcal{c}_{pc}$, so that Equation 11.52 can be written

$$q_{c,\min} = \frac{Q}{\rho_c \mathcal{c}_{pc}[T_{cf} - T]} \tag{11.54}$$

Equations 11.52 to 11.54 can be combined as

$$q_c = q_{c,\min}\left[1 - \frac{a_{\min}}{a}\right] \tag{11.55}$$

In designing a jacket these limiting values can provide a rapid feel for the numbers involved.

Example 11.7

In example 11.4 a heat load of $Q = -36,000$ BTU/hr was computed for $T = 100°$F. Suppose that cooling water is available at 65°F. How much heat transfer area and coolant is required?

$$T_{cf} = 65°F, \qquad \rho_c = 62.4 \text{ lb}_m/\text{ft}^3, \qquad \underline{c}_{pc} = 1.0 \text{ BTU/lb}_m \text{ °F}$$

A reasonable value of h is about 200 BTU/hr ft °F. From Equations 11.53 and 11.54,

$$a_{\min} = \frac{-36,000}{200 \times [65 - 100]} = 5.1 \text{ ft}^2$$

$$q_{c,\min} = \frac{-36,000}{62.4 \times 1.0 \times [65 - 100]} = 16.5 \text{ ft}^3/\text{hr}$$

The ultimate balance between heat transfer area and coolant flow is in part an economic one, trading off capital outlay versus operating costs, and of course a is always confined within certain limits since the volume of the tank will be determined by the requirements of the process taking place inside the tank. Suppose we take $a = 25 \text{ ft}^2 = 4.9 a_{\min}$. Then from Equation 11.55

$$q_c = \frac{16.5}{1 - [1/4.9]} = 21.4 \text{ ft}^3/\text{hr}$$

11.10   HEAT TRANSFER TO A COIL

Another design commonly used for heat transfer in a tank-type system is to flow coolant through a coiled pipe submerged in the tank, as shown in Figure 11.13. In this case, while the temperature in the tank remains the same, the temperature of the coolant varies along the length of the coil. Thus, we have a situation analogous to the tubular reactor in Section 7.9, in which we have to take spatial variation of the characterizing variable into account by using a very small control volume.

The control volume is shown in Figure 11.14. The control volume is a length of pipe Δx long with diameter D. Thus, x and $x + \Delta x$ correspond to inlet and outlet, respectively. The energy equation is then

$$U|_{t+\Delta t} = U|_t + \rho_c q_c H_c|_x \Delta t - \rho_c q_c H_c|_{x+\Delta x} \Delta t + Q' \Delta t \tag{11.56}$$

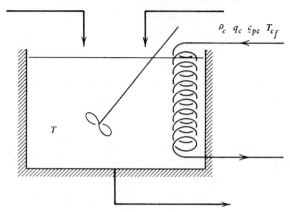

FIGURE 11.13   Cooling by means of a coil.

Q' refers to the rate of heat transfer within the control volume of length Δx. At steady state $U|_{t+\Delta t} = U|_t$. Furthermore

$$a = \pi D\,\Delta x$$

$$Q' = h\pi D\,\Delta x[T - T_c]$$

so Equation 11.56 becomes

$$\rho_c q_c \frac{H_c|_{x+\Delta x} - H_c|_x}{\Delta x} = \pi h D[T - T_c]$$

and in the limit as $\Delta x \to 0$

$$\rho_c q_c \frac{d\underline{H}_c}{dx} = \pi h D[T - T_c] \tag{11.57}$$

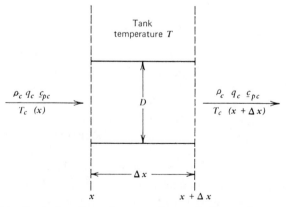

FIGURE 11.14   Control volume for a cooling coil. Properties are assumed to change negligibly over a small distance Δx of the coil.

The liquid in the coil has a constant composition, so H_c depends only on T_c, and we can replace dH_c/dx with $c_{pc}[dT_c/dx]$ in Equation 11.57

$$\rho_c q_c c_{pc} \frac{dT_c}{dx} = \pi h D[T - T_c] \tag{11.58}$$

Equation 11.58 is separable and can be integrated:

$$\frac{dT_c}{T_c - T} = -\frac{\pi h D}{\rho_c q_c c_{pc}} dx$$

$$T_c = T + [T_{cf} - T]e^{-\pi h Dx/\rho_c q_c c_{pc}} \tag{11.59}$$

Now, the rate at which heat is transferred between the tank and coil over a small distance of the coil dx is

$$Q' = \pi h D[T - T_c]\, dx = \pi h D[T - T_{cf}]e^{-\pi h Dx/\rho_c q_c c_{pc}}\, dx$$

The *total* rate of heat transfer is the summation (integration) over all small distances:

$$Q = \int_0^L \pi h D[T - T_{cf}]e^{-\pi h Dx/\rho_c q_c c_{pc}}\, dx$$

$$= \rho_c q_c c_{pc}[T - T_{cf}][1 - e^{-\pi h DL/\rho_c q_c c_{pc}}]$$

or, since the surface area $a = \pi DL$,

$$Q = \rho_c q_c c_{pc}[T - T_{cf}][1 - e^{-ha/\rho_c q_c c_{pc}}] \tag{11.60}$$

Notice that Equation 11.60 is identical to Equation 11.49 of Section 11.8 when q_c is replaced by V_1/t. The analogy between batch and tubular system carries over to heat transfer as well.

The minimum area and flow rate are readily established as identical to the values computed in Section 11.9 for a jacketed vessel. Equation 11.60 can then be rearranged to the form

$$\frac{q_c}{q_{c,\min}} = \frac{1}{1 - e^{-\{[a/a_{\min}]/[q_c/q_{c,\min}]\}}} \tag{11.61}$$

For the example used in the preceding section $a_{\min} = 5.1$ ft². Suppose the coil is made up of pipe with a diameter of $1/2$ in. $= 1/24$ ft. Then

$$5.1 \text{ ft}^2 = \pi \left[\frac{1}{24}\right] L_{\min}$$

$$L_{\min} = 39.0 \text{ ft}$$

If we choose the same coolant flow rate, $q_c = 21.4$ ft³/hr, $q_c/q_{c,\min} = 21.4/16.5 = 1.3$, then we find from Equation 11.61 that $a/a_{\min} = 1.9$, $a = 9.8$ ft², $L = 75$ ft. At a comparable flow rate, less surface area is required in a coil than in a jacket to achieve the same heat transfer. Equivalently, for a given area, less coolant is required in the coil. The reasoning is identical to that which led to the observation in Example 7.9 that an isothermal tubular reactor requires a smaller residence time than a continuous flow stirred tank reactor to achieve the same conversion.

11.11   CONCLUDING REMARKS

This chapter deals with two distinct applications of the principle of conservation of energy. In the mixing problems the primary point to recognize is the way in which the partial molar enthalpy terms combine. The final working equations involve only the heat of solution, which is considerably easier to measure than the individual partial molar quantities. Review these sections carefully, because the concepts are central to an understanding of chemically reacting systems in the next chapter. Notice the parallel between the structure of this chapter and Chapters 4, 5, 7, and 8. In all cases the batch experiment was used to measure the key variables, after which we could consider the design equations for continuously operating systems.

The material on heat transfer is another striking example of the importance of the constitutive equation. Once the idea of an experimentally measurable heat transfer coefficient has been introduced we can solve practical problems such as the design of a heat exchanger. Return to Figure 3.8 and note how the heat transfer material fits into the logical structure outlined there, and consider the close parallel to mass transfer as developed in Chapter 8 very carefully. Determination of the heat transfer and mass transfer coefficients would seem to differ little in principle from the determination of the reaction rate coefficient discussed in Chapter 5. An essential difference, however, is that the reaction rate depends only on the chemistry and is the same in all reaction vessels. The heat or mass transfer rate depends on the detailed fluid mechanics and hence on the vessel shape and agitation.

Data necessary for the types of problems studied in this chapter are in the basic reference tables, 4.6 to 4.8. Further discussion of enthalpy-concentration diagrams and diagrams for additional systems are in

11.1   O. A. Hougen, K. M. Watson, and R. A. Ragatz, *Chemical Process Principles, Part I; Material and Energy Balances*, 2nd ed., Wiley, New York, 1959.

11.2   J. H. Perry, *Chemical Engineers' Handbook*, 4th ed., McGraw-Hill, New York, 1965.

The following texts contain discussions of basic heat transfer mechanisms and design methods for heat transfer equipment:

11.3   A. J. Chapman, *Heat Transfer*, Macmillan, New York, 1960.

11.4   J. P. Holman, *Heat Transfer*, McGraw-Hill, New York, 1963.

11.5   D. Q. Kern, *Process Heat Transfer*, McGraw-Hill, New York, 1950.

11.6   W. H. McAdams, *Heat Transmission*, McGraw-Hill, New York, 1954.

11.12   PROBLEMS

11.1   In a batch processing unit the following materials are mixed together in a large stainless steel tank. Develop an expression for T, the final temperature in the tank, if the heat of solution can be assumed negligible.
  (a) n_A moles of A at T_1
  (b) n_B moles of B at T_2
  (c) n_C moles of C at T_3
  (d) n_D moles of D at T_4

11.2   If A is ethanol, B, ethylene glycol, and C, water, find the final temperature when 100 gal of each substance is dumped into a tank and mixed. The following data are available.

Substance	Temperature	$c_{p'}$, $\dfrac{cal}{g\,°C}$	ρ, $\dfrac{g}{cm^3}$
Ethanol	30°F	0.52	0.79
Ethylene Glycol	20°F	0.55	1.13
Water	80°F	1.00	1.00

11.3   The temperature after mixing in the tank for the process described in Problem 11.2 must be 70°F. If the only way to control this is to change the temperature of the water used, what should the water temperature be?

11.4   The following operation is carried out in a well stirred vessel as a first step in a series of operations in a crystallization process.
  (a) Water at 20°C is added to an empty well-insulated tank at a flow rate of 50 lb$_m$/min.
  (b) After 20 min this flow is stopped and a second stream of water at 80°C is added at a rate of 100 lb/min for 10 min.
  Develop a mathematical description that will allow you to predict the water temperature in the tank, T, as a function of time. Prepare a plot showing how T varies with time.

11.5   If the second stream in Problem 11.4 is not water but ethylene glycol at 80°C, compute T as a function of time and compare with the water case.

11.6 An insulated vessel which initially contains 200 lb of water at 20°C accidently receives a stream of 20 lb_m/min of 100 percent H_2SO_4 at 20°C. The acid is pumped into the tank for 5 min before the error is discovered. Develop the model equations for this operation and show how the equation can be manipulated to yield the tank temperature as a function of time. Plot $T(t)$.

11.7 The operations listed in Problem 10.10 are not to be kept isothermal. Find the final temperature for each.

11.8 Find the temperature that results when 50 lb of NaOH and 100 lb of water, each at 25°C, are mixed together in a well-insulated vessel.

11.9 Compute the final temperature in a batch system when 6 ft³ of a mixture of ethyl alcohol and water (60 percent by weight alcohol) is mixed with enough pure water to produce a final product of 40 weight percent alcohol. The 60 percent mixture is available at 25°C and cool spring water is available at 15°C.

11.10 A water solution of $AlCl_3$ is to be prepared for use in a pharmaceutical process. You are to set up a safe procedure for doing this in a batch mixing vessel which is required to produce 1000 lb of a 30 weight percent $AlCl_3$ solution. Water is available at 20°C and the $AlCl_3$ is available as a solid in crystalline form. As a first step in developing an operating procedure it is necessary to have an estimate of the final temperature of the mixture. Compute this quantity and carefully state all your assumptions.

11.11 An assumption of constant heat capacity has been made to derive all the working equations in Chapter 11. Comment on the validity of this assumption by finding tabular data on $\underline{c}_p$ variation with T for the following substances:

Water	0–100°C
Methyl Alcohol	5–40°C
Ethyl Alcohol	3–40°C
Ethylene Glycol	0–20°C

11.12 Fit the best expression of the following form to your data from Problem 11.11 and derive the equivalent of Equation 11.9 for an ethanol (A) and water (B) system.

$$\underline{c}_{pi} = a_i + b_i T$$

11.13 Modify Equation 11.21 so that it can be used for a system in which a mixture in stream 1 is mixed with a pure material in stream 2.

11.14 Redo Example 11.3 using Equation 11.19 assuming that the heats of solution $\Delta \underline{H}_s$, $\Delta \underline{H}_{s1}$, and $\Delta \underline{H}_{s2}$ are temperature independent. Compare the final temperature obtained with that calculated in the example.

11.15 Determine from the enthalpy-concentration diagram for sulfuric acid and water the maximum temperature at which a 90 percent sulfuric acid solution can be added to pure water at the following temperatures without the mixture boiling. In all cases a 50 percent acid solution is to be made in a batch mixing vessel.
(a) 130°F
(b) 100°F
(c) 80°F
(d) 32°F

11.16 Determine the final temperature when 100 lb_m of 50 percent sulfuric acid at 200°F are mixed with 500 lb_m of 50 percent acid at 32°F.

11.17 Modify Equation 11.33 for the case where the heat of solution is temperature independent.

11.18 Modify Equation 11.33 for the case where heat of solution effects are negligible.

11.19 1000 lb/hr of 10 percent sulfuric acid at 70°F is mixed with 300 lb/hr of 80 percent acid at 300°F in a holding vessel which is part of a more complex process. What is the composition and temperature of the exit stream? If the temperature of the 10 percent acid is fairly constant, how high can the temperature of the 80 percent acid be before the mixture boils? If the temperature of the 80 percent acid is constant, how high can the temperature of the 10 percent acid be before the mixture boils?

11.20 10,000 lb_m/hr of a 50 percent sulfuric acid mixture must be made without the temperature exceeding 200°F. If pure water is available at 70°F in any desired quantity, what feed rates are needed when the feed has each of the following compositions:
(a) 60 percent sulfuric acid
(b) 70 percent sulfuric acid
(c) 90 percent sulfuric acid
What is the maximum temperature that each inlet stream can have?

11.21 A continuous mixing operation is carried out by adding 300 lb_m/hr of calcium chloride crystals to a pure water stream flowing at 1000 lb_m/hr. If both streams are at 25°C when they enter the mixer, what is the temperature of the exit stream? How much heat would have to be removed if the operation were to be carried out isothermally at 25°C? The heat capacity of $CaCl_2$ crystals is approximated by the equation
$$\underline{c}_p = 16.9 + 0.00386T$$
T is the temperature in degrees Kelvin and $\underline{c}_p$ is the heat capacity in cal/g-mole °C.

11.22 An enthalpy concentration diagram for the ethanol-water system can be found in the *Chemical Engineers' Handbook*, 4th ed. Using the diagram compute the final temperature and composition when 1000 lb_m/hr of water at 20°C are mixed with 500 lb_m/hr of 90 percent ethanol solution at 60°C. Compare your result with that obtained using Equation 11.33 modified for the case where $\Delta \underline{H}_s^0$, $\Delta \underline{H}_{s1}$, and $\Delta \underline{H}_{s2}^0$ are all equal to zero.

11.23 A continuous mixing process operates in the following way:

(a) At time $t = 0$ a stream containing pure A at temperature T_1 and volumetric flow rate q is introduced into a well-mixed vessel containing pure B at a temperature T_0. A and B form ideal solutions with $\Delta \underline{H}_s = 0$.

(b) Also at time $t = 0$ the mixture of A and B in the vessel is pumped from the tank at a volumetric flow rate q.

Develop the mathematical description for this situation. You may assume that Q and W_s are negligible. Your development should use the same nomenclature as that used in Section 11.7. Compare the final model equation for this process with Equation 11.37.

11.24 In the process described in Problem 11.23 ethanol at 70°C is added at 1.0 ft³/min to a tank containing 10 ft³ of water that is initially at 20°C. Using data on heat capacity and density available in the standard references and the model equation from Problem 11.23 compute the exit stream temperature, T, as a function of time. How long will it take to achieve a tank temperature that is within 5 percent of the final steady state value?

11.25 A cylindrical preheat tank in a semibatch processing unit has a capacity of 1000 ft³. The tank has a length to diameter ratio of unity and is completely jacketed. T_c, the temperature of the heating material in the jacket is kept constant at 212°F by steam which condenses in the jacket. The material to be heated can be assumed to have the properties of water. Prepare a plot showing how the tank temperature, T, will vary with time for a batch operation in which the tank contents are raised from 70°F to 170°F. Q, the rate at which heat is supplied to the tank, may be represented as

$$Q = ha[T_c - T]$$

and h has a value of 200 BTU/hr ft² °F.

11.26 The heating operation described in Problem 11.25 must be carried out continuously, producing 10 ft³/min of material heated from 70°F to 170°F. Find the size of the jacketed vessel that will accomplish this task.

CHAPTER 12

Energy Balances for Reacting Liquid Systems

12.1 INTRODUCTION

In Chapters 5 and 7 we developed the mathematical models for ideal tank and tubular reactors which operate isothermally in the liquid phase. In this chapter we extend our development to nonisothermal operation and add the energy balance to the mass balances and reaction rate terms developed previously. We begin with a discussion of the heat of reaction and batch reactor experiments. We then develop the models for flow reactors and discuss reactor behavior.

12.2 HEAT OF REACTION

When a chemical reaction occurs there is a rearrangement of the molecular structure of the reactants to form a product or products. This molecular rearrangement entails energy changes within the molecules. The macroscopic manifestation of these molecular energy effects is the observed temperature increase or decrease as a reaction proceeds in an insulated laboratory vessel. In writing the energy equation for a reacting system we shall see that this behavior is associated with the *heat of reaction*, ΔH_R, which is defined in the following way:

Let the reactants be denoted as $A_1, A_2, A_3, \ldots$, and the products as $D_1, D_2, D_3, \ldots$ for the reaction

$$\alpha_1 A_1 + \alpha_2 A_2 + \alpha_3 A_3 + \cdots \rightleftarrows \delta_1 D_1 + \delta_2 D_2 + \delta_3 D_3 + \cdots$$

$$(12.1)$$

$\alpha_1, \alpha_2, \alpha_3, \ldots, \delta_1, \delta_2, \delta_3, \ldots$ are the stoichiometric coefficients for the reaction. For example, in the reaction

$$H_2SO_4 + (C_2H_5)_2SO_4 \rightleftarrows 2C_2H_5SO_4H$$

which we studied in Examples 5.1 and 5.4, we have $H_2SO_4 = A_1$, $(C_2H_5)_2SO_4 = A_2$, $C_2H_5SO_4H = D_1$, $\alpha_1 = \alpha_2 = 1$, $\delta_1 = 2$. In the reaction mixture of $A_1, A_2, A_3, \ldots, D_1, D_2, D_3, \ldots$ the partial molar enthalpies are $\tilde{H}_{A_1}, \tilde{H}_{A_2}, \tilde{H}_{A_3}, \ldots, \tilde{H}_{D_1}, \tilde{H}_{D_2}, \tilde{H}_{D_3}, \ldots$. The heat of reaction is defined as

$$\Delta H_R = \delta_1 \tilde{H}_{D_1} + \delta_2 \tilde{H}_{D_2} + \delta_3 \tilde{H}_{D_3} + \cdots - \alpha_1 \tilde{H}_{A_1}$$
$$- \alpha_2 \tilde{H}_{A_2} - \alpha_3 \tilde{H}_{A_3} - \cdots \tag{12.2}$$

Thus, in the sulfuric acid-diethyl sulfate reaction, we have for the heat of reaction

$$\Delta H_R = 2\tilde{H}_{C_2H_5SO_4H} - \tilde{H}_{H_2SO_4} - \tilde{H}_{(C_2H_5)_2SO_4}$$

The heat of reaction ΔH_R is most commonly measured in units of BTU/lb-mole or calories/g-mole.

12.3   BATCH REACTOR

Consideration of the batch reactor both demonstrates nicely the role that the heat of reaction plays in process behavior and provides the experimental means for its measurement. For definiteness we will consider the specific case studied in Section 5.6

$$A + B \rightleftharpoons nD$$

The equations for conservation of mass in a batch reactor were shown to be (Equations 5.27–5.29, 5.32)

$$\frac{dn_A}{dt} = \frac{dVc_A}{dt} = -rV \tag{12.3a}$$

$$\frac{dn_B}{dt} = \frac{dVc_B}{dt} = -rV \tag{12.3b}$$

$$\frac{dn_D}{dt} = \frac{dVc_D}{dt} = +nrV \tag{12.3c}$$

where the net reaction rate $r = r_{A-} - r_{A+}$. In writing Equations 12.3 we are, of course, describing the reactor only after the reactants have been added and we have neglected the small time during which the initial mixing takes place. This is valid if the reaction time is long compared to the few seconds required to mix and start the process. The entire semibatch operation must be taken into account in order to compute precisely any reaction that takes place while the reactants are added together. As shown in Section 5.4 (Equations 5.18 and 5.19) the mass balance Equations 12.3 can be integrated to yield a relation between the number of moles of the species which is valid for all time:

$$n_A - n_{A0} = n_B - n_{B0} = -\frac{1}{n}[n_D - n_{D0}] \tag{12.4}$$

Here, n_{A0}, n_{B0}, and n_{D0} refer to the number of moles of each of the reacting species at the start of the experiment.

The energy balance for the batch process after the reactants are put into the reactor is obtained by dropping the flow terms in Equation 10.10. Neglecting shaft work this is

$$\frac{dU}{dt} = Q \tag{12.5}$$

For the liquid system we may replace dU/dt with dH/dt with a negligible error. Thus

$$\frac{dH}{dt} = Q = \frac{dq}{dt} \tag{12.6}$$

This is integrated to yield

$$H(T, n_A, n_B, n_D) - H(T_0, n_{A0}, n_{B0}, n_{D0}) = q \tag{12.7}$$

T_0 refers to the temperature in the reactor at the start of the experiment and q is the total amount of heat added during the course of the reaction. The enthalpy is a function of composition and we have noted that explicitly here in order to keep track of the proper values. Hence the notation $H(T, n_A, n_B, n_D)$, etc.

We now follow the standard procedure which we have used consistently in applying the energy equation:

1. Use the heat capacity to calculate all enthalpies at the same temperature.
2. Express the system enthalpy in terms of the partial molar enthalpies of the components.
3. Use the mass balance to simplify the energy equation.

It is most useful to use the reactor temperature as the one at which enthalpies are evaluated. Then, from Equation 10.34,

$$H(T, n_{A0}, n_{B0}, n_{D0}) = H(T_0, n_{A0}, n_{B0}, n_{D0}) + \rho_0 V_0 \underline{c}_{p0}[T - T_0] \tag{12.8}$$

ρ_0, V_0, and $\underline{c}_{p0}$ refer to the density, volume, and heat capacity of the initial mixture, and $\underline{c}_{p0}$ has been assumed for convenience to be independent of temperature. Of course, $\rho_0 V_0 = \rho V = $ constant. If the starting mixture is approximately ideal then from Equation 10.57,

$$\text{ideal mixture: } \rho_0 V_0 \underline{c}_{p0} = n_{A0}\underline{c}_{pA} + n_{B0}\underline{c}_{pB} + n_{D0}\underline{c}_{pD} \tag{12.9}$$

Substituting Equation 12.8 into 12.7 gives

$$H(T, n_A, n_B, n_D) - H(T, n_{A0}, n_{B0}, n_{D0}) = q + \rho_0 V_0 \underline{c}_{p0}[T_0 - T] \tag{12.10}$$

The enthalpies are expressed in terms of partial molar enthalpies according to Equation 10.51:

$$H(T, n_A, n_B, n_D) = n_A \tilde{H}_A + n_B \tilde{H}_B + n_D \tilde{H}_D \tag{12.11a}$$

$$H(T, n_{A0}, n_{B0}, n_{D0}) = n_{A0} \tilde{H}_{A0} + n_{B0} \tilde{H}_{B0} + n_{D0} \tilde{H}_{D0} \tag{12.11b}$$

Here, $\tilde{H}_{A0}$ refers to the partial molar enthalpy of A in the initial mixture composition at temperature T. If we use the mass balance in the form of Equation 12.4 to solve for n_D and n_B in terms of n_A and substitute Equation 12.11 into 12.10 then, with some rearrangement, we obtain

$$\begin{aligned}
[n_{A0} - n_A][n\tilde{H}_D &- \tilde{H}_A - \tilde{H}_B] \\
&= q + \rho_0 V_0 \underline{c}_{p0}[T_0 - T] + n_{A0}[\tilde{H}_{A0} - \tilde{H}_A] + n_{B0}[\tilde{H}_{B0} - \tilde{H}_B] \\
&\quad + n_{D0}[\tilde{H}_{D0} - \tilde{H}_D]
\end{aligned} \tag{12.12}$$

The quantity on the left is the heat of reaction, defined by Equation 12.2. The last three terms on the right represent mixing terms similar to the ones considered in Section 10.7, where there are changes in the partial molar enthalpies of each species because of the changing composition. These last terms will generally be negligible. Consider, for example, the most common case in which $n_{D0} = 0$, $n_{A0} = n_{B0}$. Then $n_A = n_B$ for all time. If product D forms a nearly ideal solution in A and B then $\tilde{H}_A$ and $\tilde{H}_B$ are essentially unaffected by n_D and depend only on the mole ratio n_A/n_B, which is constant. Then $\tilde{H}_A = \tilde{H}_{A0}$, $\tilde{H}_B = \tilde{H}_{B0}$, and all three terms vanish. We will assume henceforth that these terms are negligible compared to $\Delta \underline{H}_R$. If they are not negligible then there is usually no convenient way to separate them from the measurement of $\Delta \underline{H}_R$ and they will introduce an error into the determination of the heat of reaction, but that problem is beyond the scope of our discussion here. Equation 12.12 then becomes, finally,

$$[n_A - n_{A0}]\Delta \underline{H}_R = -q + \rho_0 V_0 \underline{c}_{p0}[T - T_0] \tag{12.13}$$

The batch reactor experiment to determine $\Delta \underline{H}_R$ is generally known as a *calorimeter experiment* and is a common one in physical chemistry laboratory courses. The two simplest ways to carry out a calorimeter experiment are adiabatically and isothermally. In both, the reaction is allowed to run to completion and the difference $n_{A0} - n_A$ is measured. In the adiabatic experiment the calorimeter is well insulated ($q = 0$) and the adiabatic temperature change from start to finish is measured. Then, from Equation 12.13,

$$\textit{adiabatic}: \quad \Delta \underline{H}_R = \frac{\rho_0 V_0 \underline{c}_{p0}[T_{ad} - T_0]}{n_A - n_{A0}} \tag{12.14}$$

T_{ad} is the final adiabatic temperature. A typical temperature-time plot for an adiabatic calorimeter experiment is shown in Figure 12.1. In an isothermal calorimeter heat is added or removed so that the final temperature and the

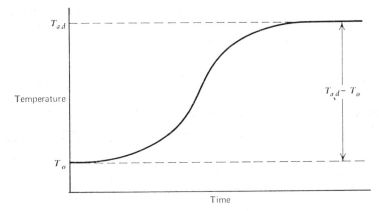

FIGURE 12.1   Temperature change due to reaction versus time in an adiabatic calorimeter.

initial temperature are equal. $\mathbf{q}$ is measured experimentally and $\Delta \underset{\sim}{H}_R$ is computed from the relation

$$isothermal: \quad \Delta \underset{\sim}{H}_R = \frac{-\mathbf{q}}{n_A - n_{A0}} \qquad (12.15)$$

An *exothermic reaction* is one in which there is an adiabatic temperature increase or, equivalently, heat must be removed to maintain $T = T_0$. Since $n_A < n_{A0}$, it then follows from Equation 12.14 or 12.15 that for an adiabatic reaction $\Delta \underset{\sim}{H}_R$ is negative ($\mathbf{q}$ is positive)

$$exothermic: \Delta \underset{\sim}{H}_R < 0 \qquad (12.16a)$$

Similarly, for an *endothermic reaction* in which there is an adiabatic temperature decrease, or in which heat must be added for isothermal operation, $\Delta \underset{\sim}{H}_R$ is positive ($\mathbf{q}$ is negative)

$$endothermic: \Delta \underset{\sim}{H}_R > 0 \qquad (12.16b)$$

$\Delta \underset{\sim}{H}_R$ is a function of temperature. Data are generally tabulated at standard temperatures such as $T° = 25°C$. If we were to repeat the steps in the above analysis but, instead of referring all enthalpies to temperature T, as in Equation 12.8, we were to evaluate all enthalpies at the reference temperature $T°$, we would obtain, in place of Equation 12.13,

$$[n_A - n_{A0}]\Delta \underset{\sim}{H}_R(T°) = -\mathbf{q} + \rho_0 V_0 \underset{\sim}{c}_{p0}[T - T_0]$$
$$+ \rho V[T° - T][\underset{\sim}{c}_{p0} - \underset{\sim}{c}_p] \qquad (12.17)$$

c_p is the heat capacity per unit mass of the final mixture. An expression for $\Delta \underset{\sim}{H}_R$ at any temperature can thus be obtained by comparing Equations 12.13 and 12.17, noting that in the batch system $\rho V = \rho_0 V_0$:

$$\Delta H_R(T) = \Delta H_R(T^\circ) + \frac{\rho V[T^\circ - T][c_p - c_{p0}]}{n_A - n_{A0}} \tag{12.18}$$

For ideal systems we have for the $A + B \rightarrow nD$ example

$$\Delta \underset{\sim}{H}_R(T) = \Delta \underset{\sim}{H}_R(T^\circ) + [nc_{pD} - c_{pA} - c_{pB}][T - T^\circ] \tag{12.19}$$

and for the more general reaction defined by Equation 12.1,

$$\Delta \underset{\sim}{H}_R(T) = \Delta \underset{\sim}{H}_R(T^\circ) + [\delta_1 c_{pD_1} + \delta_2 c_{pD_2} + \cdots - \alpha_1 c_{pA_1}$$
$$- \alpha_2 c_{pA_2} + \cdots][T - T^\circ] \tag{12.20}$$

In liquid systems c_p and c_{p0} will rarely differ by much, so $\Delta \underset{\sim}{H}_R$ may often be taken as independent of temperature without serious error.

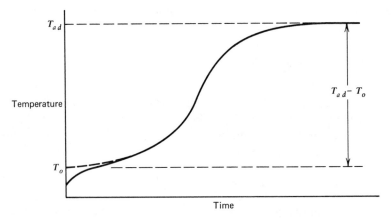

FIGURE 12.2   Temperature change versus time in an adiabatic calorimeter. The initial rise is due to mixing, the subsequent rise to reaction. T_0 is estimated by extrapolation.

It is important to emphasize that the whole analysis of this section presumes that the reactants can be mixed quickly relative to the time required for reaction. If there are enthalpy changes upon mixing the reactants, we would expect to see a rapid initial temperature change resulting from the heat of mixing. The initial temperature T_0 would thus differ from the temperature of the reactants, but T_0 could still be found by extrapolation, as shown in Figure 12.2. Equation 12.14 for $\Delta \underset{\sim}{H}_R$ can therefore still be used. Finally, it is a straightforward matter to develop equations for $\Delta \underset{\sim}{H}_R$ in which the assumption of constant heat capacities is not made. Indeed, in any equation it is only necessary to replace

$$\underline{c}_p[T_2 - T_1] \quad \text{by} \quad \int_{T_1}^{T_2} \underline{c}_p(T)\, dT$$

to obtain the appropriate relations.

12.4 TEMPERATURE EQUATION

In the context of the previous section and as preparation for subsequent work it is useful to derive an equation for the rate of change of temperature in the batch reactor. Our starting point is again the energy equation

$$\frac{dH}{dt} = Q \tag{12.6}$$

Since H depends on T, n_A, n_B, and n_D we may apply the chain rule and write (Equation 10.46)

$$\frac{dH}{dt} = \rho V \underline{c}_p \frac{dT}{dt} + \tilde{H}_A \frac{dn_A}{dt} + \tilde{H}_B \frac{dn_B}{dt} + \tilde{H}_D \frac{dn_D}{dt} = Q \tag{12.21}$$

From Equations 12.3 we can replace the molar rates of change by the reaction rate

$$\rho V \underline{c}_p \frac{dT}{dt} + [\tilde{H}_A + \tilde{H}_B - n\tilde{H}_D][-rV] = Q \tag{12.22}$$

The heat of reaction is introduced from its definition, Equation 12.2, to give

$$\rho V \underline{c}_p \frac{dT}{dt} = [-\Delta \underline{H}_R]rV + Q \tag{12.23}$$

Equation 12.23 must be solved simultaneously with Equations 12.3 for the concentrations and with a constitutive equation relating the rate to concentrations and temperature.

Equation 12.23 provides a means of evaluating the assumptions made in passing from Equation 12.12 to 12.13, in which a number of enthalpy differences were neglected. We assume that, despite the changing concentration, the mixture heat capacity $\underline{c}_p$ and the heat of reaction, $\Delta \underline{H}_R$, remain constant. Then, noting that $rV = -dn_A/dt$, $Q = dq/dt$, Equation 12.23 can be written

$$\underline{c}_p, \Delta \underline{H}_R = \text{constant:} \quad \frac{d}{dt}[\rho V \underline{c}_p T] = \frac{d}{dt}[\Delta \underline{H}_R n_A + q]$$

and, integrating both sides,

$$\rho V \underline{c}_p[T - T_0] = \Delta \underline{H}_R[n_A - n_{A0}] + q$$

which is Equation 12.13. Thus, these two conditions are sufficient to ensure that the partial molar enthalpy terms neglected in Equation 12.13 are unimportant. The assumption of constant $\underline{c}_p$ is often a good one and experimentally verifiable. Concentration dependence of $\Delta \underline{H}_R$ is difficult to obtain and rarely available, so it is usually consistent to assume constant $\Delta \underline{H}_R$ as well since we have no better information.

12.5 CALCULATION OF HEATS OF REACTION FROM TABULAR DATA

If the heat of reaction, $\Delta \underline{H}_R$, has not been measured for a particular reaction it is often possible to compute the value from tabular data. One such procedure is outlined here, together with the assumptions involved. The tabulated data used are those shown in Table 12.1 for the *Heat of Formation*

TABLE 12.1   Heat of formation of liquids and solids at 25°C, kcal/mole

n-pentane, C_5H_{12}	−41.36
n-hexane, C_6H_{14}	−47.52
2,3-dimethylbutane, C_6H_{14}	−49.58
n-octane, C_8H_{18}	−59.74
benzene, C_6H_6	11.72
toluene, C_7H_8	2.87
ethanol, C_2H_6O	−66.35
ethylene glycol, $C_2H_6O_2$	−107.91
ethylene oxide, $(CH_2)_2O$ (8°C)	−10.0
HNO_3	−41.35
H_2O	−68.32
H_2SO_4	−193.69
KNO_3	−118.08
KOH	−102.02

of compound i, $\Delta \underline{H}_{Fi}$, at 25°C. The heat of formation is defined as the heat of reaction when a compound is formed from its elements in their natural state by the (possibly nonexistent) reaction

$$\text{Elements} \rightarrow \text{Compound } i$$

The relation which we shall derive, in the absence of measurable mixing enthalpies, is

$$\Delta \underline{H}_R = \delta_1 \Delta \underline{H}_{FD_1} + \delta_2 \Delta \underline{H}_{FD_2} + \cdots$$
$$- \alpha_1 \Delta \underline{H}_{FA_1} - \alpha_2 \Delta \underline{H}_{FA_2} - \cdots \quad (12.24)$$

where $\delta_1, \delta_2, \ldots, \alpha_1, \alpha_2, \ldots$ are the stoichiometric coefficients in the reaction

$$\alpha_1 A_1 + \alpha_2 A_2 + \alpha_3 A_3 + \cdots \rightleftharpoons \delta_1 D_1 + \delta_2 D_2 + \delta_3 D_3 + \cdots$$

(12.1)

The chemical equation for the formation of one mole of species i from the elements is

$$\varepsilon_i^1 E_1 + \varepsilon_i^2 E_2 + \varepsilon_i^3 E_3 + \cdots \rightarrow i$$

(12.25)

Here, i refers to $A_1, A_2, A_3, \ldots, D_1, D_2, D_3, \ldots$ $E_1, E_2, E_3, \ldots$ are the elements in their natural state, and ε_i^k is the stoichiometric coefficient of element k in the reaction to form compound i. The heat of formation is then

$$\Delta \underset{\sim}{H}_{Fi} = \underset{\sim}{H}_i - \varepsilon_i^1 \underset{\sim}{H}_{E_1} - \varepsilon_i^2 \underset{\sim}{H}_{E_2} - \varepsilon_i^3 \underset{\sim}{H}_{E_3} - \cdots$$

(12.26)

We assume that there are no mixing effects and pure component enthalpies are used. In the reaction mixture for Equation 12.1 we will first assume ideal behavior, so $\tilde{H}_i = \underset{\sim}{H}_i$. Then combining Equation 12.2 for $\Delta \underset{\sim}{H}_R$ and Equation 12.26 for $\Delta \underset{\sim}{H}_F$ we obtain

$$\begin{aligned}
\Delta \underset{\sim}{H}_R &= \delta_1 \tilde{H}_{D_1} + \delta_2 \tilde{H}_{D_2} + \cdots - \alpha_1 \tilde{H}_{A_1} - \alpha_2 \tilde{H}_{A_2} - \cdots \\
&= \delta_1 \underset{\sim}{H}_{D_1} + \delta_2 \underset{\sim}{H}_{D_2} + \cdots - \alpha_1 \underset{\sim}{H}_{A_1} - \alpha_2 \underset{\sim}{H}_{A_2} - \cdots \\
&= \delta_1 [\Delta \underset{\sim}{H}_{FD_1} + \varepsilon_{D_1}^1 \underset{\sim}{H}_{E_1} + \varepsilon_{D_1}^2 \underset{\sim}{H}_{E_2} + \cdots] \\
&\quad + \delta_2 [\Delta \underset{\sim}{H}_{FD_2} + \varepsilon_{D_2}^1 \underset{\sim}{H}_{E_1} + \varepsilon_{D_2}^2 \underset{\sim}{H}_{E_2} + \cdots] + \cdots \\
&\quad - \alpha_1 [\Delta \underset{\sim}{H}_{FA_1} + \varepsilon_{A_1}^1 \underset{\sim}{H}_{E_1} + \varepsilon_{A_1}^2 \underset{\sim}{H}_{E_2} + \cdots] \\
&\quad - \alpha_2 [\Delta \underset{\sim}{H}_{FA_2} + \varepsilon_{A_2}^1 \underset{\sim}{H}_{E_1} + \varepsilon_{A_2}^2 \underset{\sim}{H}_{E_2} + \cdots] - \cdots
\end{aligned}$$

(12.27)

Consider the terms multiplying $\underset{\sim}{H}_{E_1}$:

$$\underset{\sim}{H}_{E_1} [\delta_1 \varepsilon_{D_1}^1 + \delta_2 \varepsilon_{D_2}^1 + \cdots - \alpha_1 \varepsilon_{A_1}^1 - \alpha_2 \varepsilon_{A_2}^1 - \cdots]$$

The sum $\delta_1 \varepsilon_{D_1}^1 + \delta_2 \varepsilon_{D_2}^2 + \cdots$ represents the total number of atoms of element 1 on the right-hand side of Equation 12.1, the sum $\alpha_1 \varepsilon_{A_1}^1 + \alpha_2 \varepsilon_{A_2}^1 + \cdots$ the total number of atoms on the left. These must be equal for the chemical equation to balance, so the coefficient of $\underset{\sim}{H}_{E_1}$ is simply equal to zero. Similarly, the coefficients of $\underset{\sim}{H}_{E_2}, \underset{\sim}{H}_{E_3}, \ldots$ also equal zero and Equation 12.27 reduces to Equation 12.24, expressing the heat of reaction in terms of heats of formation.

Notice that this result could have been obtained by adopting the convention that *the enthalpy of an element in its natural state at the reference temperature is zero*. The enthalpy of the element in its natural state will always drop out of any formulation so we could simplify the algebra here and elsewhere by setting it to zero at the start. This convention is universally applied, and the reason that it works is evident from Equation 12.27.

Example 12.1

Compute $\Delta \underset{\sim}{H}_R$ for the reaction between ethylene oxide (A_1) and water (A_2) to form ethylene glycol (D_1):

$$\begin{array}{cccc}
H_2C\!-\!CH_2 + H_2O & \rightarrow & H_2C\!-\!CH_2 \\
\diagdown\diagup & & \mid \quad \mid \\
O & & OH \quad OH
\end{array}$$

From Table 12.1,

ethylene glycol: $\Delta \underset{\sim}{H}_{FD_1} = -107.9 \text{ kcal/mole}$

ethylene oxide: $\Delta \underset{\sim}{H}_{FA_1} = -10.0$

water: $\Delta \underset{\sim}{H}_{FA_2} = -68.3$

$\Delta \underset{\sim}{H}_R = [-107.9] - [-10.0] - [-68.3] = -29.6 \text{ kcal/mole}$

The reaction is exothermic, since $\Delta \underset{\sim}{H}_R$ is negative.

   If there is a measurable heat of solution for any of the compounds involved in a reaction, the procedure derived above must be modified slightly, since the heat of formation will now be that of the compound in solution. Heat of solution data must be available if this effect is to be taken into account.

Example 12.2

Solutions of 6.25 mole percent HNO_3 (1 mole HNO_3 to 15 moles H_2O, 18.9 percent acid by weight) and 6.25 mole percent KOH (17.2 percent base by weight) react to form a 3.23 mole percent solution of KNO_3 (1 mole KNO_3 to 31 moles H_2O, 15.3 percent salt by weight). Estimate the heat of reaction.

   The reaction in solution is

$$HNO_3 + KOH \rightarrow KNO_3 + H_2O$$

Heats of solution of pure liquid HNO_3 and H_2O and crystalline KOH and KNO_3 are obtained from Table 12.1:

HNO_3 liquid: $\Delta \underset{\sim}{H}_F = -41.35 \text{ kcal/mole}$

KOH crystal: $\Delta \underset{\sim}{H}_F = -102$

KNO_3 crystal: $\Delta \underset{\sim}{H}_F = -118$

H_2O liquid: $\Delta \underset{\sim}{H}_F = -68.3$

From Figure 10.6 it can be seen that the potassium hydroxide solution is at infinite dilution, and the nitric acid and final salt solutions are nearly so. The heats of solution are

$HNO_3 + 15H_2O$: $\Delta \underset{\sim}{H}_s = -7.3 \text{ kcal/mole}$

$KOH + 15H_2O$: $\Delta \underset{\sim}{H}_s = -12.6$

$KNO_3 + 30H_2O$: $\Delta \underset{\sim}{H}_s = +7.5$

There should be no further solution effects when the dilute acid and base are mixed, so the heat of formation in the reaction solution may be approximated by the sum of heat of formation and heat of solution:

HNO_3 solution: $\Delta H_F = -41.35 - 7.3 = -48.7$ kcal/mole

KOH solution: $\Delta H_F = -102 - 12.6 = -114.6$

KNO_3 solution: $\Delta H_F = -118 + 7.5 = -110.5$

H_2O liquid: $\Delta H_F = -68.3$

$\Delta H_R = [-110.5] + [-68.3] - [-48.7] - [-114.6]$

$= -15.5$ kcal/mole

Example 12.3

Solutions of 25 mole percent HNO_3 (1 mole HNO_3 to 3 moles H_2O, 53.8 percent acid by weight) and 3.58 mole percent KOH (1 mole KOH to 27 moles H_2O, 10.33 percent base by weight) are reacted to form a 3.23 mole percent solution of KNO_3. Estimate the heat of reaction.

The heat of reaction is calculated using the concentrations of the species in the reactor. The reaction mixture in this example, 1 mole HNO_3, 1 mole KOH, and 30 moles H_2O is identical to that in the previous example, so the heat of reaction ΔH_R, must have the same value, -15.5 kcal/mole. The feed concentration of HNO_3 is not at infinite dilution, so that when the feed streams are mixed the acid is diluted with an accompanying heat of solution. This heat of solution can be computed by noting that the heat of solution for HNO_3 and $3H_2O$ is found from Figure 10.6 to be $\Delta H_s = -5.6$ kcal/mole. In an infinite solution ΔH_s is -7.3 kcal/mole, so that when the acid and base are mixed together to form the infinitely dilute solution there will be a heat of solution of -1.7 kcal/mole. A calorimeter experiment will measure the *sum* of this heat of solution and the heat of reaction and give an experimental value of $-15.5 - 1.7 = -17.2$ kcal/mole. If the solution effect is not accounted for and the measured value is assumed to be the heat of reaction, there will be an error of more than 10 percent in the value of ΔH_R. In a rapid reaction like one between an acid and a base the mixing and reaction effects will not generally be distinct as in Figure 12.2.

Heats of formation can be obtained for hydrocarbons from the tabulated *heat of combustion*, which is the heat of reaction of the oxidation reaction

$$C_nH_m + \frac{1}{2}\left[2n + \frac{m}{2}\right]O_2 \rightarrow nCO_2 + \frac{m}{2}H_2O$$

For example, the tabulated heat of combustion of liquid *n*-octane, C_8H_{18}, with gaseous oxygen to form liquid water and gaseous CO_2 is -1307.53

kcal/mole. The fact that gases are involved has no effect on the definition and calculation of heats of formation and reaction. Then

$$\Delta H_R = 8\Delta H_{F\ CO_2} + 9\Delta H_{F\ H_2O} - 12.5\Delta H_{F\ O_2} - \Delta H_{F\ C_8H_{18}}$$

The heat of formation of the element oxygen is zero. The heats of formation of gaseous CO_2 and liquid H_2O are

$$\Delta H_{F\ CO_2} = -94.05 \text{ kcal/mole}$$

$$\Delta H_{F\ H_2O} = -68.32$$

Thus

$$-1307.53 = 8[-94.05] + 9[-68.32] - 12.5[0] - \Delta H_{F\ C_8H_{18}}$$

$$\Delta H_{F\ C_8H_{18}} = -59.75 \text{ kcal/mole}$$

This computation is generally not necessary for hydrocarbons of fewer than ten carbons, where ΔH_F is usually tabulated separately. Since all combustion reactions are exothermic heats of combustion are tabulated (incorrectly) without the minus sign. Thus, the heat of combustion of *n*-octane is tabulated as 1307.53 kcal/mole.

12.6   TEMPERATURE DEPENDENCE OF THE REACTION RATE

The rate at which a chemical reaction occurs is strongly dependent on temperature. The temperature dependence of the parameters in the constitutive equation relating rate and concentrations is most easily determined by performing a series of isothermal batch experiments at various temperatures and evaluating the rate expression at each temperature according to the procedure developed in Chapter 5. For example, if the reaction

$$A + B \rightleftarrows 2D$$

has a rate

$$r = k_1 c_A c_B - k_2 c_D^2$$

as in Example 5.4, then k_1 and k_2 can be found at each temperature, T. *It is found experimentally for nearly all reaction systems that a plot of log k versus* $1/T$, *where T is absolute temperature, will be linear with a negative slope.* Typical data are shown in Figure 12.3 for the decomposition of sodium dithionate reaction discussed in Section 5.9.

The temperature dependence of the rate constant is usually written in the form

$$k = k_0 e^{-E/RT} \tag{12.28}$$

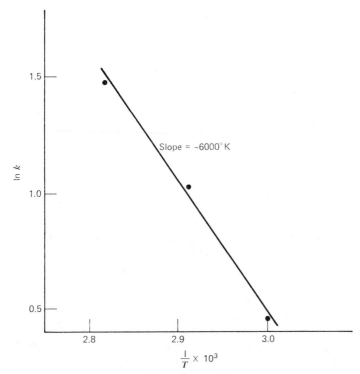

FIGURE 12.3 Decomposition of sodium dithionate, $r = kc_{S_2O_4^=}c_{H^+}$. Natural logarithm of k, liter/mole-sec, versus reciprocal temperature, $^\circ K$. $k = 1.1 \times 10^8$ exp $[-6000/T]$. [Data of Rinker et al., *Ind. Eng. Chem. Fundamentals*, **4**, 282 (1964).]

where R is the gas constant and E has dimensions of energy per mole. This *Arrhenius* form is in agreement with the data shown in Figure 12.3, since

$$\ln k = \ln k_0 - \left[\frac{E}{R}\right]\frac{1}{T}$$

For the dithionate reaction the slope of $\ln k$ versus $1/T$ is approximately $-6000^\circ K$, so E is approximately 12,000 cal/g-mole for this reaction. The *activation energy* E is typically in the range 10,000–30,000 calories/g-mole. Equation 12.28 can be deduced from physical chemical arguments, but theoretical predictions of k_0 and E are generally unsatisfactory and experimental measurement is always required.

12.7 CONCENTRATION-TEMPERATURE-TIME DEPENDENCE IN THE BATCH REACTOR

We are now in a position to treat the batch reactor completely. For the reaction $A + B \rightleftarrows nD$ the equations of conservation of mass and energy are given by Equations 12.3 and 12.23:

$$\frac{dVc_A}{dt} = -rV \tag{12.3a}$$

$$\frac{dVc_B}{dt} = -rV \tag{12.3b}$$

$$\frac{dVc_D}{dt} = nrV \tag{12.3c}$$

$$\rho V \underline{c}_p \frac{dT}{dt} = [-\Delta \underset{\sim}{H}_R]rV + Q \tag{12.23}$$

For given concentrations and temperature at time zero and for a specified rate constitutive equation these equations can be integrated to give the concentrations and temperature as functions of time. In fact, the equations simplify considerably in seeking a solution. Only one concentration equation is really required, since c_B and c_D can be related to c_A and the initial values of the concentration of A, B, and D. Furthermore, the concentration and temperature are related through Equation 12.13, so that ultimately only a single equation needs to be integrated.

To illustrate this point without algebraic complexity we will consider the slightly simpler case of an adiabatic reactor ($Q = 0$) with an irreversible reaction whose rate expression is, using Equation 12.28,

$$r = k_0 e^{-E/RT} c_A c_B$$

if $c_{A0} = c_{B0}$ then $c_A = c_B$ at all times. We will take the density as linear in concentrations and the rate of change of density with concentration proportional to the molecular weight, as discussed in Section 5.8. Then the volume is constant and Equations 12.3 and 12.23 simplify to

$$\frac{dc_A}{dt} = -k_0 e^{-E/RT} c_A{}^2 \tag{12.29}$$

$$\frac{dT}{dt} = \frac{[-\Delta \underset{\sim}{H}_R]}{\rho \underline{c}_p} k_0 e^{-E/RT} c_A{}^2 \tag{12.30}$$

$$c_A = c_{A0}, \qquad T = T_0 \quad \text{at} \quad t = 0$$

Assume further that $\Delta \underset{\sim}{H}_R$ and $\rho \underline{c}_p$ are constants. (Neither assumption is necessary but the algebra is less complicated and any error introduced is usually negligible.) Then Equations 12.29 and 12.30 combine to

$$\frac{dT}{dt} = -\frac{[-\Delta \underset{\sim}{H}_R]}{\rho \underline{c}_p} \frac{dc_A}{dt}$$

$$T - T_0 = -\frac{[-\Delta \underset{\sim}{H}_R]}{\rho \underline{c}_p} [c_A - c_{A0}] \tag{12.31}$$

Equation 12.31 is a special case of Equation 12.13. Equation 12.29 can be written entirely in terms of the concentration:

$$\frac{dc_A}{dt} = -k_0 c_A{}^2 \exp\left\{-\frac{E}{R\left[T_0 + \dfrac{-\Delta H_R}{\rho \underline{c}_p}[c_{A0} - c_A]\right]}\right\}$$ (12.32)

Equation 12.32 is separable (Section 15.9). The solution can be expressed in terms of an integral

$$\int_{c_{A0}}^{c_A} \frac{dc}{c^2 \exp\left[-\dfrac{E}{R\left[T_0 + \dfrac{-\Delta H_R}{\rho \underline{c}_p}[c_{A0} - c]\right]}\right]} = -k_0 t$$ (12.33)

The integral must be evaluated numerically.

Example 12.4

The reaction $A + B \rightarrow$ products has the following physical and chemical properties:

$$\Delta \underline{H}_R = -38{,}000 \text{ BTU/lb-mole of } A$$

$$E = 18{,}500 \text{ cal/g-mole}$$

$$k_0 = 10^9 \text{ ft}^3/\text{lb-mole min}$$

$$\rho = 62.4 \text{ lb}_m/\text{ft}^3 = \text{constant}$$

$$\underline{c}_p = 1.0 \text{ BTU/lb}_m{}^\circ\text{F} = \text{constant}$$

Initial conditions are $c_{A0} = c_{B0} = 0.5$ lb-mole/ft³, $T_0 = 200°F$. Estimate the temperature and concentration variation with time.

According to Equation 12.31 the temperature will rise monotonically from T_0 to some limiting value as c_A varies from c_{A0} to zero. It is convenient to calculate the limiting temperature directly as a check on the subsequent calculations. Setting $c_A = 0$ in Equation 12.31 we obtain

$$t \rightarrow \infty: \quad T = T_0 + \frac{[-\Delta \underline{H}_R]c_{A0}}{\rho \underline{c}_p}$$

$$= 200 + \frac{[38{,}000][0.5]}{[62.4][1.0]} = 504°F$$

The activation energy is given in cgs units, for which the gas constant is $R = 1.99$ cal/g-mole°K. For consistency the temperature must be expressed in

°K, using Equation 12.31 as follows:

$$T = T_0 + \frac{[-\Delta H_R]}{\rho \underline{c}_p} [c_{A0} - c_A] = 200 + 304 - 609c_A, \text{°F}$$

$$= 200 + 304 - 609c_A + 460, \text{°R}$$

$$= 964 - 609c_A, \text{°R}$$

$$= \frac{1}{1.8} [964 - 609c_A], \text{°K}$$

$$T = 536 - 338c_A, \text{°K}$$

Equation 12.33 can then be written

$$t = -\frac{1}{k_0} \int_{c_{A0}}^{c_A} \frac{dc}{c^2 \exp\left[-\dfrac{E}{RT}\right]}$$

$$= -10^{-9} \int_{0.5}^{c_A} \frac{dc}{c^2 \exp\left\{-\dfrac{18{,}500}{1.99[536 - 338c]}\right\}}$$

or

$$t = 10^{-9} \int_{c_A}^{0.5} \frac{dc}{c^2 \exp\left[-\dfrac{17.2}{1 - 0.631c}\right]}$$

The integral cannot be evaluated in terms of known functions, but for each value of c_A the integration can be carried out numerically using the trapezoidal rule discussed in Section 15.6 or some other numerical method. The function

$$\phi(c) = \frac{10^{-9}}{c^2 \exp\left[-\dfrac{17.2}{1 - 0.631c}\right]}$$

is plotted verus c in Figure 12.4. Since

$$t = \int_{c_A}^{0.5} \phi(c) \, dc$$

the value of t corresponding to a given c_A is computed from the integral. For example, for $c_A = 0.4$, the value of t is the shaded area in Figure 12.4. Table 12.2 contains the computed concentration-time relations and the corresponding temperatures, and these are plotted in Figure 12.5. The concentration is read on the left ordinate and temperature on the right. A single curve suffices since, according to Equation 12.31, $c_{A0} - c_A$ is always proportional to $T - T_0$.

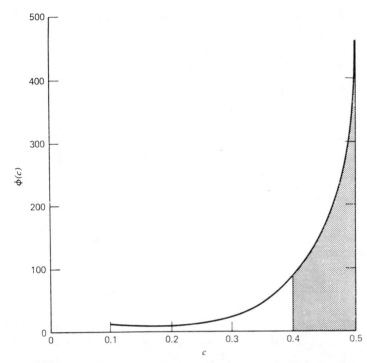

FIGURE 12.4 $\phi(c)$ versus c. The time for the concentration to drop from 0.5 to 0.4 equals the area under the curve between 0.4 and 0.5.

TABLE 12.2 Concentration and temperature versus time for the reaction $A + B \rightarrow$ products in an adiabatic batch reactor. Parameters are given in Example 12.4

c_A, lb-mole/ft³	T, °F	t, min
0.5	200	0
0.4	261	18.9
0.3	322	23.9
0.2	382	25.5
0.1	443	26.7
0	504	∞

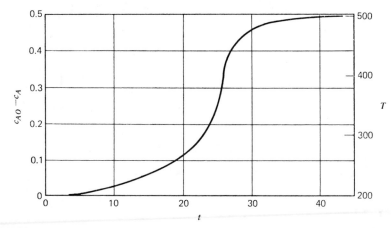

FIGURE 12.5   Concentration versus time and temperature versus time in an adiabatic batch reactor.

12.8   CONTINUOUS FLOW STIRRED TANK REACTOR

The continuous flow stirred tank reactor was discussed in considerable detail in Chapter 7, where design calculations were carried out for steady state operation. The total design problem requires that the energy equation be taken into account, and we shall derive that equation here. The flow configuration is shown in Figure 12.6. In order to keep the algebra to a minimum the feed streams are shown being mixed just prior to entering the reactor, so there is a single feed to the reactor. The flow in and out is maintained at volumetric rate q, and the standard assumptions are made concerning the density (Section 5.8) so that the volume remains constant under these conditions. For the reaction $A + B \rightleftarrows nD$ the equations of conservation of mass

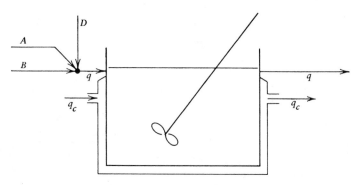

FIGURE 12.6   Reaction $A + B \rightleftarrows nD$ in a jacketed CFSTR.

are then

$$\frac{dn_A}{dt} = V\frac{dc_A}{dt} = q[c_{Af} - c_A] - rV \tag{12.34a}$$

$$\frac{dn_B}{dt} = V\frac{dc_B}{dt} = q[c_{Bf} - c_B] - rV \tag{12.34b}$$

$$\frac{dn_D}{dt} = V\frac{dc_D}{dt} = q[c_{Df} - c_D] + nrV \tag{12.34c}$$

The equation of conservation of energy applied to the tank is

$$\frac{dU}{dt} = \rho_f q \underline{H}_f(T_f) - \rho q \underline{H}(T) + Q - W_S$$

or, letting $dU/dt \simeq dH/dt$ for the liquid system,

$$\frac{dH}{dt} = \rho_f q \underline{H}_f(T_f) - \rho q \underline{H}(T) + Q - W_S \tag{12.35}$$

The enthalpies can all be evaluated at the reactor temperature by use of Equation 10.34

$$\underline{H}_f(T_f) = \underline{H}_f(T) + \underline{c}_{pf}[T_f - T] \tag{12.36}$$

$\underline{c}_{pf}$ is the heat capacity of the feed. Furthermore, we have the following identities (Equation 10.53)

$$\rho_f \underline{H}_f(T) = c_{Af}\tilde{H}_{Af}(T) + c_{Bf}\tilde{H}_{Bf}(T) + c_{Df}\tilde{H}_{Df}(T) \tag{12.37}$$

$$\rho \underline{H}(T) = c_A\tilde{H}_A(T) + c_B\tilde{H}_B(T) + c_D\tilde{H}_D(T) \tag{12.38}$$

From Equation 10.46, which is an application of the chain rule,

$$\frac{dH}{dt} = \rho \underline{c}_p V \frac{dT}{dt} + \tilde{H}_A\frac{dn_A}{dt} + \tilde{H}_B\frac{dn_B}{dt} + \tilde{H}_D\frac{dn_D}{dt} \tag{12.39}$$

Using Equations 12.34 for the rates of change of compositions Equation 12.35 then becomes

$$\rho V \underline{c}_p \frac{dT}{dt} + \tilde{H}_A[qc_{Af} - qc_A - rV] + \tilde{H}_B[qc_{Bf} - qc_B - rV]$$

$$+ \tilde{H}_D[qc_{Df} - qc_D + nrV]$$

$$= \rho_f q \underline{c}_{pf}[T_f - T] + qc_{Af}\tilde{H}_{Af} + qc_{Bf}\tilde{H}_{Bf} + qc_{Df}\tilde{H}_{Df}$$

$$- qc_A\tilde{H}_A - qc_B\tilde{H}_B - qc_D\tilde{H}_D + Q - W_S$$

Some of the same terms appear on the left and right, such as $qc_A\tilde{H}_A$. These can be cancelled and the equation rearranged to give

$$\rho V \underline{c}_p \frac{dT}{dt} = \rho_f q \underline{c}_{pf}[T_f - T]$$

$$- [n\tilde{H}_D - \tilde{H}_A - \tilde{H}_B]rV + q\{c_{Af}[\tilde{H}_{Af} - \tilde{H}_A]$$

$$+ c_{Bf}[\tilde{H}_{Bf} - \tilde{H}_B] + c_{Df}[\tilde{H}_{Df} - \tilde{H}_D]\} + Q - W_S$$

$$(12.40)$$

The term $n\tilde{H}_D - \tilde{H}_A - \tilde{H}_B$ is simply the heat of reaction, $\Delta \underline{H}_R$, defined by Equation 12.2. We have discussed previously how to obtain the numerical value of this term by experiment or from calculation using tabular data. The term involving $\tilde{H}_{Af} - \tilde{H}_A$ and so on, represents the enthalpy change upon mixing the feed stream with the reactor contents. As was discussed following Equation 12.12, these terms will generally be small compared to $\Delta \underline{H}_R$ and will vanish in an ideal solution. We shall neglect them here. Thus, the energy equation becomes

$$\rho V \underline{c}_p \frac{dT}{dt} = \rho_f q \underline{c}_{pf}[T_f - T] + [-\Delta \underline{H}_R]rV + Q - W_S \quad (12.41)$$

If the reactor is cooled by a jacket, as shown in Figure 12.6, with coolant temperature T_c, then Equation 11.50 for the jacket temperature can be written

$$\rho_c V_c \underline{c}_{pc} \frac{dT_c}{dt} = \rho_c q_c \underline{c}_{pc}[T_{cf} - T_c] - Q \quad (12.42)$$

where subscript "c" refers to coolant and we have made use of the fact that $Q_c = -Q$. From Equation 11.45, the rate of heat transfer is expressed in terms of area and heat transfer coefficient as

$$Q = -ha[T - T_c] \quad (12.43)$$

By setting time derivatives to zero Equations 12.34, 12.41 to 12.43, together with the reaction rate expression, provide the design equations for the steady state continuous flow reactor. Notice that the design problem decomposes itself into two parts. At any given temperature the equations of conservation of mass are those treated in Chapter 7, and thus the full discussion there is relevant. We must, however, specify the operating temperature of the reactor before any design is considered complete and this superimposes an additional set of considerations on those discussed in Chapter 7. We examine the essential features of this problem by considering the steady state design.

12.9 STEADY STATE CFSTR

The equations for the continuous flow stirred tank reactor have an interesting mathematical structure which has a significant effect on both the design and the actual operation of the reactor. The structure manifests itself nicely with a minimum of mathematical complication by considering the pseudo first-order reaction $A \rightarrow D$, as discussed in Section 5.5. In that case we have

$$r = kc_A = k_0 e^{-E/RT} c_A$$

Equations 12.34a and 12.41 to 12.43 then become

$$V \frac{dc_A}{dt} = q[c_{Af} - c_A] - k_0 V e^{-E/RT} c_A \tag{12.44}$$

$$\rho V \underline{c}_p \frac{dT}{dt} = \rho_f q \underline{c}_{pf} [T_f - T] + [-\Delta \underline{H}_R] k_0 V e^{-E/RT} c_A$$

$$- ha[T - T_c] - W_S \tag{12.45}$$

$$\rho_c V_c \underline{c}_{pc} \frac{dT_c}{dt} = \rho_c q_c \underline{c}_{pc} [T_{cf} - T_c] + ha[T - T_c] \tag{12.46}$$

It is convenient to define the residence time for the tank, $\theta = V/q$, and for the cooling jacket, $\theta_c = V_c/q_c$, as well as the groups $J = -\Delta \underline{H}_R / \rho \underline{c}_p$, $K = ha/\rho_c q_c \underline{c}_{pc}$, and $u = \rho_c q_c \underline{c}_{pc} K/[1 + K]\rho q \underline{c}_p$. The essential features of the problem are retained if we make the very reasonable approximations $\rho_f = \rho$, $\underline{c}_{pf} = \underline{c}_p$. The working equations are thus

$$\theta \frac{dc_A}{dt} = c_{Af} - c_A - k_0 \theta e^{-E/RT} c_A \tag{12.47}$$

$$\theta \frac{dT}{dt} = T_f - T + J k_0 \theta e^{-E/RT} c_A$$

$$- u[1 + K][T - T_c] - W_S/\rho q \underline{c}_p \tag{12.48}$$

$$\theta_c \frac{dT_c}{dt} = T_{cf} - T_c + K[T - T_c] \tag{12.49}$$

At steady state, neglecting the shaft work term for simplicity ($W_S = 0$), we have the basic design equations:

$$0 = c_{Af} - c_A - k_0 \theta e^{-E/RT} c_A \tag{12.50}$$

$$0 = T_f - T + J k_0 \theta e^{-E/RT} c_A - u[1 + K][T - T_c] \tag{12.51}$$

$$0 = T_{cf} - T_c + K[T - T_c] \tag{12.52}$$

This is simplified by solving Equation 12.52 for T_c and substituting into Equation 12.51, leaving only two equations to solve simultaneously, Equation 12.50 for concentration and Equation 12.53 for temperature,

$$0 = T_f - T + Jk_0\theta e^{-E/RT}c_A + u[T_{cf} - T] \tag{12.53}$$

We can rearrange these two equations into a form more suitable for determining the nature of the solution. We multiply Equation 12.50 by J and add to Equation 12.53 to obtain

$$0 = J[c_{Af} - c_A] + T_f - T + u[T_{cf} - T]$$

or

$$c_A = c_{Af} + \frac{1}{J}[T_f + uT_{cf}] - \frac{1+u}{J}T \tag{12.54}$$

Equation 12.50 can be rearranged to

$$c_A = \frac{c_{Af}}{1 + k_0\theta e^{-E/RT}} \tag{12.55}$$

Equations 12.54 and 12.55 each allow us to plot c_A versus T. The intersection of the two curves defines the point at which both equations are satisfied simultaneously, providing the solution.

Equation 12.55 has the form shown in curve I in Figure 12.7. As $T \to 0$, $c_A \to c_{Af}$ with a zero slope, while as $T \to \infty$, c_A approaches $c_{Af}/[1 + k_0\theta]$

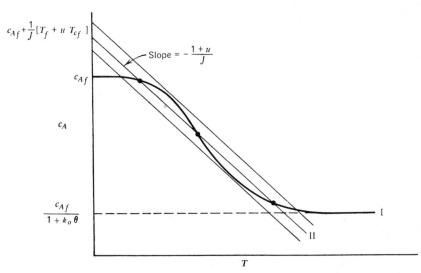

FIGURE 12.7 c_A versus T. Line I is Equation 12.55, line II is Equation 12.48. The intersections of the two lines define possible steady state solutions for the CFSTR.

asymptotically. On the other hand, Equation 12.54 is a straight line with slope $-[1 + u]/J$, shown as line II in Figure 12.7. The intersection of lines I and II defines the pair c_A, T which provides the simultaneous solution to the reactor equations. *Depending on the values of the parameters there can be as many as three different solutions to these equations.*

It might seem somewhat surprising at first to find that there can be more than one concentration-temperature pair which provides a solution to the reactor design equations. On reflection it is not really so strange. The algebraic equations are nonlinear, and even the simplest nonlinear algebraic equations often have more than one real solution. (Consider $x^2 - 1 = 0$, which has *two* solutions, $x = +1$ and $x = -1$.) In physical problems we can often discard one or more solutions on the grounds that they do not make physical sense—a negative temperature, for example. That simple approach will not work here, since it is evident from Figure 12.7 that all three possible temperatures are positive and lie within physically possible limits, as do the concentrations. Examples 16.1 and 16.3 consider a particular case with $u = 0$, $\theta = 0.3$ hr, $c_{Af} = 0.5$ moles/ft^3, $T_f = 311°$K, $k_0 = 10^9$ min^{-1}, $E = 1.85 \times 10^4$ cal/g-mole, $\Delta H_R = -3.8 \times 10^4$ BTU/lb-mole, and the physical properties of water. Solutions are found at $c_A = 0.499$, $T = 311.4$; $c_A = 0.257$, $T = 393.4$; and $c_A = 0.008$, $T = 477.5$.

The parameters for the example were chosen to produce the three-solution case. In practical applications it is not uncommon to have such a case and we need to be aware of the possibility because the questions which are raised are serious and extremely interesting. How can we be certain that a system designed to operate at $T = 478°$K and high conversion will not operate after being started up at $T = 311°$K and negligible conversion instead? If a process is operating at design conditions and there is a momentary change in conditions—for example, the cooling water temperature rises by a few degrees—will the system return to design conditions or will it go to some other possible solution to the equation?

Figure 12.8 shows results of an experimental study of the reaction

$$2Na_2S_2O_3 + 4H_2O_2 \rightarrow Na_2S_3O_6 + Na_2S_2O_4 + 4H_2O$$

in an adiabatic continuous flow stirred tank reactor. Batch data were used to determine the reaction rate and heat of reaction, leading to the computed curve of T versus θ shown in the figure. For residence times in the range of approximately 7–18 seconds there are three possible steady state temperatures. Two were found experimentally, one at high temperature, one at low. The intermediate steady state solution to the equations could not be achieved in practice. We shall show subsequently that whenever the slope of line I in Figure 12.7 is more negative than that of line II, the steady state cannot be attained, though the converse is not necessarily true.

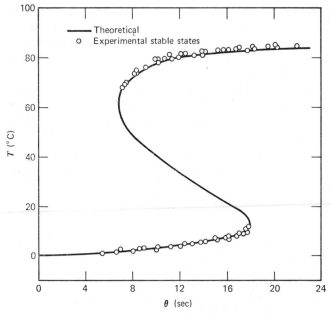

FIGURE 12.8   Experimental steady states in a CFSTR as a function of θ. Two different steady states can be obtained at the same feed conditions for θ in the range 7–18 sec. The theoretical curve, showing three possible steady states, was computed from batch reactor data. [Data of Vejtasa and Schmitz, *A.I.Ch.E.J.*, **16**, 410 (1970), reproduced by permission.]

12.10   UNATTAINABILITY OF A STEADY STATE

The question of whether a steady state solution to a set of nonlinear equations is attainable requires some care, but insight into *unattainability* can be obtained rather easily. In any real system there are fluctuations in all streams, so a process is always fluctuating somewhat about its steady state. If we can find *any* fluctuation such that a small deviation from the steady state will become larger, then it is clear that we have an inherently unstable situation that cannot be achieved in practice. We can do this for the intermediate steady state of the reactor by considering a *special fluctuation* in which c_A and T are algebraically related to one another through the relation which holds at steady state, Equation 12.50. Temperature and coolant temperature will be assumed to vary according to Equation 12.52. The other transient equation, 12.48, can be used to examine the nonsteady state behavior under these *special conditions*. In this case, for any temperature the system lies along line I in Figure 12.7 and the temperature satisfies the following transient equation, obtained after combining Equations 12.48,

12.50, and 12.52 and neglecting W_s

$$\theta \frac{dT}{dt} = T_f - T + J[c_{Af} - c_A] + u[T_{cf} - T] \qquad (12.56)$$

Line II is defined by Equation 12.54. If we let the subscripts I and II refer to lines I and II, respectively, at a given temperature, then Equation 12.56 can be combined with Equation 12.54 to give

$$\theta \frac{dT}{dt} = J[c_{AII} - c_{AI}] \qquad (12.57)$$

Consider the intermediate steady state in Figure 12.7. If we move to a slightly lower temperature, then $c_{AII} - c_{AI}$ is negative, $dT/dt < 0$, and the temperature decreases further! Similarly, at a slightly higher temperature $c_{AII} - c_{AI} > 0$, $dT/dt > 0$, and the temperature increases further. Clearly the intermediate steady state cannot be maintained, since certain small fluctuations which follow line I will reinforce themselves and move further away. This is equivalent to the observation regarding slopes of lines I and II made at the end of the preceding section. A similar argument applied to the high and low temperature steady states would suggest that they are inherently stable and can be maintained, since small fluctuations of the type described here will become smaller. Such logic cannot be used to prove stability, however, since there might be some other disturbance, unaccounted for here, which would grow rather than die out. The complete quantitative analysis will be considered in Chapter 13.

12.11 QUALITATIVE TRANSIENT BEHAVIOR

The graphical approach adopted in the preceding sections can be extended somewhat to yield information about the general transient behavior of the stirred tank reactor and specifically about the question of starting the reactor up in such a way that the desired operating conditions will be obtained. The method is adequately revealed with a minimum of manipulation by considering the adiabatic reactor, $Q = 0$, and neglecting shaft work. Equations 12.47 and 12.48 can then be rearranged to isolate c_A,

$$\text{Equation 12.47:} \quad c_A = \frac{c_{Af}}{1 + k_0\theta e^{-E/RT}} - \frac{\theta \dfrac{dc_A}{dt}}{1 + k_0\theta e^{-E/RT}} \qquad (12.58)$$

$$\text{Equation 12.48:} \quad c_A = \frac{[T - T_f]e^{E/RT}}{Jk_0\theta} + \frac{\theta e^{E/RT} \dfrac{dT}{dt}}{Jk_0\theta} \qquad (12.59)$$

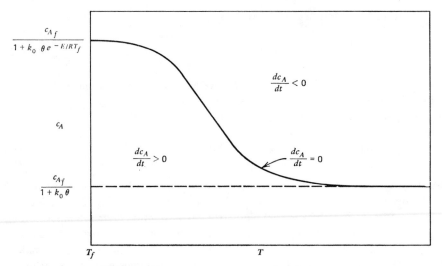

FIGURE 12.9 Regions of positive and negative values of dc_A/dt in the c_A-T plane. Line I, with $dc_A/dt = 0$, is computed from Equation 12.58.

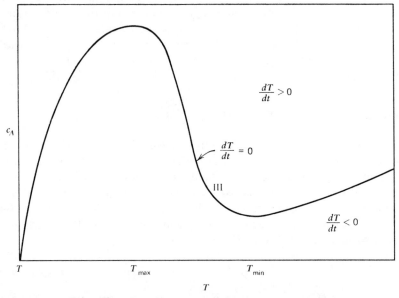

FIGURE 12.10 Regions of positive and negative values of dT/dt in the c_A-T plane. Line III, with $dT/dt = 0$, is computed from Equation 12.59. $T_{\max}$ and $T_{\min}$ are computed from Equation 12.60.

When dc_A/dt is zero, Equation 12.58 leads to curve I for c_A versus T, redrawn in Figure 12.9. This line divides the $c_A - T$ plane into two regions. Above the line, where $c_A > c_{Af}/[1 + k_0\theta e^{-E/RT}]$, it follows from Equation 12.58 that $dc_A/dt < 0$ and c_A is decreasing with time. Similarly, below the line $dc_A/dt > 0$ and c_A is increasing with time.

We can do the same with Equation 12.59. When $dT/dt = 0$ we obtain curve III for c_A versus T, shown in Figure 12.10. By differentiating c_A with respect to T and setting the derivative to zero, it is established that the maximum and minimum in curve III occur at

$$T_{\max,\min} = \frac{E \pm E^{1/2}[E - 4RT_f]^{1/2}}{2R} \tag{12.60}$$

According to Equation 12.60, a maximum and minimum will occur as long as $E > 4RT_f$. This will generally be the case, since E will normally exceed 10,000 calories/mole, while R is approximately 2 cal/mole °K and T would not usually reach 1000°K. When c_A lies above line III it follows from Equation 12.59 that $dT/dt > 0$ and T increases with time. For c_A below line III $dT/dt < 0$ and T decreases in time.

Figure 12.11 shows lines I and III superimposed. The curves are shown intersecting three times, but note that by adjusting J, $k_0\theta$, and T_f they could be shifted relative to one another so that only one intersection is possible. Since these lines represent $dc_A/dt = 0$ and $dT/dt = 0$ their intersection

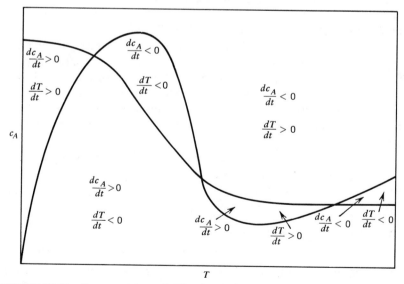

FIGURE 12.11 Superposition of Figures 12.9 and 12.10 showing regions of positive and negative dc_A/dt and dT/dt in the c_A-T plane.

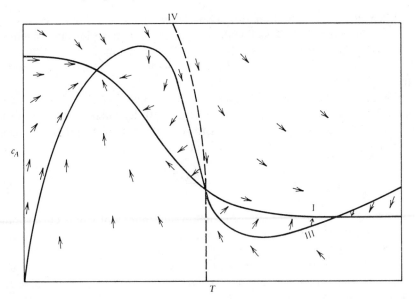

FIGURE 12.12   Directions of change of c_A and T computed from the algebraic signs of dc_A/dt and dT/dt.

corresponds to the steady state. The two lines divide the $c_A - T$ plane into six regions, and in each of these the algebraic sign of dc_A/dt and dT/dt is shown. In Figure 12.12 the inequalities are replaced by arrows. For example, an arrow pointing to the left and up means "decreasing T, increasing c_A," or $dT/dt < 0$, $dc_A/dt > 0$, and so on. By following the arrows we can determine how c_A and T will change as time goes on. Notice that the arrows move away from the middle steady state, indicating its unattainability. By following the arrows we find that when to the right of the dashed line IV, known as a *separatrix*, we will always go to the high temperature steady state; to the left we will always go to the low temperature state. Thus we know what conditions must prevail when we start the reactor in order to reach the desired operating point.

The method of solution shown in Figure 12.12 can be made quantitative by noting that the slopes of the arrows represent the rate of change of c_A with T at any point in the $c_A - T$ plane, sometimes known as a *phase plane*. From the chain rule, Section 15.4,

$$\frac{dc_A}{dT} = \frac{dc_A/dt}{dT/dt} = \frac{c_{Af} - c_A - k_0\theta e^{-E/RT}c_A}{T_f - T + Jk_0\theta e^{-E/RT}c_A} \tag{12.61}$$

Thus, for any value of c_A and T we can compute the slope. This is known as the *method of isoclines*. The method is a particularly simple way of determining the temperature-composition behavior, but it gives no information about the time involved for the process to occur. Time dependence can only be obtained by integration of the differential Equations 12.47 and 12.48. Figures 12.13 and 12.14 show the temperature and concentration as functions of time for the reactor parameters used in Example 16.1 with an initial reactor concentration $c_0 = 0.3$ lb-moles/ft^3 and initial temperatures $T_0 = 390°K$ and $T_0 = 400°K$, respectively. The equations were integrated numerically using the technique described in Section 17.12. Notice that in one case the low temperature steady state is obtained while the high temperature state occurs for the other. The corresponding curves in the concentration-temperature plane are shown in Figure 12.15.

12.12   THE DESIGN OF CFSTR SYSTEMS

It was shown in Chapter 7 that the optimal design of a continuous flow stirred tank reactor could not be achieved without considering the effect of the reactor on the economics of the total process. In that discussion we implicitly assumed that the reactor temperature was known and nothing was said about the selection of the best operating temperature. Although a complete discussion of that problem is well outside the scope of this introductory text we can, using the developments of the previous sections, present some of the basic logic needed to determine the optimal operating temperature for a stirred tank reactor system.

Equations 12.50 to 12.52 are the basic design equations for the simple reaction $A \rightarrow D$. Batch data provides us with values of k_0, E, ΔH_R and $\underline{c}_p$ and we will assume that the inlet conditions c_{Af}, T_f and T_{cf} are specified and not subject to our control. For the reaction being considered, therefore, we can control the operating temperature only by the conversion of A and the heat exchanger design. The simplest design results if a heat exchanger is not needed, and this case should always be considered first in any analysis. There is a disadvantage to adiabatic operation only if it leads to excessively high temperatures, requiring expensive high pressure equipment; if it leads to very low temperatures, requiring an excessively large reactor; or if the temperature changes are such that undesirable side reactions not previously accounted for will occur.

For any single reaction, then, our strategy is clear. We design for adiabatic operation and then consider the economic consequences of either heating or cooling the reactor. A cooling system can usually be justified when the adiabatic design leads to temperatures that are unattainable with reasonable materials of construction, when the adiabatic design leads to excessively high working pressures, or when the adiabatic design causes a loss of raw material by production of unwanted by-products. A heating system for the reactor can only be justified if its capital and operating costs are less than the capital and operating costs for the larger reactor which will result if an adiabatic design

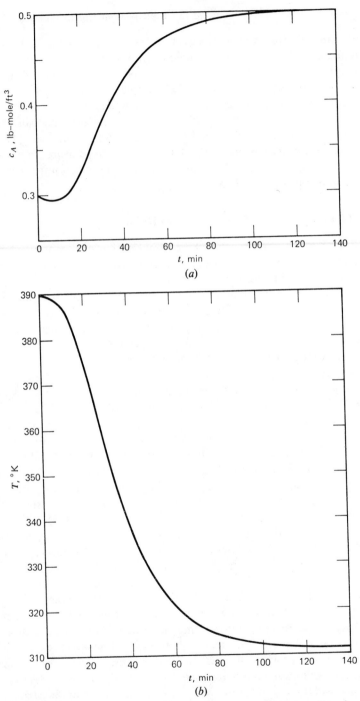

FIGURE 12.13 Concentration (*a*) and temperature (*b*) versus time for a CFSTR with initial conditions $c_{A0} = 0.3$ lb-mole/ft, $T_0 = 390°$K. Note the initial decrease in concentration.

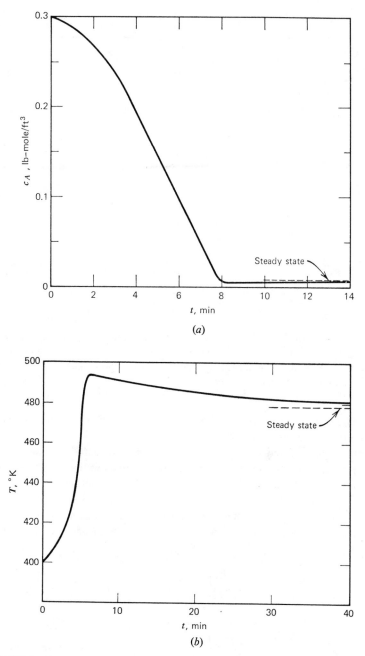

FIGURE 12.14 Concentration (a) and temperature (b) versus time for a CFSTR with initial conditions $c_{A0} = 0.3$, $T_0 = 400°K$. Note how both concentration and temperature overshoot the ultimate steady state. Compare the time scales on these figures with those in Figure 12.13.

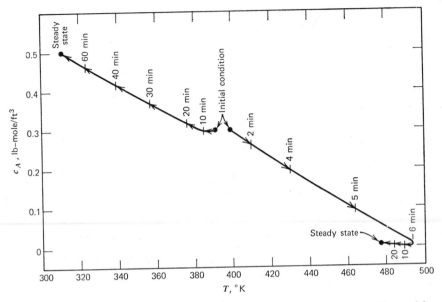

FIGURE 12.15   Concentration-temperature phase plane plot for CFSTR with $c_{A0} = 0.3$, $T_0 = 390°K$ and $400°K$. Times required to reach various points along the path are shown in the figure. Note the more rapid approach to the high temperature-high conversion steady state.

is used. The same comments would apply to a preheater for the feed streams if feed temperatures are allowed to vary.

For more complicated reactions such as those considered in Section 7.6 the effect of temperature on the product distribution as well as on the reactor size must be considered. A feel for the structure of this general problem can be obtained by examining the specific case represented by Equations 7.45 to 7.47. In these steady state equations for the product distribution of R, S, and T, the temperature dependence is contained in the ratios of the rate constants

$$\frac{k_2}{k_1} = \frac{k_{20}}{k_{10}} \exp\left[\frac{E_1 - E_2}{RT}\right]$$

$$\frac{k_3}{k_1} = \frac{k_{30}}{k_{10}} \exp\left[\frac{E_1 - E_3}{RT}\right]$$

If E_1, E_2, and E_3 are not equal the product distribution at any value of x_A will be affected by the operating temperature, T. Then an exchanger that either heats or cools the reactor may be economically justified on the basis of allowing a more favorable product mix of R, S, and T to be made. If $E_1 = E_2 = E_3$ (which is the case for the ethylene glycol reaction), the operating temperature has no effect and the optimal T may be found as discussed for the single reaction.

12.13 TUBULAR REACTOR

The energy equation for the tubular reactor is developed following the approach in Sections 7.5 and 11.10. The mass balance at steady state was found to be

$$\frac{4q}{\pi D^2} \frac{dc_A}{dz} = -r \tag{12.62}$$

The energy equation similarly becomes

$$\rho q c_p \frac{dT}{dz} = \frac{\pi D^2}{4} [-\Delta H_R] r + \pi D h [T_c - T] \tag{12.63}$$

The manipulations in Section 12.7 for the batch reactor can be carried over directly to these design equations for the tubular reactor. Design of the cooling jacket will determine the manner in which T_c varies with z, and an additional equation for T_c will generally be required. The analysis of the tubular reactor is more complicated than that of the continuous flow stirred tank because the temperature in the tubular device varies with distance along the reactor unless a cooling system is employed to maintain isothermal conditions.

Because of the mathematical complications the design of nonisothermal tubular systems is beyond the scope of this text, but it is possible to make some general remarks paralleling our discussion of stirred tank systems:

1. An adiabatic design ($Q = 0$) should always be considered first, since this will result in the simplest design.
2. The economic feasibility of either heating or cooling the reactor must be determined by balancing heat exchanger costs against reactor costs without an exchanger.
3. For complicated reactions, the effect of the temperature profile on the product distribution must be considered along with the reactor size problem. This analysis can get quite complicated for tubular systems and its structure must often be investigated using a computer. An approximation of isothermal behavior often allows some useful design decisions to be made, however, and this approach should be used before excessive programming is carried out.

12.14   CONCLUDING REMARKS

The heat of reaction is the most important new idea introduced in this chapter. Be certain that you understand the way in which it is defined (Equation 12.2) and how it can be determined for a particular reaction by experiment or by use of tabulated heats of formation. Compare the development of the reactor equations with those for mixing in the previous chapter. If you completely understand the way in which the partial molar enthalpy is used to arrive at these final working equations, then you will never have any difficulty with more complex situations.

Notice that the derivation of the mass balance is unchanged from Chapters 5 and 7, where we had assumed isothermal conditions. This mass balance must always be part of the more complete description. Because of the way in which temperature enters the rate expression, the manipulation of the model equations is considerably more difficult, and elementary integration methods will rarely be adequate to determine model behavior. Consider Section 12.9 carefully. On first reading it presents a real challenge to one's physical intuition and dramatically demonstrates the importance of examining model behavior.

Heat of formation and combustion data are tabulated in the basic reference tables, 4.6 to 4.8. The engineering texts on reactor design cited in Chapter 7 are fundamental references for this chapter as well.

12.15   PROBLEMS

12.1   Develop the complete mathematical description in terms of first-order differential equations in T and c_i for the following liquid phase reactions taking place in a nonisothermal batch reactor.

(a) $A \rightarrow R$

$$r_{A-} = r_{R+} = kc_A$$
$$k = k_0 e^{-E/RT}$$

(b) $A \rightleftarrows R$

$$r_{A-} = r_{R+} = k_1 c_A$$
$$r_{A+} = r_{R-} = k_2 c_R$$
$$k_1 = k_{10} e^{-E_1/RT}$$
$$k_2 = k_{20} e^{-E_2/RT}$$

(c) $A \rightarrow B \rightarrow C$

$$r_{A-} = r_{B+} = k_1 c_A$$
$$r_{B-} = r_{C+} = k_2 c_B$$
$$k_1 = k_{10} e^{-E_1/RT}$$
$$k_2 = k_{20} e^{-E_2/RT}$$

(d) $A + B \rightarrow R$

$$A + B \rightarrow S$$
$$r_{R+} = k_1 c_A c_B$$
$$r_{S+} = k_2 c_A c_B$$
$$k_1 = k_{10} e^{-E_1/RT}$$
$$k_2 = k_{20} e^{-E_2/RT}$$

Assume that there is only reactant present at time $t = 0$. Designate initial temperatures T_0 and clearly show how ΔH_R is obtained in the model equation containing dT/dt.

12.2 Using Equation 12.20 and a source of heat capacity data, determine the coefficient of $[T - T^\circ]$ for the following reaction
$$CH_2OCH_2 + H_2O \rightarrow CH_2OH - CH_2OH \quad c_{pCH_2OCH_2} = 0.44\,BTU/lb_m\,^\circ F.$$

12.3 (a) Using a heat of formation table find ΔH_R for each of the reactions below.

(b) Using a heat of combustion table find ΔH_R and compare with the result obtained in part (a) if applicable.

(i) $C_6H_6 + HNO_3 \longrightarrow C_6H_5NO_2 + H_2O$

(ii) $C_2H_5OH + CH_3COOH \longrightarrow CH_3C\overset{O}{\underset{OC_2H_5}{\diagup}} + H_2O$

(iii) $CH_3-C\overset{O}{\underset{Cl}{\diagup}} + H_2O \longrightarrow CH_3-C\overset{O}{\underset{OH}{\diagup}} + HCl$

(iv) $\begin{matrix} CH_3-C\overset{O}{\diagdown} \\ \quad\quad O + H_2O \longrightarrow 2CH_3-C\overset{O}{\underset{OH}{\diagup}} \\ CH_3-C\diagup \\ \quad\quad O \end{matrix}$

(v) $C-C\overset{O}{\underset{OH}{\diagup}} + NH_4OH(aq) \longrightarrow CH_3-C\overset{O}{\underset{NH_2}{\diagup}} + 2H_2O$

(vi) $C_6H_6 + CH_3OH \longrightarrow C_6H_5CH_3 + H_2O$

$C_6H_5CH_3 + CH_3OH \longrightarrow C_6H_4(CH_3)_2 + H_2O$

$C_6H_4(CH_3)_2 + CH_3OH \longrightarrow C_6H_3(CH_3)_3 + H_2O$

(vii) $CH_3CN + C_2H_5OH + H_2O \longrightarrow$

$$CH_3C \overset{\displaystyle O}{\underset{\displaystyle OC_2H_5}{\big\backslash}} + NH_4OH \quad (aq)$$

(viii) $H-\overset{\displaystyle H}{\underset{}{C}}\!\!-\!\!\overset{\displaystyle H}{\underset{}{C}}\!\!-\!\!H + H_2O \longrightarrow H-\overset{\displaystyle H}{\underset{\displaystyle OH}{C}}\!\!-\!\!\overset{\displaystyle H}{\underset{\displaystyle OH}{C}}\!\!-\!\!H$

(ix) $CH_2{=}C{=}O + CH_3C\overset{\displaystyle O}{\underset{\displaystyle OH}{\big\backslash}} \longrightarrow \quad \overset{\displaystyle CH_3-C}{\underset{\displaystyle CH_3-C}{}} \xrightarrow{C_2H_5OH}$

$$CH_3C\overset{\displaystyle O}{\underset{\displaystyle OH}{\big\backslash}} + CH_3C\overset{\displaystyle O}{\underset{\displaystyle OC_2H_5}{\big\backslash}}$$

12.4 For the reaction

urea + formaldehyde → monomethylolurea

Smith [*J. Phys. & Colloid Chem.*, **51**, 369 (1947)] has reported the specific reaction rate constant as a function of temperature:

T, °C	30	40	50	60
k, liters/ mole sec	5.5×10^{-5}	11.8×10^{-5}	24.5×10^{-5}	50.1×10^{-5}

Check the validity of the Arrhenius relation and determine the activation energy, E.

12.5 For the formation of glucose from cellulose discussed in Problem 5.6 rate constant k_1 shows the following temperature dependence at an HCl concentration of 0.055 moles/liter:

T, °C	160	170	180	190
k_1, min^{-1}	0.00203	0.00568	0.0190	0.0627

What is the activation energy, E?

12.6 Approximately what value must the energy of activation, E, have if the rule of thumb that "the rate of a reaction doubles for each 10°C rise in temperature" holds?

12.7 For reaction (a) in Problem 12.1 carried out adiabatically ($Q = 0$), develop an expression for the time to reach a given value of c_A in terms of an integral involving only c_A and constants.

12.8 If an isothermal experiment for the following reaction is to be carried out show how Q must vary as a function of time.

$$A + B \rightarrow nD$$

12.9 If the isothermal experiment referred to in Problem 12.8 is carried out in a jacketed reactor, how should the coolant conditions be varied with time?

12.10 (a) Repeat Example 12.4 for isothermal operation at 200°F and plot Q as a function of time using the parameters given. ($V = 10$ ft³)
(b) Compare the concentration of A obtained after 20 min of operation at 200°F with that obtained for the same time in the adiabatic case.

12.11 Compute the number of BTUs that would have to be added to the system described in Example 12.4 if the concentration of A is to be reduced to 0.1 lb-moles/ft³ in 10 min with *isothermal* operation. At what temperature must the reactor operate?

12.12 Modify the development leading to Equations 12.29 and 12.30 for nonadiabatic operation in a reactor with
(a) a jacket
(b) a coil
Present the complete set of differential equations needed and comment on their solution.

12.13 The following data are available for the reaction $A \rightarrow R$.

$$r_{A-} = kc_A$$

$k = k_0 e^{-E/RT}$, $k_0 = 10^9$ min⁻¹, $E = 18,500$ cals/g-mole, $\Delta H_R = -38,000$ BTU/lb-mole of A reacted (may be assumed constant with respect to T). Assume all physical properties are those of water and essentially constant, and take the molecular weight of A as 18. The volume of an adiabatic CFSTR in which this reaction is taking place is 30 ft³ and the steady state flow rate to and from the reactor is 100 ft³/hr. Find c_A and T, the exit concentration and temperature, if c_{Af} is 0.50 lb-moles/ft³, $T_f = 100$°F.

12.14 If c_A must be 0.1 lb-mole/ft³, find T and the volume of the reactor. All other conditions are the same as in Problem 12.13.

12.15 Compute the volume of a reactor needed for the situation described in Problem 7.17 if the reaction is carried out at 150°C. The energy of activation, E, is 18,500 calories/g-mole.

12.16–12.20 This process description applies to the following five problems.

The catalytic gas phase reaction between naphthalene and oxygen to produce phthalic anhydride is thought to take place as follows:

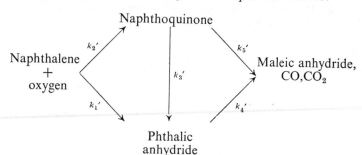

This reaction has been studied by DeMaria, Longfield, and Butler, *Ind. Eng. Chem.*, **53**, 259 (1961) in a fluidized bed reactor with two different catalysts designated "*A*" and "*B*." A surprisingly good way to model the fluidized bed reactor is to assume that it behaves like a continuous flow stirred tank reactor in which the same procedure and nomenclature used for liquid systems can be applied.

Laboratory analysis has also shown that it is feasible to assume that the following kinetic scheme will describe the process:

$$A \rightarrow R \rightarrow S$$

$$r_{A-} = k_1 c_A$$

$$r_{R-} = k_2 c_R$$

In this kinetic scheme A is the naphthalene, R is the phthalic anhydride, and S is combustion products of no commercial value. This description only applies when oxygen is in excess, but in commercial production such is always the case. The following data apply:

$$k_1 = k_{10}e^{-E_1/RT} \qquad k_2 = k_{20}e^{-E_2/RT}$$

Catalyst A	Catalyst B
$E_1 = 42,400$ cal/g-mole	20,600 cal/g-mole
$E_2 = 20,200$	46,200
$k_1 = 2.6 \times 10^{-4}$ sec^{-1} at 224°C	13 sec^{-1} at 497°C
$k_2 = k_1$ at 224°C	k_1 at 497°C

12.16 Write the steady state mass and energy balance equations for operation in a continuous flow stirred tank and tubular plug flow reactor.

12.17 Compute reactor selectivity, s, and yield, y, in a continuous flow stirred tank reactor of residence time $V/q = 1$ sec for reactor temperatures $T = 224°C$, $300°C$, $400°C$, and $450°C$ for each catalyst, and plot s and y versus T.

$$s = c_R/[c_{Af} - c_A] \qquad y = c_R/c_{Af}$$

(Compare Problem 7.11.)

12.18 Select the reactor operating temperature as that point where you estimate there is a maximum in the yield curve. Pick two holding times different from 1 sec and attempt to improve the yield. Which catalyst is more effective?

12.19 How does the selectivity affect the reactor design? Are there any differences between catalyst A and B?

12.20 How would you select the optimal holding time for the phthalic anhydride reaction? Does a solution to Problem 7.12 aid your decision making?

12.21 The reaction rate constants for the reaction described in Problem 7.24 have been measured and reported by Laidler (*Chemical Kinetics*, 2nd ed., McGraw-Hill, 1965) as

$$k_1 = 3 \times 10^6 \exp\left[\frac{-16{,}700}{RT}\right] \text{ liters/g-mole sec}$$

$$k_2 = 6 \times 10^{12} \exp\left[\frac{-34{,}000}{RT}\right] \text{ sec}^{-1}$$

$$k_3 = 10^{10} \exp\left[\frac{-25{,}000}{RT}\right] \text{ liters/g-mole sec}$$

The energies of activation; 16,700, 34,000, and 25,000 are all kcals/g-mole.

Compute c_R, c_B, and c_D if this reaction takes place in a CFSTR operating at 158°F. Ninety percent of the cyclopentadiene is to be converted and it enters the reactor with a temperature of 100°F. The composition of the inlet stream is 20 weight percent cyclopentadiene (A), 70 weight percent isopentane, which can be considered to be an inert material, and 10 weight percent $t - 1, 3$ pentadiene.

CHAPTER 13

Transient Process Behavior

13.1 INTRODUCTION

We have regularly considered the time dependence of the process models developed throughout the text. An understanding of time dependence is crucial in the design and analysis of experiments and in appreciating how a process will operate relative to its steady state design specifications. We have generally emphasized the former, as in Chapters 5 and 8. Here we focus, in the context of the stirred tank reactor, on the actual operation of a process that has been designed to provide specified behavior. We specifically consider the effect of feed variations on reactor conversion and the design of a simple control system to maintain a constant output in the face of fluctuating feed conditions. In so doing we stress the prediction and design aspects of the analysis process and we develop our skills in determining model behavior for second-order systems.

13.2 REACTOR EQUATIONS

The equations describing the pseudo first-order reaction $A \rightarrow D$ in a jacketed stirred tank reactor were derived in Section 12.9 as

$$\theta \frac{dc_A}{dt} = c_{Af} - c_A - k_0 \theta e^{-E/RT} c_A \tag{13.1}$$

$$\theta \frac{dT}{dt} = T_f - T + J k_0 \theta e^{-E/RT} c_A - u[1 + K][T - T_c] \tag{13.2}$$

$$\theta_c \frac{dT_c}{dt} = T_{cf} - T_c + K[T - T_c] \tag{13.3}$$

Here we have neglected shaft work, defined $\theta = V/q$, $J = -\Delta H_R/\rho \underline{c}_p$, $K = ha/\rho_c q_c \underline{c}_{pc}$, $u = \rho_c q_c \underline{c}_{pc} K/[1 + K]\rho q \underline{c}_p$, and the subscript c refers to the jacket. It is convenient to make one simplification to minimize the manipulations without seriously affecting the basic structure of the problem. We assume that the residence time θ_c of the jacket is small compared to that of the reactor (small volume or large flow rate), and we formally let $\theta_c \to 0$. (A rigorous justification is outlined in Problem 13.1.) In that case, Equation 13.3 simplifies to an algebraic equation,

$$0 = T_{cf} - T_c + K[T - T_c] \tag{13.4}$$

which can be solved for T_c and substituted into Equation 13.2. The resulting equation for reactor temperature is then

$$\theta \frac{dT}{dt} = T_f - T + Jk_0\theta e^{-E/RT}c_A - u[T - T_{cf}] \tag{13.5}$$

Equations 13.1 and 13.5, together with conditions at time $t = 0$, determine the transient behavior. We have already discussed the steady state solution of these equations.

The distinction between linear and nonlinear equations is discussed in Sections 17.1 and 17.2. Equations 13.1 and 13.5 involve the exponential $e^{-E/RT}$ as well as the product $c_A e^{-E/RT}$, and are therefore nonlinear. Hence it is unlikely that solutions in terms of simple functions can be obtained. Numerical integration of the type used in Section 12.11 can be employed to obtain specific answers for a given set of parameters, but broad conclusions with general validity usually require analytical expressions. The clue to the necessary approximations which will lead to a meaningful set of equations having simple analytical solutions is contained in the discussion at the end of Section 2.4. We observed there that small excursions from a designated value would be closely approximated by a linear function. We can apply that notion here to great advantage.

The physical concession that we have to make in order to obtain a tractable mathematical problem is a willingness to consider only temperatures and concentrations that are close to the design steady state. We will subsequently see the precise mathematical statement of this assumption and why it is necessary. At steady state, Equations 13.1 and 13.5 become

$$0 = c_{Afs} - c_{As} - k_0\theta e^{-E/RT_s}c_{As} \tag{13.6}$$

$$0 = T_{fs} - T_s + Jk_0\theta e^{-E/RT_s}c_{As} - u_s[T_s - T_{cfs}] \tag{13.7}$$

It is helpful to our physical understanding, as well as being mathematically convenient, to consider the steady state as the reference point and to measure quantities in terms of their difference from the steady state value. Thus, let $x(t)$ be the difference between the actual concentration of A at any time and the steady state value

$$x(t) = c_A(t) - c_{As} \tag{13.8a}$$

Similarly, let $y(t)$ be the temperature difference

$$y(t) = T(t) - T_s \tag{13.8b}$$

We wish to consider the case where the feed concentration and temperature may vary, so we define the fluctuations as $x_f(t)$ and $y_f(t)$, respectively,

$$x_f(t) = c_{Af}(t) - c_{Afs} \tag{13.8c}$$

$$y_f(t) = T_f(t) - T_{fs} \tag{13.8d}$$

To keep the discussion simple we will assume that flow rates q and q_c remain constant, as does the coolant feed temperature, T_{cf}. We will relax these last assumptions in a subsequent section.

Equations 13.1 and 13.5 can be rewritten in terms of the fluctuation variables x and y as

$$\theta \frac{d[c_{As} + x]}{dt} = \theta \frac{dx}{dt}$$

$$= c_{Afs} + x_f - c_{As} - x - k_0 \theta e^{-E/R[T_s+y]}[c_{As} + x] \tag{13.9}$$

$$\theta \frac{d[T_s + y]}{dt} = \theta \frac{dy}{dt}$$

$$= T_{fs} + y_f - T_s - y + Jk_0\theta e^{-E/R[T_s+y]}[c_{As} + x]$$
$$- u[T_s + y - T_{cf}] \tag{13.10}$$

We have made use of the fact that c_{As} and T_s are constant, so $dc_{As}/dt = dT_s/dt = 0$. These equations involve a number of steady state terms, and they can be rearranged and simplified somewhat by substitution of Equations 13.6 and 13.7 for the steady state. This gives

$$\theta \frac{dx}{dt} = x_f - x - k_0\theta e^{-E/R[T_s+y]}x$$

$$- k_0\theta c_{As}\{e^{-E/R[T_s+y]} - e^{-E/RT_s}\} \tag{13.11}$$

$$\theta \frac{dy}{dt} = y_f - y + Jk_0\theta e^{-E/R[T_s+y]}x$$

$$+ Jk_0\theta c_{As}\{e^{-E/R[T_s+y]} - e^{-E/RT_s}\} - uy \tag{13.12}$$

At this point we make use of the idea that we are close to the steady state and that x and y are therefore small in magnitude. The exponential can be expressed in terms of a Taylor series (Equation 15.20) about $y = 0$ as

$$e^{-E/R[T_s+y]} = e^{-E/RT_s} + \frac{E}{RT_s^2} e^{-E/RT_s}y$$

$$+ \frac{1}{2}\frac{E}{RT_s^4}\left[\frac{E}{R} - 2T_s\right]e^{-E/RT_s}y^2 + \cdots$$

Equation 13.11 is then

$$\theta\frac{dx}{dt} = x_f - x - k_0\theta e^{-E/RT_s}x - \frac{k_0\theta E}{RT_s^2}e^{-E/RT_s}xy$$

$$- \frac{k_0\theta E}{RT_s^4}\left[\frac{E}{2R} - T_s\right]e^{-E/RT_s}xy^2 + \cdots$$

$$- \frac{k_0\theta c_{As}Ee^{-E/RT_s}}{RT_s^2}y$$

$$- \frac{k_0\theta c_{As}E}{RT_s^4}\left[\frac{E}{2R} - T_s\right]e^{-E/RT_s}y^2 + \cdots \qquad (13.13)$$

Equation 13.13 is, of course, completely equivalent to the original mass balance equation, 13.1. It is in a form, however, more suitable for approximation. *We assume that x and y are sufficiently small in magnitude that y^2, xy, etc., are negligible in comparison to x and y.* We can obtain very useful results by this approximation, but note what a severe physical restriction it is! Now, however, Equation 13.13 becomes

$$\theta\frac{dx}{dt} = x_f - x - k_0\theta e^{-E/RT_s}x - \frac{k_0\theta c_{As}Ee^{-E/RT_s}}{RT_s^2}y \qquad (13.14a)$$

Similarly, the temperature equation, 13.12, becomes

$$\theta\frac{dy}{dt} = y_f - y + Jk_0\theta e^{-E/RT_s}x + \frac{Jk_0\theta c_{As}Ee^{-E/RT_s}}{RT_s^2}y - uy$$

$$(13.15a)$$

or, regrouping,

$$\theta \frac{dx}{dt} = -[1 + k_0\theta e^{-E/RT_s}]x - \left[\frac{k_0\theta c_{As}Ee^{-E/RT_s}}{RT_s^2}\right]y + x_f$$

(13.14b)

$$\theta \frac{dy}{dt} = +[Jk_0\theta e^{-E/RT_s}]x - \left[1 + u - \frac{Jk_0\theta c_{As}Ee^{-E/RT_s}}{RT_s^2}\right]y + y_f$$

(13.15b)

These equations are now linear, and the coefficients of x and y are known constants whose values depend on the steady state near which the calculations are to be carried out. By restricting ourselves to concentrations and temperatures near the steady state we have obtained equations that can be treated analytically by the methods outlined in Chapter 17.

For application it is convenient to convert the two linear first-order differential equations to a single second-order equation for either concentration or temperature. Equations 13.14 and 13.15 are of the form of Equations 17.13. It is shown in Section 17.10 that we may replace these equations with either of the following equivalent second-order equations:

$$\frac{d^2x}{dt^2} + a_1 \frac{dx}{dt} + a_0x = f_1(t)$$

(13.16)

$$\frac{d^2y}{dt^2} + a_1 \frac{dy}{dt} + a_0y = f_2(t)$$

(13.17)

$$a_1 = \frac{1}{\theta}\left\{2 + u + k_0\theta e^{-E/RT_s}\left[1 - \frac{Jc_{As}E}{RT_s^2}\right]\right\}$$

(13.18a)

$$a_0 = \frac{1}{\theta^2}\left\{1 + u + k_0\theta e^{-E/RT_s}\left[1 + u - \frac{Jc_{As}E}{RT_s^2}\right]\right\}$$

(13.18b)

$$f_1(t) = \frac{1}{\theta^2}\left[1 + u - \frac{Jk_0\theta c_{As}Ee^{-E/RT_s}}{RT_s^2}\right]x_f$$

$$+ \frac{1}{\theta}\frac{dx_f}{dt} - \frac{k_0c_{As}Ee^{-E/RT_s}}{\theta RT_s^2}y_f$$

(13.19a)

$$f_2(t) = \frac{1}{\theta^2}[1 + k_0\theta e^{-E/RT_s}]y_f + \frac{1}{\theta}\frac{dy_f}{dt} + \frac{1}{\theta}Jk_0e^{-E/RT_s}x_f$$

(13.19b)

For the reactor parameters used in Examples 16.1 and 16.4 the values of a_1, a_0, $f_1(t)$ and $f_2(t)$ at each of the three steady states are as follows:

$$c_{As} = 0.499, \qquad T_s = 311.4: \quad a_1 = 6.57 \qquad a_0 = 10.80$$

$$f_1(t) = 10.78x_f + 3.33 \frac{dx_f}{dt} - 9.92 \times 10^{-4} y_f$$

$$f_2(t) = 11.13y_f + 3.33 \frac{dy_f}{dt} + 7.01x_f$$

$$c_{As} = 0.257, \qquad T_s = 393.4: \quad a_1 = -6.71 \qquad a_0 = -33.46$$

$$f_1(t) = -44.00x_f + 3.33 \frac{dx_f}{dt} - 0.163y_f$$

$$f_2(t) = 21.65y_f + 3.33 \frac{dy_f}{dt} + 3.56 \times 10^3 x_f$$

$$c_{As} = 0.008, \qquad T_s = 477.5: \quad a_1 = 188.3 \qquad a_0 = 616.6$$

$$f_1(t) = -64.11x_f + 3.33 \frac{dx_f}{dt} - 0.222y_f$$

$$f_2(t) = 691.8y_f + 3.33 \frac{dy_f}{dt} + 2.30 \times 10^5 x_f$$

13.3 ATTAINABILITY OF A STEADY STATE

In Section 12.10 we began to explore the question of whether a computed steady state could in fact be attained for operation. We developed a graphical criterion for unattainabilty, and we are now in a position to deal with the problem fully. We presume that disturbances in the feed have moved the concentration and temperature an amount x_0 and y_0, respectively, from steady state design conditions. For $t > 0$ we may presume for simplicity that no further disturbances enter ($x_f = y_f = 0$). Equation 13.16 describing the reactor response then becomes

$$\frac{d^2x}{dt^2} + a_1 \frac{dx}{dt} + a_0 x = 0 \qquad (13.20)$$

If $x(t)$ goes to zero with time, the steady state design is stable and can be maintained. If $x(t)$ does not return to zero with time, the steady state cannot be attained.

Equation 13.20 is a linear, second-order equation with constant coefficients. The solution procedure is outlined in Section 17.3. We assume a solution of the form

$$x = Ce^{mt}$$

Substitution into Equation 13.20 leads to

$$Ce^{mt}[m^2 + a_1m + a_0] = 0$$

and, since the exponential cannot vanish, we obtain the characteristic equation

$$m^2 + a_1m + a_0 = 0 \tag{13.21}$$

This has roots m_1 and m_2 given by

$$m_1 = -\frac{a_1}{2} + \left[\frac{a_1^2}{4} - a_0\right]^{1/2} \tag{13.22a}$$

$$m_2 = -\frac{a_1}{2} - \left[\frac{a_1^2}{4} - a_0\right]^{1/2} \tag{13.22b}$$

Notice that

$$m_1 + m_2 = -a_1 \tag{13.23}$$

$$m_1m_2 = +a_0 \tag{13.24}$$

The general solution is

$$x(t) = C_1e^{m_1t} + C_2e^{m_2t} \tag{13.25}$$

The behavior of the concentration and temperature following time zero is governed entirely by the character of m_1 and m_2. It is convenient initially to distinguish between the cases when m_1 and m_2 are real and when they are complex, depending on the sign of the discriminant, $a_1^2 - 4a_0$.

Case I: $a_1^2 - 4a_0 > 0$, m_1 and m_2 Real

In this case there are three possibilities: two positive roots, two negative roots, and one positive and one negative root. A term e^{mt} will decay to zero if m is negative and will grow without bound if m is positive. Thus, if both m_1 and m_2 are negative, $x(t)$ will go to zero, as shown in Figure 13.1. If *either* m_1 or m_2 is positive, the solution will grow to $\pm\infty$, as indicated in the figure. The steady state is stable and can be maintained despite small disturbances *if and only if* m_1 and m_2 are both negative. Clearly, when one or both roots is positive, the conversion does not really go to infinity. After a short period of time the assumption that x^2 is small compared to x will no longer be valid and Equation 13.20 will no longer describe the physical situation.

Equations 13.23 and 13.24 provide a simple criterion for determining when the roots are both negative. The sum of two negative numbers is

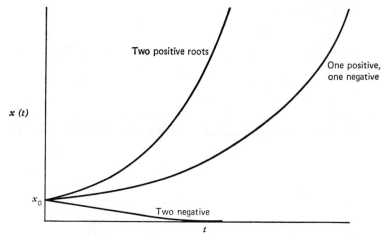

FIGURE 13.1 Transient response of a linear homogeneous second-order system with real characteristic roots, $a_1^2 - 4a_0 > 0$. $x(t)$ goes to zero if and only if both roots are negative.

negative, so, from Equation 13.23, $-a_1$ must be negative, or $a_1 > 0$. The product of two negative numbers is positive, so from Equation 13.24 we have $a_0 > 0$. From Equations 13.18 these can be expressed in terms of the reactor parameters as

$$\theta a_1 = 2 + u + k_0 \theta e^{-E/RT_s} \left[1 - \frac{Jc_{As}E}{RT_s^2} \right] > 0 \qquad (13.26a)$$

$$\theta^2 a_0 = 1 + u + k_0 \theta e^{-E/RT_s} \left[1 + u - \frac{Jc_{As}E}{RT_s^2} \right] > 0 \qquad (13.26b)$$

Notice that for the adiabatic reactor, $u = 0$, the second condition implies the first, since $\theta a_1 = \theta^2 a_0 + 1$. This is not true in general.

Case II: $a_1^2 - 4a_0 < 0$, m_1 and m_2 Complex

This case can occur only if $a_0 > 0$. m_1 and m_2 are now

$$m_1 = -\frac{a_1}{2} + i \sqrt{a_0 - \frac{a_1^2}{4}} \qquad m_2 = -\frac{a_1}{2} - i \sqrt{a_0 - \frac{a_1^2}{4}}$$

where $i^2 = -1$. Then Equation 13.25 becomes

$$x(t) = C_1 e^{-a_1 t/2} e^{i \sqrt{a_0 - a_1^2/4}\, t} + C_2 e^{-a_1 t/2} e^{-i \sqrt{a_0 - a_1^2/4}\, t}$$

Following Section 17.4 we can use the equality

$$e^{i \sqrt{a_0 - a_1^2/4}\, t} = \cos \sqrt{a_0 - \frac{a_1^2}{4}}\, t + i \sin \sqrt{a_0 - \frac{a_1^2}{4}}\, t$$

and rewrite $x(t)$ in the form

$$x(t) = e^{-a_1 t/2}\left[K_1 \cos\sqrt{a_0 - \frac{a_1{}^2}{4}}\, t + K_2 \sin\sqrt{a_0 - \frac{a_1{}^2}{4}}\, t\right] \quad (13.27)$$

Here $K_1 = C_1 + C_2$, $K_2 = i[C_1 - C_2]$. Then if $a_1 > 0$, the response is damped, as shown in Figure 13.2(a). If $a_1 < 0$ the concentration fluctuation grows away from zero in an oscillatory manner, as shown in Figure 13.2(b).

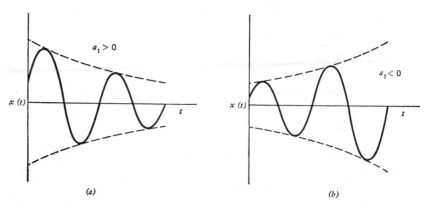

(a)   (b)

FIGURE 13.2   Transient response of a linear homogeneous second-order system with complex characteristic roots, $a_1 - 4a_0{}^2 < 0$. $x(t)$ goes to zero in an oscillatory manner if and only if both roots have negative real parts [(a) $a_1 > 0$], grows in an oscillatory manner if a positive real part [(b), $a_1 < 0$].

Thus, the design steady state is attainable and stable if and only if $a_1 > 0$. Since oscillatory behavior (complex roots) requires $a_0 > 0$, we are led to the same conditions as in Case I, $a_1 > 0$, $a_0 > 0$, which are expressed in terms of reactor parameters in Equations 13.26.

13.4 SLOPE CONDITION

In Section 12.10 we used specialized arguments to conclude that a steady state solution will be unattainable in practice whenever the slope of line I in Figure 12.7 is more negative than that of line II. This slope condition can be related to the criteria developed in the preceding section as follows:

Line I is a plot of c_A versus T defined by Equation 12.55:

$$c_{AI} = \frac{c_{Af}}{1 + k_0 \theta e^{-E/RT}} \quad (12.55)$$

We have used the subscript I to denote "from line I." The slope is formed from differentiating with respect to T

$$\frac{dc_{AI}}{dT} = - \frac{k_0\theta e^{-E/RT}c_{Af}E}{RT^2[1 + k_0\theta e^{-E/RT}]^2}$$

At steady state $c_{AI} = c_{As}$, $T = T_s$, and Equation 12.55 can be substituted to obtain the slope at the intersection of lines I and II:

$$\text{steady state:} \quad \frac{dc_{AI}}{dT} = - \frac{k_0\theta E}{RT_s^2} \frac{c_{As}e^{-E/RT_s}}{1 + k_0\theta e^{-E/RT_s}} \tag{13.28}$$

Line II is the straight line defined by Equation 12.54,

$$c_{AII} = c_{Af} + \frac{1}{J}[T_f + uT_{cf}] - \frac{1+u}{J}T \tag{12.54}$$

The slope is then

$$\frac{dc_{AII}}{dT} = - \frac{1+u}{J} \tag{13.29}$$

Comparisons of Equations 13.28 and 13.29 with Equation 13.26b shows that the condition $a_0 > 0$ is identical to the condition that $dc_{AI}/dT > dc_{AII}/dT$ at steady state. Thus, the middle steady state must have $a_0 < 0$ and can never be attained in practice. The high and low temperature steady states always have $a_0 > 0$. Unless the reactor is adiabatic ($u = 0$) it is still necessary to check that $a_1 > 0$ to ensure that the steady state is a stable one. It is possible to build a reactor with conditions $a_0 > 0$, $a_1 < 0$. A nice example of this is to be found for the reaction of acetyl chloride and water to form acetic acid and hydrochloric acid in a continuous flow reactor in a paper by Baccaro, Gaitonde, and Douglas in the *AIChE Journal*, **16**, 249 (1970).

13.5 STEP RESPONSE

The simple linear model equations for describing reactor behavior when concentrations and temperatures are near the steady state enables us to compute the effects of operating changes quite readily. Suppose, for example, that the process is operating at a stable steady state when the feed temperature changes by a constant amount y_f. The change in effluent concentration can then be calculated.

From Equations 13.16 and 13.19 the concentration change $x(t) = c_A - c_{As}$ is described by the equation

$$\frac{d^2x}{dt^2} + a_1\frac{dx}{dt} + a_0x = -\frac{k_0c_{As}Ee^{-E/RT_s}}{\theta RT_s^2}\, y_f = \text{constant} \qquad (13.30)$$

We have set $x_f = 0$ in the function $f_1(t)$. The solution of this nonhomogeneous equation is obtained using the methods outlined in Section 17.7. The general solution is a sum of a homogeneous and particular solution

$$x(t) = x_h(t) + x_p(t)$$

The homogeneous solution was discussed in Section 13.3 and is

$$x_h(t) = C_1e^{m_1t} + C_2e^{m_2t}$$

where m_1 and m_2 are given by Equation 13.22. The particular solution can be found from the method of undetermined coefficients as in Example 17.12 to be

$$x_p = \frac{k_0c_{As}Ee^{-E/RT_s}y_f}{a_0\theta RT_s^2}$$

Thus, the general solution is

$$x(t) = C_1e^{m_1t} + C_2e^{m_2t} - \frac{k_0c_{As}Ee^{-E/RT_s}y_f}{a_0\theta RT_s^2} \qquad (13.31)$$

The constants C_1 and C_2 are found from the initial conditions. At $t = 0$ the reactor is at steady state, so $x(0) = y(0) = 0$. From Equation 13.14 it follows that $dx/dt = 0$ at $t = 0$ when $x(0)$, $y(0)$, and x_f all equal zero. Thus

$$x(0) = C_1 + C_2 - \frac{k_0c_{As}Ee^{-E/RT_s}y_f}{a_0\theta RT_s^2} = 0$$

$$\frac{dx}{dt}(0) = m_1C_1 + m_2C_2 = 0$$

Upon solving for C_1 and C_2 we obtain

$$c_A(t) - c_{As} = x(t) = -\frac{k_0c_{As}Ee^{-E/RT_s}y_f}{a_0\theta RT_s^2}$$

$$\times \left[1 + \frac{m_2}{m_1 - m_2}e^{m_1t} - \frac{m_1}{m_1 - m_2}e^{m_2t}\right] \qquad (13.32)$$

The concentration change in time then appears as shown in Figure 13.3(a) for m_1 and m_2 real or Figure 13.3(b) for m_1 and m_2 complex.

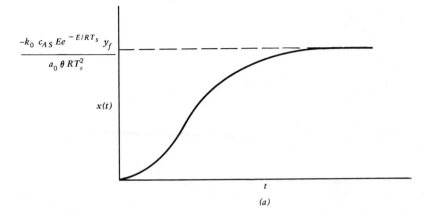

$$\frac{-k_0 \, c_{AS} \, E e^{-E/RT_s} \, y_f}{a_0 \, \theta \, R T_s^2}$$

$x(t)$

t

(a)

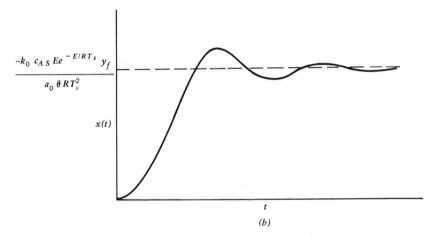

$$\frac{-k_0 \, c_{AS} \, E e^{-E/RT_s} \, y_f}{a_0 \, \theta \, R T_s^2}$$

$x(t)$

t

(b)

FIGURE 13.3 Concentration change of a CFSTR calculated from linearized equations for a constant step change in feed temperature. (a) Real characteristic roots. (b) Complex characteristic roots.

13.6 ANALOGIES

We have noted on numerous occasions that different physical phenomena can be described by the same mathematical equation as long as the proper meaning is ascribed to the coefficients. Models of mass transfer and heat transfer between phases, for example, have the same mathematical form. It is often possible to exploit such similarities in mathematical structure by using knowledge of one physical phenomenon to infer the behavior of another, and these analogies play a prominent role in engineering design.

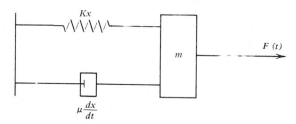

FIGURE 13.4   Mass-spring-dashpot system. Spring force $= -Kx$, viscous damping force $= -\mu\, dx/dt$.

We need not restrict ourselves just to phenomena that occur commonly in chemical engineering situations. Consider, for example, the mechanical system shown in Figure 13.4. It follows from principles discussed in a basic physics course that the displacement of the mass from equilibrium, x, is determined by the equation

$$\frac{d^2x}{dt^2} + \frac{\mu}{m}\frac{dx}{dt} + \frac{K}{m}x = \frac{1}{m}F(t) \tag{13.33}$$

K is the spring constant, μ the viscous damping coefficient, m the mass, and $F(t)$ the external force applied to the mass. Equation 13.33 is identical in form to Equation 13.16 for the reactor concentration. K/m corresponds to a_0. Notice that the spring constant must be positive to return the system toward equilibrium. The sum of kinetic and potential energy of the mass can be written

$$E = \frac{m}{2}\left[\frac{dx}{dt}\right]^2 + \tfrac{1}{2}Kx^2$$

Then

$$\frac{1}{m}\frac{dE}{dt} = \frac{dx}{dt}\frac{d^2x}{dt^2} + \frac{K}{m}\frac{dx}{dt}x = -\frac{\mu}{m}\left[\frac{dx}{dt}\right]^2$$

The energy of the mass must decrease to zero as equilibrium is approached, so stability requires $dE/dt < 0$, or $\mu/m > 0$. Thus, for the stable mechanical system we must have $a_1 > 0$. Conservation of energy requires, of course, that the dashpot heat up because of the frictional work.

For the resistance-inductance-capacitance circuit shown in Figure 13.5, the equation for voltage drop x across the capacitance is

$$\frac{d^2x}{dt^2} + \frac{1}{RC}\frac{dx}{dt} + \frac{1}{LC}x = \frac{1}{RC}\frac{d\mathscr{E}}{dt} \tag{13.34}$$

R is the resistance, L the inductance, C the capacitance, and $\mathscr{E}$ the voltage source. This, too, is identical to the reactor equation. Electrical analogies

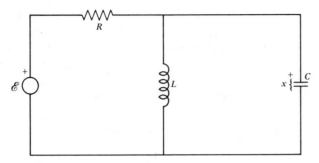

FIGURE 13.5   *R-L-C* circuit.

are particularly convenient, since by construction of an electrical circuit with an equation corresponding to some other physical situation it is possible to follow the circuit voltage with time and determine the behavior of the process of interest. This is the basic principle by which *analog computers* operate. It is necessary, of course, to match the parameters of the electrical circuit with the analogous parameters in the real process. Mechanical and fluid analog computers are also known, though much less widely employed.

13.7 CONTROL SYSTEM DESIGN

In Section 13.5 we computed the change in effluent concentration in the face of a change in feed temperature. In most processes it is desirable to maintain the output as constant as possible despite fluctuating inputs. To this end it is necessary to design control systems that change certain of the operating conditions to maintain the desired output. In Section 4.3 we examined the problem of designing a level controller for a tank. Here we consider the design of the control system for a stirred tank reactor.

The basic reactor equations are 13.1 and 13.5, repeated here:

$$\theta \frac{dc_A}{dt} = c_{Af} - c_A - k_0\theta e^{-E/RT}c_A \tag{13.1}$$

$$\theta \frac{dT}{dt} = T_f - T + Jk_0\theta e^{-E/RT}c_A - u[T - T_{cf}] \tag{13.5}$$

We suppose that the concentration or temperature of the feed stream might fluctuate, but we will continue to assume for simplicity that T_{cf} is constant. The first problem is what to measure. It is likely that we wish to keep c_A constant, but concentrations are difficult to measure rapidly with accuracy. Thus, in most cases we will be forced to measure the reactor temperature and from it infer concentration deviations.

Next we need to establish the quantity that we will manipulate. In this case we will probably vary the coolant flow rate q_c or, equivalently, u. When the concentration c_A falls too low we will increase u, raise the rate of cooling and thus drop the reactor temperature and decrease the reaction rate. This will tend to drive c_A back up. Similarly, if c_A gets too high, we will decrease u. The control system is shown schematically in Figure 13.6.

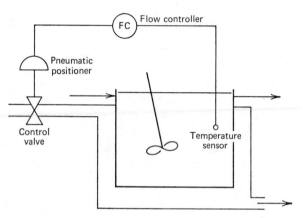

FIGURE 13.6   Feedback control system for CFSTR. Coolant flow rate is adjusted by measuring reactor temperature.

The design equations are derived by linearization in exactly the manner that we obtained Equations 13.14 and 13.15. Set $x = c_A - c_{As}$, $x_f = c_{Af} - c_{Afs}$, $y = T - T_s$, $y_f = T_f - T_{fs}$, and define a new variable, $w = u - u_s$, where u_s is computed at steady state conditions. When these variables are substituted into Equations 13.1 and 13.5 we obtain, after linearization,

$$\theta \frac{dx}{dt} = -[1 + k_0\theta e^{-E/RT_s}]x - \left[\frac{k_0\theta c_{As}E e^{-E/RT_s}}{RT_s^2}\right]y + x_f$$

$$(13.35)$$

$$\theta \frac{dy}{dt} = [Jk_0\theta e^{-E/RT_s}]x - \left[1 + u_s - \frac{Jk_0\theta c_{As}E e^{-E/RT_s}}{RT_s^2}\right]y$$

$$- [T_s - T_{cf}]w + y_f \quad (13.36)$$

Equation 13.35 is identical to Equation 13.1, while Equation 13.36 differs from 13.5 only by the extra term $-[T_s - T_{cf}]w$.

The control strategy is to increase u when we wish to decrease the temperature. Thus, w and y should have the same algebraic sign. The simplest scheme is *proportional control*, in which the change in u is taken as proportional to the change in T

$$w = Ky$$

$$(13.37)$$

In this way we expend a large control effort when temperature excursions are large, little effort for small temperature changes. Notice that the control action is not based on direct observation of the concentration fluctuation, which is the quantity in which we are really interested. Substitution of Equation 13.37 into 13.36 gives the final working equation for temperature

$$\theta \frac{dy}{dt} = [Jk_0\theta e^{-E/RT_s}]x$$
$$- \left\{1 + u_s + K[T_s - T_{cf}] - \frac{Jk_0\theta Ee^{-E/RT_s}}{RT_s^2}\right\}y + y_f$$

(13.38)

Combination of Equations 13.35 and 13.38 into a single second-order equation for concentration then gives

$$\frac{d^2x}{dt^2} + A_1\frac{dx}{dt} + A_0x = F_1(t)$$

(13.39)

$$A_1 = a_1 + \frac{1}{\theta}K[T_s - T_{cf}]$$

$$A_0 = a_0 + \frac{1}{\theta^2}K[T_s - T_{cf}][1 + k_0\theta e^{-E/RT_s}]$$

$$F_1(t) = f_1(t) + \frac{1}{\theta^2}K[T_s - T_{cf}]x_f$$

a_1, a_0, and $f_1(t)$ are defined by Equations 13.18 and 13.19.

The main effect of the control system is to increase the coefficients of the characteristic equation and therefore to make the roots more negative. Thus, *the transient response is quickened*. We can obtain a quantitative measure of the effect of the control by reconsidering the step response problem of Section 13.5. In that case, for $x_f = 0$, $y_f = $ constant, the solution analogous to Equation 13.32 is

$$x(t) = \frac{-k_0c_{As}Ee^{-E/RT_s}y_f}{\left\{a_0 + \frac{1}{\theta^2}K[T_s - T_{cf}][1 + k_0\theta e^{-E/RT_s}]\right\}RT_s^2}$$
$$\times \left[1 + \frac{M_2}{M_1 - M_2}e^{M_1t} - \frac{M_1}{M_1 - M_2}e^{M_2t}\right]$$

(13.40)

M_1 and M_2 are roots of the characteristic equation

$$M^2 + A_1M + A_0 = 0$$

(13.41)

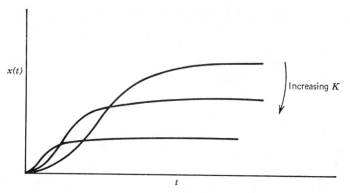

FIGURE 13.7   Concentration change of a CFSTR with real characteristic roots for a constant step change in feed temperature at various levels of proportional feedback control.

The transient response for various values of the *feedback gain K* is shown in Figure 13.7. The choice of K can be made on the basis of the maximum expected value of $|y_f|$ and the maximum allowable value of $|x(t)|$, much like in Chapter 4. Notice that for sufficiently large K, the ultimate steady state $x(\infty)$ computed from Equation 13.41 can be made as close to zero as desired.

It is also interesting to compare the response of the effluent concentration to a change in feed temperature, computed above, with the response to a step change in feed concentration. In that case $y_f = 0$, $x_f = $ constant, and

$$F_1(t) = \frac{1}{\theta^2}\left\{1 + u + K[T_s - T_{cf}] - \frac{Jk_0\theta c_{As}Ee^{-E/RT_s}}{RT_s^2}\right\}x_f$$

The homogeneous solution is the same, $C_1e^{M_1t} + C_2e^{M_2t}$, and the particular solution is found as before. At time zero we have $x(0) = y(0) = 0$. The initial condition for dx/dt is obtained from Equation 13.35

$$t = 0: \quad \frac{dx}{dt}(0) = \frac{1}{\theta}x_f$$

Using these initial conditions to evaluate the constants the solution is then found to be

$$x(t) = \frac{\{1 + u + K[T_s - T_{cf}] - [Jk_0\theta c_{As}Ee^{-E/RT_s}/RT_s^2]\}x_f}{a_0\theta^2 + K[T_s - T_{cf}][1 + k_0\theta e^{-E/RT_s}]}$$

$$\times \left\{1 + \frac{M_2}{M_1 - M_2}e^{M_1t} - \frac{M_1}{M_1 - M_2}e^{M_2t}\right\}$$

$$+ \frac{x_f}{\theta[M_1 - M_2]}[e^{M_1t} - e^{M_2t}] \qquad (13.42)$$

In this case K appears in both numerator and denominator, so $x(t)$ can never be made to go to zero for constant nonzero x_f. For large K we obtain

$$\lim_{K \to \infty} x(\infty) = \frac{x_f}{1 + k_0 \theta e^{-E/RT_s}}$$

Thus, there is no advantage in making the gain excessively large. This inability to eliminate a feed concentration fluctuation completely is a consequence of the structure that we have adopted for the control system. Furthermore, the whole strategy is built around temperatures, despite the fact that concentration is the quantity of real interest.

The linear structure of the equations and the ability to design simple electrical and mechanical devices to carry out the control is retained if we take w to have certain more general dependences on y than Equation 13.37. Most control systems used in process applications fall in the general category of *three-mode controls*

$$w = K_1 y + K_2 \frac{dy}{dt} + K_3 \int_0^t y(\tau) \, d\tau \tag{13.43}$$

The properties introduced by the *derivative* and *integral modes* are explored in the problems. It should be mentioned that the reactor example studied here is unique in one important respect. Although it might not be economically advantageous to make the proportional gain K excessively large, there is no harm in doing so. For situations other than the one considered here it is possible to make a system *unstable* by making the gain too large. Thus, there will usually be an upper bound on the control parameters. This, too, is left to the problems.

13.8 CONCLUDING REMARKS

There are no new physical principles introduced in this chapter. The chapter is a test, however, of your facility in reading mathematics. The material is closely referenced to Chapter 17 on differential equations, and you should consult that chapter or a text on differential equations if you are having any difficulty. Aside from the use of linear second-order differential equations to express model behavior the one new concept is that of linearizing the model equations for operation close to the steady state.

This chapter contains an introduction to the area generally known as process dynamics and control. Most texts in this area assume that the reader is skillful in model development and emphasize the consequences of mathematical manipulation of model equations. Basic texts written for chemical engineers are:

13.1 P. S. Buckley, *Techniques of Process Control*, Wiley, New York, 1964.

13.2 D. R. Coughanowr and L. B. Koppel, *Process Systems Analysis and Control*, McGraw-Hill, New York, 1965.

13.3   J. M. Douglas, *Process Dynamics and Control*, Prentice-Hall, Engle-wood Cliffs, 1972.

13.4   P. Harriott, *Process Control*, McGraw-Hill, New York, 1964.

13.5   D. D. Perlmutter, *Introduction to Process Control*, Wiley, New York, 1965.

There is an extensive discussion of control instrumentation in Section 22 of the *Chemical Engineers' Handbook*:

13.6   J. H. Perry, *Chemical Engineers' Handbook*, 4th ed., McGraw-Hill, New York, 1963.

Analogies like those mentioned in Section 13.6 are the basis of the treatment in

13.7   J. L. Shearer, A. T. Murphy, and H. H. Richardson, *Introduction to System Dynamics*, Addison-Wesley, Reading, Mass., 1967.

Application to physiological problems is the subject of

13.8   J. H. Milsum, *Biological Control Systems Analysis*, McGraw-Hill, New York, 1966.

An introduction to optimal control theory is contained in the first few chapters of

13.9   M. M. Denn, *Optimization by Variational Methods*, McGraw-Hill, New York, 1969.

The broad implications of feedback in qualitative and sometimes philo-sophical terms are surveyed in a collection of *Scientific American* articles published as a book:

13.10   *Automatic Control*, Simon and Schuster, New York, 1955.

13.9   PROBLEMS

13.1   In passing from Equation 13.3 to 13.4 we indicated that when θ_c is small compared to θ we can neglect the term $\theta_c[dT_c/dt]$ in the equation for T_c. This problem takes you through the steps of showing this rigorously.

(a) The "natural time" for the temperature and concentration response is θ. Let $\tau = t/\theta$, $\varepsilon = \theta_c/\theta[1 + K]$. Show that the solution to Equation 13.3 is

$$T_c(\tau) = T_{c0}e^{-\tau/\varepsilon} + \int_0^\tau \left[\frac{1}{\varepsilon} e^{-\lambda/\varepsilon}\right] \frac{KT(\tau - \lambda) + T_{cf}}{1 + K} \, d\lambda$$

T_{c0} is the value of T_c at time zero. You may wish to refer to Section 15.10.

(b) Take ε to be a small number, say $\varepsilon = 0.05$. Plot $[1/\varepsilon]e^{-\lambda/\varepsilon}$ versus λ. Notice that the function is different from zero only for λ between zero and about 5ε. Similarly, for τ greater than about 5ε, $e^{-\tau/\varepsilon} \simeq 0$.

(c) We can now write

$$T_c(\tau) \simeq \int_0^{5\varepsilon} \left[\frac{1}{\varepsilon} e^{-\lambda/\varepsilon} \right] \frac{KT(\tau - \lambda) + T_{cf}}{1 + K} \, d\lambda$$

Justify the statement that $T(\tau - \lambda) \simeq T(\tau)$ for $0 \le \lambda \le 5\varepsilon$. Thus, write

$$T_c \simeq \frac{KT + T_{cf}}{1 + K} \int_0^{5\varepsilon} \frac{1}{\varepsilon} e^{-\lambda/\varepsilon} \, d\lambda$$

(d) Carry out the integration and show that this leads to Equation 13.4.

13.2 Batch growth of microbial populations can sometimes be described approximately by the equation

$$\frac{dx}{dt} = k_1 x (1 - k_2 x)$$

where x is the total mass of the bioorganism.

(a) Show that the equation can be put into the form $dy/d\tau = y[1 - y]$ and solve for $y(\tau)$. Plot the solution for the following initial conditions:

$$y(0) = 0, 0.05, 0.50, 0.95, 1.00$$

(b) What are the steady states of the system? Are they stable or unstable? Explain physically.

(c) *Linearize* the equation for y about *each* steady state and solve. Plot each linearized solution together with the exact solution for the three intermediate initial conditions in part (b). What conclusions can you draw?

13.3 Consider an *adiabatic* continuous flow stirred tank reactor described by Equations 13.1 and 13.2 with $u = 0$.

(a) Show that when the conditions at $t = 0$ satisfy the *special relation*

$$Jc_A(0) + T(0) = Jc_{Af} + T_f$$

a single first-order differential equation for T is obtained.

(b) Manipulate the equation for T into the form

$$t = \int_{T(0)}^{T(t)} \frac{dT}{F(T)}$$

Using the reactor parameters from Example 16.1 use the trapezoidal rule (Section 15.6) to plot T versus t for the following initial conditions:

(i) $T(0) = 300°K$ $c_A(0) = 0.533$ moles/ft³
(ii) $T(0) = 350$ $c_A(0) = 0.385$
(iii) $T(0) = 390$ $c_A(0) = 0.267$
(iv) $T(0) = 395$ $c_A(0) = 0.252$
(v) $T(0) = 425$ $c_A(0) = 0.163$
(vi) $T(0) = 480.2$ $c_A(0) = 0$

(c) For cases (i), (iii), (iv), and (vi) in part (b), which are near a steady state, use the linearized Equations 13.17 and the parameters given at the end of Section 13.2 to obtain $T(t)$. How does the linear solution compare with the solution to the nonlinear equation?

(d) For cases (ii) and (v) in part (b) compare the nonlinear solution to the solution of the linearized equation. For parameters for the linearized equation use the steady state both above and below the initial temperature.

13.4 Develop numerical methods analogous to Equations 17.18 and 17.19 for the coupled differential equations

$$\frac{dx}{dt} = F(x, y) \qquad x(0) = x_0$$

$$\frac{dy}{dt} = G(x, y) \qquad y(0) = y_0$$

13.5 (a) Using either your result from Problem 13.4 or a library routine from your computing center repeat some of the calculations in part (c) of Problem 13.3 by *numerically* solving the linearized reactor equations, 13.14 and 13.15. Comparison of the numerical and analytical solutions to the linearized equations provides a check on your numerical method. A desk calculator can be used, but slide rule accuracy is not sufficient.

(b) Obtain numerical solutions of the *nonlinear* adiabatic reactors, Equations 13.1 and 13.2 with $u = 0$ for the initial conditions given by cases (ii) and (v) in Problem 13.3, part (b). Check your calculations by comparison with the numerical integration in part(b) of Problem 13.3.

(c) If you are using a computer program, choose some other initial conditions·which do not satisfy $Jc_A(0) + T(0) = Jc_{Af} + T_f$ and integrate the nonlinear equations. Plot the results in the "phase plane," as in Figure 12.15, and compare with calculations using the method of isoclines (Equation 12.61). Your computing center might have a library program for phase plane plotting.

13.6 The irreversible reaction $A \rightarrow$ products has a reaction rate $r_{A-} = k_0 e^{-E/RT} c^n$. Derive the linearized transient equations for an adiabatic continuous flow stirred tank reactor operating close to the steady state. Check your result by setting $n = 1$ and comparing with Equations 13.14 and 13.15 for $u = 0$.

13.7 Consider the control problem in Section 13.7. Show that for *proportional-plus-integral* control

$$\omega = K_1 y + K_3 \int_0^t y(\tau) \, d\tau$$

the temperature fluctuation $y(t)$ will go to a steady state value of zero for constant x_f and y_f with K_1 finite.

13.8 For the control problem in Section 13.7 suppose that you wished to use *proportional-plus-derivative* control

$$w = K_1 y + K_2 \frac{dy}{dt}$$

Deduce the effect of the constant K_2 for the derivative mode.

13.9 Consider the control problem in Section 13.7. Show that a concentration fluctuation resulting from a step change in feed concentration can never be eliminated by any feedback control system which adjusts only coolant flow rate based on temperature deviation.

13.10 For the control problem described in Section 13.7 suppose that u is maintained constant but that the flow rate q through the reactor is adjusted. Thus, take the residence time, θ, as the control variable and set

$$\tau = \frac{V}{q} - \theta_s$$

Obtain the appropriate linearized equations and determine the effect of a proportional controller in which τ is proportional to y.

13.11 A controlled process is described by the equation

$$\frac{d^3 x}{dt^3} + 6 \frac{d^2 x}{dt^2} + 11 \frac{dx}{dt} + 6x = u(t)$$

$$u(t) = -Kx$$

Show that when $K > 60$ the system becomes unstable. (Hint: the characteristic equation can be solved simply in terms of K. Use the formula for a cubic in the *Chemical Engineers' Handbook* or an algebra book. If you wish, assume $K \gg 1$ to simplify the relations.)

CHAPTER 14

Simple Gas Systems

14.1 INTRODUCTION

Until this point the text has dealt almost entirely with liquid systems. In a liquid system the density is relatively insensitive to changes in pressure and temperature. This allows us to neglect certain terms in the energy equation and to consider the effect of temperature in the mass balance only in the reaction rate term. Thus, it has been possible to go rather deeply into the development and use of models for certain reaction and separation processes without undue complication.

The density of a gas depends strongly on temperature and pressure. The ideas that we have developed for liquids can be applied *without change* to gases and to mixed gas-liquid systems, but because of the density dependence the descriptions are of necessity more complex. Detailed study of such systems is traditionally undertaken in courses in thermodynamics. In this single chapter we shall demonstrate the essential continuity of the analysis process in passing from liquid to gas and present some of the important constitutive equations used for gaseous systems. In addition to providing an introduction to gas behavior this chapter will serve to review the basic principles of analysis introduced earlier.

14.2 NONREACTING BATCH SYSTEMS

Batch experiments with gases are carried out in vessels usually designed to withstand pressures above atmospheric. The vessels generally are rigid with a fixed volume, but some key experiments have been carried out in variable volume devices fitted with a piston to allow for expansion or contraction. The laws of conservation of mass and energy were stated without regard to whether we had liquids, gases, or solids and can readily be applied to yield the basic model equations for batch gas systems:

$$\frac{d\rho V}{dt} = 0 \tag{14.1}$$

$$\frac{dU}{dt} = Q - W_s \tag{14.2}$$

Of course, these equations are the same as for a liquid. We also follow our previous procedure by expressing Equation 14.2 in terms of the enthalpy, using Equation 10.8

$$\frac{dH}{dt} - \frac{dpV}{dt} = Q - W_s$$

Equation 10.46, an application of the chain rule, is used finally to yield

$$\rho V \underline{c}_p \frac{dT}{dt} + \left[V - T\frac{\partial V}{\partial T}\bigg)_{p,n_i} \right] \frac{dp}{dt} + \tilde{H}_1 \frac{dn_1}{dt} + \cdots$$

$$+ \tilde{H}_S \frac{dn_S}{dt} - \frac{dpV}{dt} = Q - W_s \tag{14.3}$$

For nonreacting batch systems, Equation 14.3 simplifies, since $dn_1/dt = dn_2/dt \cdots = dn_S/dt = 0$:

$$\rho V \underline{c}_p \frac{dT}{dt} + \left[V - T\frac{\partial V}{\partial T}\bigg)_{p,n_i} \right] \frac{dp}{dt} - \frac{dpV}{dT} = Q - W_s \tag{14.4}$$

In Equation 14.4 we have retained two terms on the left-hand side that were too small to be of significance in batch liquid systems.

An alternate form of the energy balance can be written using Equations 14.2 and 10.51

$$\rho V \underline{c}_V \frac{dT}{dt} + \left[T\frac{\partial p}{\partial T}\bigg)_{V,n_i} - p \right] \frac{dV}{dt} = Q - W_s \tag{14.5}$$

It turns out to be convenient in dealing with gas systems to have both forms of the energy balance available for further manipulations, some of which are more easily done with one form than with the other.

Equations 14.4 and 14.5 involve characterizing variables density, volume, temperature, and pressure. Density and volume are related by knowing molecular weights and number of moles of each species, so we can alternatively characterize the system by temperature, pressure, number of moles, and either volume or density. We know from experience that these variables are not all independent. If we specify the number of moles, volume, and temperature of a gas, for example, the pressure is determined. Indeed, in

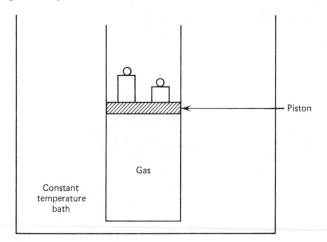

FIGURE 14.1   Boyle's experiment. As temperature is changed weights are added to the piston to maintain a constant volume.

Equations 14.4 and 14.5 we need to compute rates of change of one characterizing variable with respect to a second while holding the others fixed. Thus, *we need a constitutive equation relating these variables.*

Experimental studies for this gas constitutive equation (sometimes, unfortunately, called an equation of state) date to the seventeenth century. The most important is that of Boyle, published in 1660 in his book *The Spring of Air*. Boyle's apparatus is shown in Figure 14.1. At constant temperature and at moderate pressures he studied the volume variation in a batch system with changing pressure. According to Equation 14.1 the total mass, ρV, is constant. Over the range studied Boyle found that the product of pressure and volume is constant,

$$pV = \text{constant}$$

A related set of experiments by Charles in the eighteenth century established that in a batch system at constant pressure the volume varies linearly with temperature, and that for each rise of one degree centigrade the volume increases by 1/273 of the volume occupied at 0°C.

These experimental observations have been found over the years to apply to most gases over at least some range of temperature and pressure. Together they lead to what has become known as the *ideal gas law*, or *perfect gas law:*

$$pV = nRT \tag{14.6}$$

n is the number of moles and R the *gas constant*. R is found experimentally to be nearly the same for all gases over the range of conditions where Equation 14.6 applies. The value of R for p, V, and T expressed in a variety of units is shown in Table 14.1. Notice that T must be measured in absolute

TABLE 14.1 Values of the gas constant, R, in various units.

p	V	T	n	R	
$\dfrac{lb_f}{ft^2}$	ft^3	$°R$	lb-mole	1545	$\dfrac{ft^3\ lb_f}{ft^2\ \text{lb-mole}\ °R}$
atm	cm^3	$°K$	g-mole	82.06	$\dfrac{\text{atm}\ cm^3}{\text{g-mole}\ °K}$
atm	ft^3	$°R$	lb-mole	0.73	$\dfrac{\text{atm}\ ft^3}{\text{lb-mole}\ °R}$
$psi = \dfrac{lb_f}{in^2}$	ft^3	$°R$	lb-mole	10.73	$\dfrac{\text{psi}\ ft^3}{\text{lb-mole}\ °R}$
mm Hg	liter	$°K$	g-mole	62.36	$\dfrac{\text{mmHg liter}}{\text{lb-mole}\ °K}$
atm	ft^3	$°K$	lb-mole	1.314	$\dfrac{\text{atm}\ ft^3}{\text{g-mole}\ °K}$
mm Hg	ft^3	$°K$	lb-mole	999	$\dfrac{\text{mmHg}\ ft^3}{\text{lb-mole}\ °K}$

pV	T	n	R	
BTU	$°R$	lb-mole	1.987	$\dfrac{\text{BTU}}{\text{lb-mole}\ °R}$
calories	$°K$	g-mole	1.987	$\dfrac{\text{calorie}}{\text{g-mole}\ °K}$
ft lb_f	$°R$	lb-mole	1.544×10^3	$\dfrac{ft\ lb_f}{\text{lb-mole}\ °R}$
$\dfrac{g\ cm^2}{sec^2}$	$°K$	g-mole	8.314×10^7	$\dfrac{\text{gm}\ cm^2}{sec^2\ \text{g-mole}\ °K}$

temperature, $°R$ or $°K$, when using Equation 14.6. The ideal gas law was deduced in Example 3.3 using dimensional analysis, and some indication of the range of applicability is shown in Figure 3.9. In general, the more dilute the gas is, the better an approximation the ideal gas becomes. Small n/V corresponds to high temperature and low pressure. Gases like oxygen and nitrogen are well approximated by Equation 14.6 at room temperature and atmospheric pressure. The ideal gas equation can be derived by applying the principle of conservation of momentum at a molecular level and assuming that n/V is small. Corrections for large n/V can also be derived. Constitutive equations for gases outside the ideal range are discussed in the next section.

Some useful relations follow directly from Equations 14.1, 14.2, and 14.6. For a single gas it is convenient to express the density in terms of number of moles and volume. For ρ in mass/volume we have

$$\rho = nM_w/V = M_w p/RT \tag{14.7}$$

M_w is the molecular weight. Equation 14.1 can then be written

$$\frac{d\rho V}{dt} = \frac{d}{dt}\left[\frac{M_w pV}{RT}\right] = \frac{M_w}{R}\frac{d}{dt}\left[\frac{pV}{T}\right] = 0$$

Integration between any two times yields the familiar relationship

$$\frac{p_1 V_1}{T_1} = \frac{p_2 V_2}{T_2} \tag{14.8}$$

The energy equation can be similarly integrated. The coefficient of dp/dt in Equation 14.4 is evaluated using the constitutive equation, 14.6, as follows:

$$V - T\frac{\partial V}{\partial T}\bigg)_{p,n} = V - T\frac{\partial}{\partial T}\left[\frac{nRT}{p}\right]\bigg)_{p,n}$$

$$= V - T\frac{nR}{p} = V - V = 0$$

That is, for a gas which obeys the ideal gas constitutive equation the enthalpy depends only on temperature, as for a liquid or solid. Similarly, in Equation 14.5 the coefficient of dV/dt is zero. The batch energy equation for an ideal gas can then be written in either of the two equivalent forms

$$\rho V \underline{c}_p \frac{dT}{dt} - \frac{dpV}{dt} = Q - W_s \tag{14.9}$$

$$\rho V \underline{c}_V \frac{dT}{dt} = Q - W_s \tag{14.10}$$

Since $pV = nRT$ and n is constant in a nonreacting batch system we have

$$\frac{dpV}{dt} = nR\frac{dT}{dt}$$

and Equation 14.9 can be rewritten

$$[\rho V \underline{c}_p - nR]\frac{dT}{dt} = Q - W_s \tag{14.11}$$

Comparison of Equations 14.10 and 14.11 establishes a relation referred to in Section 10.5 for ideal gases

$$\underline{c}_p = \underline{c}_V + \frac{nR}{\rho V} \tag{14.12a}$$

Since $\rho V = M_0 = $ total mass and $n = $ number of moles, the ratio $\rho V/n = M_w$, the molecular weight. $M_w \underline{c}_p = \underline{c}_p$, $M_w \underline{c}_V = \underline{c}_V$, so

$$\underline{c}_p = \underline{c}_V + R \tag{14.12b}$$

Using the fact that $\rho V = M_0 = $ constant, Equation 14.10 or 14.11 can be integrated to give

$$M_0 \int_{T_1}^{T_2} \underline{c}_V(T)\, dT = \mathbf{q} - w \tag{14.13}$$

$\mathbf{q}$ is the total amount of heat added over the time interval and w is the work done by the system. Heat capacity data for gases are often reported in the empirical form

$$\underline{c}_V = a + bT + cT^2$$

in which case Equation 14.13 becomes

$$M_0 a[T_2 - T_1] + \frac{M_0 b}{2}[T_2^2 - T_1^2] + \frac{M_0 c}{3}[T_2^3 - T_1^3] = \mathbf{q} - w \tag{14.14}$$

For $\underline{c}_V = $ constant we have

$$M_0 \underline{c}_V[T_2 - T_1] = \mathbf{q} - w \tag{14.15}$$

Some typical problems with gas systems are illustrated in the following examples.

Example 14.1

Gas quantities are often stated in volumetric rather than molar or mass units. Since the volume of a gas depends strongly on the temperature and pressure, it is necessary to specify a standard state when volumetric units are employed. The term STP refers to a standard temperature of $0°C$ and a standard pressure of 1 atm. If the gas is assumed to behave as an ideal gas, compute the volume that one lb-mole and one gram-mole of gas occupy at STP.

R is found from Table 14.1. In units of atm ft³/lb-mole °R the value of R is 0.73. 0°C = 492°R. In units atm cm³/g-mole °K, R = 82.06. 0° C = 273°K. The computation is then as follows:

$$V = \frac{nRT}{p}$$

(a) $n = 1$ lb–mole

$$V = \frac{1 \times 0.73 \times 492}{1} = 359 \text{ ft}^3$$

(b) $n = 1$ g–mole

$$V = \frac{1 \times 82.06 \times 273}{1} = 22.4 \times 10^3 \text{ cm}^3 = 22.4 \text{ liters}$$

Example 14.2

Air, oxygen, hydrogen, nitrogen, and a variety of other gases are available in small, relatively portable pressure vessels, called cylinders. A common size is the "300 cubic foot cylinder." The 300 ft³ refers to the volume of gas at STP that the cylinder holds when it is filled at 20°C to a pressure of 2400 psig (lb/sq in. gauge, i.e., pressure in excess of atmospheric. 2400 psig = 2414.7 psi). What is the volume of the empty cylinder if the gas it contains behaves as an ideal gas?

We are concerned in this problem with a constant mass of gas that has been compressed from STP to the cylinder conditions. Although the actual operation is one in which air is pumped by a compressor to the tank, we are only concerned in this statement of the problem with a fixed mass of gas and hence a batch system. We will designate the standard state with subscript 1 and the pressurized vessel conditions with subscript 2

$$\frac{P_1 V_1}{T_1} = \frac{P_2 V_2}{T_2}$$

$$\frac{[1 \text{ atm}][300 \text{ ft}^3]}{273°K} = \frac{[24{,}147 \text{ psi}][V_2 \text{ ft}^3]}{293°K} \frac{1 \text{ atm}}{14.7 \text{ psi}}$$

$$V_2 = 1.89 \text{ ft}^3$$

Example 14.3

What will the temperature increase be when 100 ft³ of nitrogen at atmospheric pressure and 70°F is compressed adiabatically ($\mathbf{q} = 0$) by a rotary compressor using 1000 BTU?

Equation 14.15 is used to solve this problem. We compute M_0 using Equation 14.6 and an R value of 0.73 atm ft³/lb-mole °R. The molecular weight of nitrogen is 28, and $70°F = 530°R$.

$$M_0 = nM_w = M_m \frac{RT}{pV}$$

$$= \frac{28 \times 0.73 \times 530}{1 \times 100}$$

$$= 108.5 \text{ lb}_m$$

c_p for nitrogen (70°F) is readily found in any standard reference as 0.25 BTU/lb$_m$°F and relatively constant up to about 200°F. The ratio c_p/c_V is also commonly tabulated. c_V can be computed by finding $c_p/c_V = 1.404$ in the same reference. For nitrogen $c_V = 0.18$ BTU/lb$_m$°F and T_2 is computed from Equation 14.15

$$530 - T_2 = \frac{-1000}{108.5 \times 0.18}$$

$$T_2 = 581°R = 121°F$$

It is possible to express w in terms of an integral involving pdV and hence compute the volume and pressure in the final compressed state. This is a subject traditionally left to thermodynamics and we shall not deal with it here.

14.3 CONSTITUTIVE EQUATIONS FOR REAL GASES

A constitutive relationship between T, p, and ρ must be available if one is going to work with gas systems. The simplest and easiest to work with is the ideal gas law, but in practice we may frequently encounter engineering problems in which the conditions are outside the range in which the ideal gas equation accurately describes the behavior of a given gas. The deviations from ideal behavior for some gases are shown in Figure 3.9. The behavior of "real" gases has been a subject of active research over the past century and various methods have been devised to relate p, T, and ρ or p, T, and V.

One of the first attempts to describe p-V-T behavior of gases with an equation valid outside the range of ideal behavior was by van der Waals in 1873. The van der Waals equation is only a slight improvement over the ideal gas, but it nicely illustrates the analytical complexities of nonideal (i.e., real) behavior. The equation is

$$\left\{ p + a \left[\frac{n}{V} \right]^2 \right\} \{V - nb\} = nRT \tag{14.16}$$

TABLE 14.2   van der Waals constants for several gases

Gas	$a \times 10^{-6}$ $\mathrm{atm}\left[\dfrac{\mathrm{cm}^3}{\mathrm{g\text{-}mole}}\right]^2$	b $\dfrac{\mathrm{cm}^3}{\mathrm{g\text{-}mole}}$
Helium	0.0354	23.6
Hydrogen	0.245	26.5
Oxygen	1.32	32.2
Air	1.33	36.6
Nitrogen	1.35	38.3
Carbon Dioxide	3.60	42.5
Methane	2.27	42.6
Water Vapor	5.48	30.6
Chlorine	6.65	56.0

(a) To convert a to psi $[\mathrm{ft}^3/\mathrm{lb\text{-}mole}]^2$ multiply table value by 3.776×10^{-3}. (b) To convert b to $\mathrm{ft}^3/\mathrm{lb\text{-}mole}$ multiply table value of 1.60×10^{-2}

Experimental values of a and b for several gases are shown in Table 14.2. These values are determined by fitting Equation 14.16 to data using techniques like those in Chapter 6. Notice that for small n/V the van der Waals equation simplifies to the ideal gas.

The influence of the nonideal gas on integration of the energy equation can be seen by evaluating the coefficient of dV/dt in Equation 14.5. Equation 14.16 can be rearranged to solve for p

$$p = \frac{nRT}{V - nb} - a\left[\frac{n}{V}\right]^2$$

$$\left.\frac{\partial p}{\partial T}\right)_{V,n} = \frac{nR}{V - nb}$$

$$T\left.\frac{\partial p}{\partial T}\right)_{V,n} - p = a\left[\frac{n}{V}\right]^2$$

Equation 14.5 then becomes

$$M_0 \underline{c}_V \frac{dT}{dt} + \frac{an^2}{V^2}\frac{dV}{dt} = Q - W_s$$

For constant $\underline{c}_V$ this can be integrated to give

$$M_0 \underline{c}_V [T_2 - T_1] - an^2\left[\frac{1}{V_2} - \frac{1}{V_1}\right] = \mathbf{q} - w \tag{14.17}$$

In a constant volume system there is no effect, since $V_1 = V_2$, but if volume is allowed to vary, the nonideal behavior will lead to different results from the ideal gas.

Three of the commonly used constitutive equations for real gases are listed in Table 14.3. Notice that all reduce to the ideal gas in the limit of dilute

TABLE 14.3 Constitutive equations for real gases

Redlich-Kwong

$$\left[p + \cfrac{a}{T^{1/2} \dfrac{V}{n}\left[\dfrac{V}{n} + b\right]} \right] \left[\frac{V}{n} - b \right] = RT \qquad a = 0.4278 \,\frac{R^2 T_c^{2.5}}{p_c}$$

$$b = 0.0867 \,\frac{RT_c}{p_c}$$

Virial

$$pV = nRT\left\{ 1 + B\left[\frac{n}{V}\right] + C\left[\frac{n}{V}\right]^2 \cdots \right\}$$

Benedict-Webb-Rubin

$$\frac{pV}{n} = RT + \beta\left[\frac{n}{V}\right] + \sigma\left[\frac{n}{V}\right]^2 + \eta\left[\frac{n}{V}\right]^4 + \omega\left[\frac{n}{V}\right]^6$$

behavior. In general, the more constants an equation contains the wider the range of p-V-T data that it fits and the more difficult it is to use without machine computation. Some equations are more easily solved for p in terms of n, V, and T, others for V in terms of n, p, and T. All are completely empirical except the virial equation, where the coefficients can be evaluated in principle using techniques of statistical mechanics. We will not go into the relative accuracy or usefulness of the various equations here. Texts on thermodynamics deal with this subject in great detail.

It is found that for nearly all gases the deviation from ideal behavior can be correlated quite well on a single plot using *reduced variables*. The reduced temperature, pressure, and volume are defined as the ratio of temperature to critical temperature, pressure to critical pressure, and volume to critical volume

$$T_R = \frac{T}{T_c} \qquad p_R = \frac{p}{p_c} \qquad V_R = \frac{V}{V_c} \tag{14.18}$$

The critical conditions are those at which the gas and liquid have the same physical properties. T_c and p_c for several gases are shown in Table 14.4. The correlation valid for most gases is shown in Figure 14.2 as deviation from

TABLE 14.4   Critical temperature and pressure for several gases

Gas	T_c °K	p_c atm
Air	132.5	37.2
Carbon Dioxide CO_2	304.2	72.9
Chlorine Cl_2	417.0	76.1
Ethane C_2H_6	305.4	48.2
Methane CH_4	190.7	45.8
Nitrogen N_2	126.2	33.5

ideality versus p_R at various values of T_R. The deviation from ideality is expressed in terms of the *compressibility*, z,

$$z = pV/nRT \tag{14.19}$$

This approach is known as the *law of corresponding states*. Notice that at pressures of several atmospheres and temperatures above room temperature most common gases will have values of p_R much less than one and values of

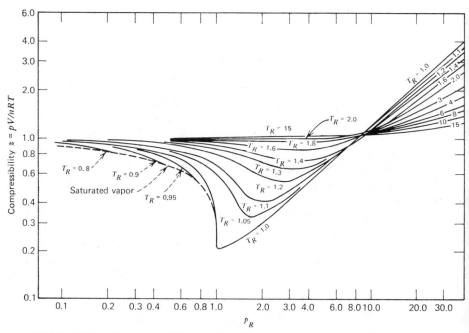

FIGURE 14.2   Compressibility as a function of reduced pressure at various values of reduced temperature. From Hougen and Watson, *Industrial Chemical Calculations*, 2nd Ed., Wiley, New York, 1936. Reproduced by permission.

T_R above one. According to Figure 14.2 the deviation from ideality ($z = 1$) will not be great under these conditions. Most of the constitutive equations in Table 14.3 can be formulated in reduced variables and applied with the same set of constants to a large number of gases. The van der Waals equation can be written in reduced form

$$\left\{ p_R + 3\left[\frac{1}{V_R}\right]^2 \right\}\{3V_R - 1\} = 8T_R \tag{14.20}$$

The critical properties, particularly the volume, are difficult to obtain accurately, but some measurements do exist and a variety of estimation procedures are available. For example, for some liquids the normal boiling temperature is approximately two-thirds the critical temperature.

Example 14.4

One hundred pounds of methane is to be stored in a 10 cubic foot vessel at a temperature of 80°F. Compute the pressure in the tank by the following methods

(a) Ideal gas law $pV = nRT$
(b) van der Waals equation

$$p = \frac{nRT}{V - nb} - \frac{n^2 a}{V^2}$$

(c) Compressibility chart
   Using Tables 14.1 to 14.3 the following information is obtained

$$T_c = 190.7°K$$

$$p_c = 45.8 \text{ atm}$$

$$R = 0.73\,\frac{\text{atm ft}^3}{\text{lb-mole °R}} = 1.314\,\frac{\text{atm ft}^3}{\text{lb-mole °K}}$$

$$a = 2.27\times10^6 \text{ atm}\left[\frac{\text{cm}^3}{\text{g-mole}}\right]^2 = 5.811\times10^3 \text{ atm}\left[\frac{\text{ft}^3}{\text{lb-mole}}\right]^2$$

$$b = 42.6\,\frac{\text{cm}^3}{\text{g-mole}} = 0.682\,\frac{\text{ft}^3}{\text{lb-mole}}$$

The molecular weight of CH_4 is 16.04 and thus

$$n = \frac{100}{16.04} = 6.23 \text{ lb-moles}$$

$$T = 80°F = 540°R = 300°K$$

(a) Ideal gas law

$$p = \frac{6.23 \times 1.314 \times 300}{10}$$

$$= 245.6 \text{ atm}$$

(b) van der Waals equation

$$p = \frac{6.23 \times 1.314 \times 310}{10 - 4.249} - \frac{6.23^2 \times 5.811 \times 10^2}{10^2}$$

$$= 201.5 \text{ atm}$$

(c) Compressibility chart

$$T_R = \frac{300}{190.7} = 1.57$$

$$p_R = \frac{p}{45.8}$$

$$z = \frac{10p}{6.23 \times 1.314 \times 300} = 0.187 p_R$$

Plotting this expression on Figure 14.2 will yield a value of z where the line intersects with $T_R = 1.57$.

$$z = 0.85$$

$$p = 208 \text{ atm}$$

14.4 SEMIBATCH OPERATION—FILLING A PRESSURE VESSEL

In this section we develop the mathematical description of the process of filling a vessel with a gas. The analogous liquid system was dealt with in Chapters 2 and 4. The physical situation is shown schematically in Figure 14.3 and we will assume that it is possible to deliver a constant mass flow rate of gas, $\rho_f q$, to the tank at a temperature T_f and pressure p_f. To develop the mathematical description we refer, as we did for liquid systems, to the logical procedure developed in Figure 3.4. The primary fundamental variable of concern is, of course, the mass of gas in the vessel, which we can represent by the characterizing variables ρV. If we choose the vessel

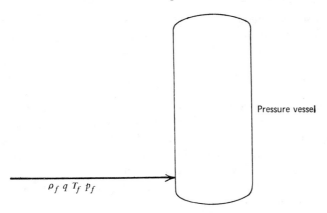

FIGURE 14.3 Filling a vessel with a gas at constant mass flow rate.

itself as the control volume it adequately meets the test that ρ have a constant value throughout. Application of the law of conservation of mass then yields

$$\frac{d\rho V}{dt} = \rho_f q \tag{14.21}$$

Since the gas always occupies the entire vessel, V is a constant and Equation 14.21 can be integrated.

$$\rho = \rho_0 + \frac{\rho_f q}{V} [t - t_0] \tag{14.22}$$

Here ρ_0 is the initial density of any gas in the vessel at temperature T_0 and pressure p_0.

The density of a gas system is, of course, very much dependent on the temperature and pressure of the system. This is not the case for liquids, and in our previous analysis of the various liquid systems we always made the reasonable assumption that the liquid density was either constant or dependent only on composition. Furthermore, the process of a liquid flow to a tank is isothermal, and since there are no heat effects associated with the liquid filling process it is possible to determine the system behavior using only mass as a fundamental dependent variable. This may well not be true for gas systems, and in fact even with a limited previous knowledge of the behavior of gas systems we would not expect the process to be isothermal. Since Equation 14.22 contains ρ, which we know from Equation 14.7 to be dependent upon the characterizing variables T and p and since we cannot assume isothermal behavior, it is obvious that we must deal with the energy of the system as a second fundamental dependent variable. Equation 10.10 can be simplified for the system we are considering to yield

$$\frac{dU}{dt} = \rho_f q \underline{H}_f + Q \tag{14.23}$$

We have made the reasonable assumptions that the kinetic and potential energy effects are negligible in comparison with thermal effects, and restricted our analysis to the case where no shaft work is done on or by the system.

To determine the system behavior Equation 14.23 must be rewritten in terms of characterizing variables which can be either measured or are tabulated. Following our procedure with batch systems we will first work with the equation in terms of the enthalpy.

$$\frac{dU}{dt} = \frac{dH}{dt} - \frac{dpV}{dt} = \rho_f q \underline{H}_f + Q$$

The enthalpy change of the gas in the tank with time, dH/dt, can be expressed in terms of T, p, and the system composition by using Equation 10.46

$$\frac{dH}{dt} = \rho V \underline{c}_p \frac{dT}{dt} + \left[V - T \frac{\partial V}{\partial t} \right)_{p, n_i} \right] \frac{dp}{dt}$$

$$+ \tilde{H}_1 \frac{dn_1}{dt} + \cdots + \tilde{H}_s \frac{dn_s}{dt}$$

If we assume a single gas in the system, then there is only one species, $\tilde{H}_1[dn_1/dt] = \underline{H}[dn/dt]$, and the energy balance becomes

$$\rho V \underline{c}_p \frac{dT}{dt} + \left[V - T \frac{\partial V}{\partial t} \right)_{p,n} \right] \frac{dp}{dt} + \underline{H} \frac{dn}{dt} - \frac{dpV}{dt} = \rho_f q \underline{H}_f + Q$$

$$\tag{14.24}$$

We can eliminate $\underline{H}_f$ from the energy balance if we express $\underline{H}_f$ in terms of the inlet gas temperature and pressure, T_f and p_f, and the vessel temperature and pressure, T and p. We make use of Equation 10.45 and the relationship $\underline{H} = H/\rho V$, recognizing that changes in composition need not be considered, to obtain

$$\underline{H}_f(T_f, p_f) - \underline{H}_f(T, p)$$

$$= \underline{c}_p[T_f - T] + \left[\frac{1}{\rho} - T \frac{\partial \frac{1}{\rho}}{\partial T} \right)_{p,n} \right] [p_f - p]$$

We have assumed a constant heat capacity for simplicity. Since we are dealing with a single gas $\underline{H}_f(T, p)$ is equal to $\underline{H}(T, p)$. Equation 14.24 can be simplified by making use of this fact, the mass balance (Equation 14.21), and the definition for ρ.

$$\underline{H} \frac{dn}{dt} = \underline{H} \frac{d\rho V}{dt} = \rho_f q \underline{H}_f(T, p)$$

$$\rho V \underline{c}_p \frac{dT}{dt} + \left[V - T \left(\frac{\partial V}{\partial T} \right)_{p, n} \right] \frac{dp}{dt} - \frac{dpV}{dt}$$

$$= \rho_f q \underline{c}_p [T_f - T]$$

$$+ \rho_f q \left[\frac{1}{\rho} - T \frac{\partial \frac{1}{\rho}}{\partial T} \right)_{p, n} \right] [p_f - p] + Q \quad (14.25)$$

The behavior of the system can be determined with much less mathematical complexity if ideal gas behavior is assumed. The coefficient of dp/dt has already been shown to be zero, and this is also true for the coefficient of $p_f - p$. Equation 14.25 becomes

$$\rho V \underline{c}_p \frac{dT}{dt} - \frac{dpV}{dt} = \rho_f q \underline{c}_p [T_f - T] + Q$$

A further simplification results when Equation 14.7 is used to rewrite the dpV/dt term for this constant volume system.

$$\frac{dpV}{dt} = V \frac{dp}{dt} = \frac{VR}{M_w} \frac{d}{dt} [\rho T]$$

$$= \frac{VR}{M_w} \left[T \frac{d\rho}{dt} + \rho \frac{dT}{dt} \right]$$

$$= \frac{RT}{M_w} \rho_f q + \frac{\rho VR}{M_w} \frac{dT}{dt}$$

The energy balance expressed entirely in terms of characterizing variables thus becomes

$$\rho V \left[\underline{c}_p - \frac{R}{M_w} \right] \frac{dT}{dt} = \rho_f q \underline{c}_p [T_f - T] + \frac{\rho_f q RT}{M_w} + Q \quad (14.26)$$

Equation 14.26 can be equivalently expressed in terms of $\underline{c}_V$ using Equation 14.12:

$$\rho V \underline{c}_V \frac{dT}{dt} = -\rho_f q \underline{c}_V T + \rho_f q \underline{c}_p T_f + Q \quad (14.27)$$

This equation can also be derived from Equations 14.23 and 10.50 using $\underline{U}$ and $\underline{U}_f$ in the manipulation in place of $\underline{H}$ and $\underline{H}_f$.

Equation 14.27 can be integrated to yield the temperature of the gas in the vessel as a function of time, noting that ρ is a function of time given by Equation 14.22. The integration is analogous in its mathematical form to that encountered in Section 4.52. To simplify the manipulations we restrict ourself to the adiabatic case where $Q = 0$. This restriction is easily removed at the expense of some algebra in a manner similar to the procedures used in Chapter 11. The integrated form is

$$\ln \left[\frac{\gamma T_f - T}{\gamma T_f - T_0}\right] = \ln \left[\frac{V \rho_0 / \rho_f q}{\dfrac{V \rho_0}{\rho_f q} + t}\right]$$

or, rearranging to give an explicit expression for T,

$$T = \gamma T_f - \left(\frac{\gamma T_f - T_0}{1 + \dfrac{\rho_f q t}{V \rho_0}}\right) \tag{14.28}$$

where $\gamma = c_p / c_V$. If the tank is initially empty, then $\rho_0 \simeq 0$ and Equation 14.28 reduces to $T = \gamma T_f = $ constant.

Example 14.5

Nitrogen is delivered to an evacuated insulated tank at a constant mass flow rate at temperature 70°F. What is the temperature of gas in the tank during filling?

For nitrogen $\gamma = 1.40$. $T_f = 70°F = 530°R$. Thus

$$T = 1.4 \times 530 = 742°R = 282°F$$

14.5   MIXTURES OF GASES

In studying liquid mixtures beginning in Section 4.4 we found that the primary complication was in expressing mixture properties, such as density, in terms of pure component properties. A similar situation arises for gases. The simplest constitutive equation for gas mixtures is obtained in the following way. Define the partial pressure p_i of species i in a mixture as

$$p_i \equiv y_i p \tag{14.29}$$

y_i is the mole fraction of species i in the mixture. The total pressure is then, by definition, the sum of the partial pressures, since the mole fractions sum to unity:

$$\sum p_i = \sum y_i p = p \sum y_i = p$$

For a mixture that behaves as an ideal gas we have

$$p_i = y_i p = y_i \frac{nRT}{V} = n_i \frac{RT}{V} \tag{14.30}$$

n_i is the number of moles of i in the mixture. Equation 14.30 is called *Dalton's law* and may be stated as follows:

The total pressure of a gas mixture is the sum of the pressures that each individual gas would exert if it occupied the entire volume by itself at the temperature in question.

If all of the component gases are ideal *and* the mixture is ideal, then Dalton's law is a consequence of the ideal gas. In general there is no reason to assume that a mixture will be ideal even if the components would be if present by themselves, so Dalton's law is in fact an additional constitutive assumption.

Another constitutive equation often used for gas mixtures is called *Amagat's law*. Define the partial volume v_i by

$$v_i = y_i V$$

The assumption of an ideal mixture leads to

$$v_i = n_i \frac{RT}{p} \tag{14.31}$$

and the statement that the total volume of a gas mixture is equal to the sum of the volumes of the individual gases evaluated at the temperature and pressure of the system. For ideal gases Dalton's and Amagat's "laws" lead to identical results. For nonideal systems they are contradictory and must lead to different properties for the mixture.

14.6 REACTING BATCH SYSTEMS

The development for reacting systems in this section parallels those in Chapters 5 and 12. What we shall find is that although gas systems are generally more complex than corresponding liquid systems, there is one case in which the gas can be simpler. That is in the experimental measurement of rate constants. Where in the liquid some means of inferring concentrations of reacting species as functions of time is needed it will often be adequate in the gas to simply monitor the pressure.

We shall restrict our attention to the simple pseudo first-order reaction

$$A \rightarrow \delta D$$

$$r_{A-} = kc_A \tag{14.32}$$

More general schemes can be handled at the expense of more algebra. The overall mass balance in the batch reactor is clearly the same as Equation 14.1

$$\frac{d\rho V}{dt} = 0 \qquad (14.1)$$

The component balances are the same as for a liquid system, Equations 5.3 and 5.5 without the flow terms:

$$\frac{dn_A}{dt} = \frac{dc_A V}{dt} = -r_{A-} V \qquad (14.33a)$$

$$\frac{dn_D}{dt} = \frac{dc_D V}{dt} = +r_{D+} V \qquad (14.33b)$$

For gases it is usually more convenient to use number of moles instead of concentration, and we shall do so henceforth. The intrinsic rate can be written

$$r = -r_{A-} = +\frac{r_{D+}}{\delta} = kc_A$$

in which case Equations 14.33 become

$$\frac{dn_A}{dt} = -rV = -kn_A \qquad (14.34a)$$

$$\frac{dn_D}{dt} = +\delta rV = +\delta kn_A \qquad (14.34b)$$

Clearly, combination of Equations 14.34 leads to

$$n_A - n_{A0} = -\frac{1}{\delta}[n_D - n_{D0}] \qquad (14.35)$$

where n_{A0} and n_{D0} refer to initial numbers of moles.

The principle of conservation of energy, neglecting shaft work, is

$$\frac{dU}{dt} = Q \qquad (14.36)$$

This can be rewritten in terms of the enthalpy using $U = H - pV$, and the rate of change of enthalpy is then expanded by means of Equation 10.46. The energy equation then takes the form

$$\rho V \underline{c}_p \frac{dT}{dt} + \left[V - T\left(\frac{\partial V}{\partial T}\right)_{p,n_i}\right]\frac{dp}{dt} + \tilde{H}_A \frac{dn_A}{dt}$$

$$+ \tilde{H}_D \frac{dn_D}{dt} - \frac{dpV}{dt} = Q \quad (14.37)$$

Using Equations 14.34 for component mass balances and Equation 12.12, the definition of heat of reaction, we obtain, finally,

$$\rho V \underline{c}_p \frac{dT}{dt} + \left[V - T \left(\frac{\partial V}{\partial T} \right)_{p, n_i} \right] \frac{dp}{dt} - \frac{dpV}{dt} = rV[-\Delta \underline{H}_R] + \dot{Q}$$

(14.38)

Equation 14.38 is valid for any gas, ideal or nonideal, and a careful look at the derivation will show that it is valid for any rate expression, not just Equation 14.32. It is useful at this point to assume an ideal gas. In that case the coefficient of dp/dt vanishes. The term dpV/dt can be written

$$\frac{dpV}{dt} = \frac{d}{dt}[nRT] = nR \frac{dT}{dt} + RT \frac{dn}{dt}$$

(14.39)

Also, $n = n_A + n_D$. From Equations 14.34

$$\frac{dn}{dt} = \frac{d}{dt}[n_A + n_D] = [\delta - 1]rV$$

(14.40)

Combining of Equations 14.38 to 14.40, together with 14.12 relating $\underline{c}_p$ and $\underline{c}_V$ for an ideal gas we obtain the final working form of the energy equation

$$\rho V \underline{c}_V \frac{dT}{dt} = rV\{-\Delta \underline{H}_R + [\delta - 1]RT\} + \dot{Q}$$

(14.41)

Except for the interchange of $\underline{c}_V$ and $\underline{c}_p$ this differs from the liquid case only in the addition of $[\delta - 1]RT$ to the heat of reaction.

The complete mathematical description of the reacting batch system for an ideal gas consists of Equations 14.32, 14.34, and 14.41, together with the constitutive equation. In order to determine behavior we must specify two quantities. Several important cases can be distinguished.

Case I: *T* and *V* Constant

The isothermal reactor of constant volume is the most important laboratory configuration for obtaining rate data. For the first-order reaction Equation 14.34a can be integrated directly if k is relatively independent of pressure. The result is

$$n_A = n_{A0}e^{-kt}$$

(14.42)

For simplicity take $n_{D0} = 0$. Then from Equation 14.35

$$n_D = \delta n_{A0} - \delta n_{A0}e^{-kt}$$

(14.43)

$$n = n_A + n_D = \delta n_{A0} + [1 - \delta]n_{A0}e^{-kt}$$

(14.44)

The pressure in the reactor at any time can then be computed:

$$p = \frac{nRT}{V} = \frac{\delta n_{A0}RT}{V} + \frac{n_{A0}RT}{V}[1 - \delta]e^{-kt} \tag{14.45}$$

If we call p_0 the pressure at $t = 0$ with n_{A0} moles in the reactor and p_∞ the pressure when the reaction is over and there are δn_{A0} moles in the reactor, then Equation 14.45 can be written

$$p = p_\infty + p_0[1 - \delta]e^{-kt}$$

or

$$\frac{p_\infty - p}{p_0} = [\delta - 1]e^{-kt} \tag{14.46}$$

Thus, a plot of $\ln\{[p_\infty - p]/p_0\}$ versus t will have a slope $-k$ and intercept $\delta - 1$. *The rate constant can be obtained simply by monitoring the pressure.* The rate at which heat needs to be removed to keep the reactor at constant temperature follows from Equation 14.41 by setting $dT/dt = 0$:

$$Q = -kn_{A0}e^{-kt}\{-\Delta H_R + [\delta - 1]RT\} \tag{14.47}$$

Example 14.6

The decomposition of di-t-butyl peroxide,

$$(CH_3)_3COOC(CH_3)_3 \rightarrow \text{products}$$

is believed to be first order. Data are shown in Table 14.5 for pressure versus time in a reactor at 154.6°C. Verify the first-order assumption and calculate k.

TABLE 14.5   Decomposition of di-t-butyl peroxide at 154.6°C

Time, min	pressure, mmHg	$\dfrac{p_\infty - p}{p_0}$	$\ln\left[\dfrac{p_\infty - p}{p_0}\right]$
0	$p_0 = 174$	1.83	0.60
3	193	1.72	0.54
6	211	1.62	0.48
9	229	1.52	0.42
12	244	1.42	0.35
15	259	1.34	0.29
18	274	1.26	0.23
21	287	1.18	0.17
∞	$p_\infty = 492$	0	

Source. Data of Raley, Rust, and Vaughan, *J. Am. Chem. Soc.*, **70**, 98 (1948).

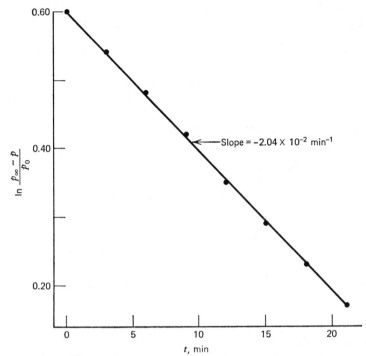

FIGURE 14.4 Calculation of first-order rate constant for decomposition of di-t-butyl peroxide at 154.6°C by following pressure change with time. [Data of Raley, Rust, and Vaughan, *J. Am. Chem. Soc.*, **70**, 88 (1948).]

The data are plotted as $\ln\{[p_\infty - p]/p_0\}$ versus t in Figure 14.4 and the linear dependence is good, as required by Equation 14.46. The value of k computed from the slope is

$$k = 2.04 \times 10^{-2}\,\text{min}^{-1} = 3.4 \times 10^{-4}\,\text{sec}^{-1}$$

Applying Equation 14.46 at $t = 0$ we obtain

$$\delta = 1 + \frac{p_\infty - p_0}{p_0} = 1 + 1.83 = 2.83$$

This nonintegral value is explained by the fact that the normal reaction product is two moles of acetone and one of ethane, a total of three moles, but there are small amounts of other decomposition products.

Case II: p and T Constant

This case would describe an isothermal reactor fitted with a piston to allow the volume to vary, as in Figure 14.1. Because k is again a constant the solutions to the mass balance equations and the heat removal rate are the same as for case I. By identical reasoning the volume can be shown to follow the relation

$$\frac{V_\infty - V}{V_0} = [\delta - 1]e^{-kt} \tag{14.48}$$

Thus, the rate coefficient can be evaluated by following the volume variation with time.

Case III: $V = $ Constant, Q Specified

The constant volume reactor with specified heat flux is the case most likely to be encountered for actual production. If pressure dependence of k can be neglected then the only coupling between mass and energy equations is through the temperature dependence of the rate coefficient

$$k = k_0 e^{-E/RT} \tag{14.49}$$

This is the same coupling as in liquid systems, and the approach to solution is almost identical, except that $-\Delta H_R$ is replaced by $-\Delta H_R + [\delta - 1]RT$.

For $\delta = 1$ the equations reduce identically to those for a liquid system. For $\delta \neq 1$ it is helpful to consider Equation 12.20 for temperature dependence of the heat of reaction

$$\Delta H_R(T) = \Delta H_R(T_0) + [\delta c_{pD} - c_{pA}][T - T_0] \tag{14.50}$$

From Equation 14.12 we have $c_p = c_V + R$. For algebraic simplicity we restrict ourselves to gases for which c_{pD} and c_{pA} are approximately equal. Thus, Equation 14.50 can be written

$$\Delta H_R(T) = \Delta H_R(T_0) + c_p[\delta - 1][T - T_0]$$

and Equation 14.41 becomes

$$\rho V c_V \frac{dT}{dt} = rV\{-\Delta H_R(T_0) + c_p[\delta - 1]T_0 - c_V[\delta - 1]T\} + Q \tag{14.51}$$

The special case $Q = 0$, c_p and c_V constant, is then solved as in Section 12.7 for liquids. Equations 14.51 and 14.34 combine to give

$$\frac{1}{-c_V[\delta - 1]T - \Delta H_R(T_0) + c_p[\delta - 1]T_0} \frac{dT}{dt} = -\frac{V}{M_0 c_V} \frac{dn_A}{dt}$$

which integrates to

$$T = T(n_A)$$

$$= -\frac{\Delta H_R}{c_V[\delta - 1]} + \frac{c_p T_0}{c_V}$$

$$+ \left\{ T_0 + \frac{\Delta H_R}{c_V[\delta - 1]} - \frac{c_p T_0}{c_V} \right\} e^{\{-c_V[\delta - 1]V/M_0 c_V\}[n_{A0} - n_A]} \tag{14.52}$$

Substitution into Equations 14.49 and 14.34 and integration leads to the relation for n_A

$$\int_{n_{A0}}^{n_A} \frac{e^{E/RT(n)} \, dn}{n} = -k_0 t \tag{14.53}$$

Equation 14.53 is analogous to Equation 12.33 for liquid systems and will require numerical integration to obtain n_A as a function of t. The temperature variation with time can then be extracted from Equation 14.52.

We shall go no further with gas systems in this text. It should be clear by now that extensions of the analysis process to other situations, such as flowing and reacting gases and liquid-vapor separation is straightforward, although ultimate working equations might be quite complex.

14.7 CONCLUDING REMARKS

This chapter covers a great deal of ground. It is possible to do so because there is only one major change introduced by considering gas systems, that is, the necessity of including constitutive equations for the density in terms of temperature, pressure, and composition. Sections 14.2, 14.3, and 14.5 are the key sections to ensure understanding. The equations for gases are more complicated than for liquid systems, but they can be used in exactly the same way. It is quite easy to go back to each chapter and to rederive the equations for comparable situations involving gas systems or, in the case of Chapters 8 and 9, for liquid-gas systems.

Many standard chemistry texts have good introductory sections on gas constitutive relations. See, for example,

14.1 J. V. Quagliana and L. M. Vallarino, *Chemistry*, 3rd ed., Prentice-Hall, Englewood Cliffs, 1969.

Steady state gas problems are discussed in many introductory chemical engineering texts, such as

14.2 D. Himmelblau, *Basic Principles and Calculations in Chemical Engineering*, 2nd ed., Prentice-Hall, Englewood Cliffs, 1967.

14.3 O. A. Hougen, K. M. Watson, and R. A. Ragatz, *Chemical Process Principles, Part I: Material and Energy Balance*, 2nd ed., Wiley, New York, 1954.

See also the thermodynamics texts cited in Chapter 10 and the kinetics and reactor design texts cited in Chapters 5 and 7.

14.8 PROBLEMS

14.1 A rather simple experiment can be performed to determine the approximate molecular weight of an unknown gas, if a container of known volume is filled with the gas, weighed, and the pressure and

temperature recorded. Estimate the molecular weight of 0.100 g
of gas occupying a 100 ml vessel at a temperature of 180°C and a
pressure of 810 mm of Hg.

14.2 Air for SCUBA (self contained underwater breathing apparatus) is
contained in a "72 cubic foot" cylinder designed to be filled to 2250
psig at 70°F. These cylinders are often stored in car trunks or in the
sun. It is common practice to use the following rule of thumb to
compute the rise in pressure under these conditions; "the pressure
will increase 5 psig for each 1°F increase in temperature."
   (a) Show by plotting p versus T over the range 40°F to 200°F just
       how adequate this rule is. You may assume that Equation 14.6
       applies.
   (b) Obtain the relation for rate of change of pressure with tem-
       perature and check the rule of thumb.

14.3 Under normal conditions an average swimmer uses about 1 cubic
foot of air per minute. If the swimmer is obtaining his air supply
from a "72 cubic foot tank" (Problem 14.2) how long will the tank
supply his needs in fresh water at a depth of
   (a) 30 feet?
   (b) 70 feet?
   (c) 110 feet?
   Assume a constant water temperature of 65°F.

14.4 The Natural Gas Industries standard conditions of temperature and
pressure are 60°F and 14.7 psia. How many cubic feet will one lb-
mole of an ideal gas occupy under these conditions?

14.5 What should the design pressure of a 0.75 ft³ cylinder of oxygen be if
it must contain 2.0 lb_m of oxygen and be subjected to a temperature
of 150°C?

14.6 Six lb_m of air is stored in a 0.5 ft³ vessel which is initially at 70°F.
If this vessel is placed in a bath which supplies 2000 BTU of heat,
what gas temperature will be achieved and what pressure must the
cylinder withstand?

14.7 If the temperature of the filling operation described in Example 14.5
must be kept constant at the feed temperature, how many BTU of
heat would have to be removed? If this heat is removed with a coil
containing cooling water at 60°F, what is the minimum length of
1/4 in. copper tubing needed if h may be assumed equal to 5 BTU/hr
ft² °F? The filling rate is 10 lb_m/min.

14.8 A pressure vessel containing nitrogen is heated by means of a coil in
which steam condenses at a constant temperature T_c.
   (a) Develop the basic mathematical description for this situation
       using as a constitutive relation for Q the equation discussed in
       Section 11.10. Notice that we can keep a constant temperature

in the coil if we vary the pressure in the coil by means of a steam trap.

(b) The pressure vessel has a volume of 10 ft³ and contains 10 lb$_m$ of nitrogen at 70°F before the steam is turned on to the coil. The steam condenses at 260°F. The coil consists of 400 ft of 1/2 in. diameter copper tubing and the value of the heat transfer coefficient, h, may be assumed to be 5 BTU/hr ft² °F. Prepare a plot of the pressure and temperature in the vessel versus time.

14.9 Evaluate the coefficient of dV/dt in Equation 14.5 for a gas that obeys the Benedict-Webb-Rubin constitutive equation and derive an expression similar to Equation 14.17.

14.10 A reactor with a volume of 8 ft³ contains 90 lb$_m$ of ethane. If a pressure gauge on this reactor reads 820 psig, what is the temperature of the gas? Compute the temperature using
(a) Ideal gas law.
(b) Van der Waals equation.
(c) Compressibility chart.

14.11 If the filling process described in Section 14.4 is carried out so that the whole operation is isothermal at T_f, develop a mathematical description that will give the tank pressure, p, as a function of time. Use as constitutive equations the ideal gas law and the van der Waals equation.

14.12 Derive Equation 14.27 using $\underline{U}$ and $\underline{U}_f$ in the manipulations with Equation 14.23.

14.13 Find the coefficient of dV/dt in Equation 10.50 for a gas which obeys van der Waals' constitutive equation.

14.14 Consider the adiabatic filling operation in Section 14.4 and plot the tank temperature, T, versus time for methane using the ideal gas development. The following conditions apply:

$$T_0 = 20°C \qquad e_0 = 0.046 \text{ lb}_m/\text{sec}$$
$$V = 10 \text{ ft}^3$$
$$T_f = 10°C$$
$$\rho_{fq} = 0.1 \text{ lb}_m/\text{sec}$$

14.15 (a) Develop the basic model equations for the situation in which a gas is flowing from a pressure vessel of volume V. Before flow begins conditions in the tank can be designated by n_0, p_0, and T_0.
(b) These equations cannot be manipulated in a development which parallels the tank filling analysis in Section 14.4. Why not? Be specific in your discussion and point out the essential differences from the liquid tank-draining problem.

14.16 If the gas flows from the tank in the preceding problem through a converging nozzle, the maximum possible mass flow rate in pounds per second is given by the following relation

$$\rho_l q_l = S_c \sqrt{\gamma p \rho \left(\frac{2}{\gamma + 1}\right)^{[\gamma+1]/[\gamma-1]} g_c}$$

p and ρ are the conditions in the tank and $\gamma = \underline{c}_p/\underline{c}_v$. S_c is a constant.

Develop an expression which would allow the tank pressure, p, and the tank temperature, T, to be computed as a function of time. For an ideal gas it can be shown that $T = T_l$.

14.17 Develop an equation for p_i similar to Equation 14.30 for a gas which obeys
  (a) Van der Waals equation.
  (b) Benedict-Webb-Rubin equation.

14.18 The following mixture of gases is contained in a vessel of 5 ft³ at a temperature of 90°C

Gas	Composition in lb-moles
Methane	0.20
Ethane	0.30
Hydrogen	0.20
Nitrogen	0.30

Compute the tank pressure using Daltons "law" with
  (a) Ideal gas equation.
  (b) Van der Waals equation.
  (c) A method using a mean compressibility factor which you develop yourself.

14.19 Repeat the development in Section 14.6 leading to Equation 14.41 for the following gas phase reaction.

$$\alpha_A A + \alpha_B B \rightarrow \delta_R R \qquad r_{A-} = k c_A c_B$$

14.20 If the reaction in Problem 14.19 is carried out in a batch reactor in which T and p are kept constant show how n_A, n_B, and n_R vary with time. Describe how you would design such a reactor. How would you calculate the total volume needed? $\alpha_A n_{AO} = \alpha_B n_{BO}$, $n_{RO} = 0$.

14.21 The decomposition of nitrogen pentoxide has been reported as follows (F. Daniels, *Chemical Kinetics*, Cornell Univ. Press, 1938).

Time (seconds)	$p_{N_2O_5}$, (mm of Hg)
0	348.4
600	247
1200	185
1800	140
2400	105
3000	78
3600	58
4200	44
4800	33
5400	24
6000	18
7200	10
8400	5
9600	3
∞	0

The partial pressure data were obtained in a batch reactor at 45°C Propose a rate expression, justify the form, and compute the specific reaction rate constant from the data given.

14.22   The reversible vapor phase decomposition of hydrogen iodide is characterized by the following chemical equation

$$2HI \leftarrow H_2 + I_2$$

Postulate a constitutive equation for the rates, r_{HI-}, r_{H_2+}, and r_{I_2+} and develop the batch reactor model equations.

14.23   The following references contain data that will allow you to check the rate expressions postulated in Problem 14.22. Examine one or two of them in detail and either verify or change your rate expressions to correspond with the data. M. Bodenstein, Z. *Physik Chem.*, **13**, 56 (1894); Z. *Physik Chem.*, **22**, 1 (1897); Z. *Physik Chem.*, **29**, 295 (1898); H. A. Taylor, *J. Phys. Chem.*, **28**, 984 (1924); G. B. Kistia-kowsky, *J. Am. Chem. Soc.*, **50**, 2315 (1928).

14.24   In the presence of *p*-toluene sulfonic acid, tertiary butyl alcohol reacts to form isobutylene and water.

$$CH_3-\overset{\overset{CH_3}{|}}{\underset{\underset{CH_3}{|}}{C}}-OH \rightarrow CH_3-\overset{\overset{CH_2}{\|}}{\underset{\underset{CH_3}{|}}{C}} + H_2O$$

The batch reaction can be followed by measuring the amount of isobutylene which is produced as a vapor. The following data were obtained this way in an experiment run at 77°C.

Time (minutes)	Liters of isobutylene at 25°C and 1 atm
0	0
2	0.133
4	0.385
6	0.625
8	0.965
10	1.235
12	1.510
14	1.793
16	2.055
18	2.310
20	2.555
24	3.060
28	3.540
33	4.090
36	4.423
40	4.830
44	5.207
48	5.556

Develop a mathematical description of the batch experiment and test the following rate expression for the disappearance of tertiary butyl alcohol, A. $n_{AO} = 2.1$ g-moles.

$$r_{A-} = kc_A$$

If this expression does not adequately fit the data, can you find one which is better? (Data from H. W. Heath, Jr., M.Ch.E. Thesis, Univ. of Delaware, 1971.)

Postface

Our goal in this text has been the development of a logical process for attacking chemical engineering problems. In so doing we have focused on the modeling of specific items of equipment that can be broadly classified as tank-type or tubular devices. We have exploited the tractability of the mathematical descriptions for these systems to illustrate the entire analysis process from experiment to design. The design problems have always involved use of a constitutive relationship and we have introduced and emphasized relations for the rate of reaction, rate of mass transfer, and rate of heat transfer.

The coefficients characterizing the rates of heat and mass transfer generally need to be obtained experimentally, but we can learn much about their dependence on the system variables by studying the detailed fluid motions within a piece of equipment. This is done by applying the law of conservation of momentum together with the other conservation principles to small control volumes within the larger device. The additional constitutive relations required here relate fluid stresses to the other characterizing variables. A deeper understanding of thermodynamics is also required, based on the concept of entropy and its consequences. The same principles of analysis apply to the more complex problem, although the resulting mathematical descriptions are not usually as easy to manipulate as those that we have studied in this text.

In our experience with practical engineering problems we have found that the approach developed in this book has always provided us with the correct description for a problem, or revealed those unknown relations which prevented us from obtaining a set of working equations. Often we have found that shortcuts would have taken us to the correct answer more quickly. More often we have found that shortcuts, with which the literature abounds, have introduced subtle errors with sometimes serious consequences.

Finally, it is important to repeat that skill in analysis, although necessary, is not sufficient for the solution of real-world problems. There are many aspects of engineering that cannot be quantified. We have dealt here with the introductory aspects of those that can, and it is our hope that the reader will consider his analysis skills an important tool to be used in conjunction with whatever other skills and experience a particular situation may demand.

PART IV

Mathematical Review Notes

CHAPTER 15

Calculus

15.1 INTRODUCTION

The calculus is the most important working mathematical tool in engineering analysis. In fact, the development of the calculus in the seventeenth century by Newton, Leibniz, and others was primarily motivated by the need to establish a mathematical framework that was adequate for the description of physical phenomena. The detailed techniques of the calculus require substantial practice to develop the necessary proficiency, and it is for that reason that all science and engineering students take one or more courses in the subject. The basic principles, however, are few and simple. This chapter is intended to serve as a review and reference for those principles that are required elsewhere in the text. It is neither rigorous nor complete, and a calculus text should be consulted when needed.

15.2 FUNCTION

A *function* is a rule for performing operations on one variable to obtain values of another. For example, when we write

$$y = t^2$$

we mean "for a given value of t compute the value of y by squaring the value of t." When $t = 2$, the value of y is 4, when $t = 3$, the value of y is 9, and so on. In physical terms we might say "flow rate is a function of height," meaning that when the height is specified, we can calculate the flow rate. The calculational rule might be an equation, as above, or a table or a graph. Symbolically we write

$$y = f(t)$$

"y is a function of t" or

$$q = f(h)$$

"flow rate, q, is a function of height, h." Notice that the symbol $f(t)$ is simply a shorthand for the words "is a function of t" and does not tell us anything about the rule for computing y from t.

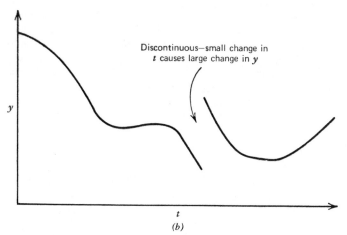

FIGURE 15.1 y plotted as a function of t. (a) Continuous function. (b) Discontinuous function.

t (or h) is referred to as the *independent variable*, or the *argument of the function*, and y (or q) is the *dependent variable*. A function is *continuous* if very small changes in the independent variable cause only small changes in the dependent variable, discontinuous otherwise (see Figure 15.1). Most (but not all) physical processes are described by continuous functions.

15.3 DERIVATIVE

The calculus is concerned primarily with changes of continuous functions. Consider the function $y = f(t)$ shown graphically in Figure 15.2. At a value t_1 of the

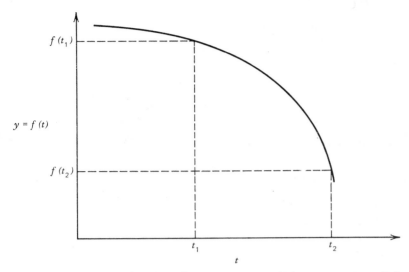

FIGURE 15.2 y as a function of t. At $t = t_1$, $y = f(t_1)$; at $t = t_2$, $y = f(t_2)$.

independent variable the function has a value calculated by $f(t_1)$, and value $f(t_2)$ at t_2. The *change* in going from t_1 and t_2 is denoted

$$\Delta_{1\to 2} f \equiv f(t_2) - f(t_1)$$

while the corresponding change in the independent variable is

$$\Delta_{1\to 2} t \equiv t_2 - t_1$$

The *average rate of change* of $f(t)$ is then

$$\text{average rate of change}_{1\to 2} = \frac{\Delta_{1\to 2} f}{\Delta_{1\to 2} t} = \frac{f(t_2) - f(t_1)}{t_2 - t_1}$$

(For simplicity we shall henceforth write only Δ instead of $\Delta_{1\to 2}$, but the latter is always understood.) Notice that the average rate of change $\Delta f/\Delta t$ is simply the slope of the straight line chord in Figure 15.3 between $f(t_1)$ and $f(t_2)$.

The average rate of change is a rather gross feature of the entire interval from t_1 to t_2. Had we chosen a different point, t_3, for example, we could write

$$\text{average rate of change}_{t_1 \to t_3} = \frac{\Delta f}{\Delta t} = \frac{f(t_3) - f(t_1)}{t_3 - t_1}$$

and the average rate of change would be the slope of the straight line chord in Figure 15.3 between $f(t_1)$ and $f(t_3)$. In general, we might consider any two values of the independent variable, t_1 and $t_1 + \Delta t$, with corresponding values of the dependent variable calculated by $f(t_1)$ and $f(t_1 + \Delta t)$. For various values of Δt the average rate of change, $\Delta f/\Delta t$, is obtained from the slope of one of the straight lines in Figure 15.4. Notice that as Δt gets smaller and smaller the slope of the

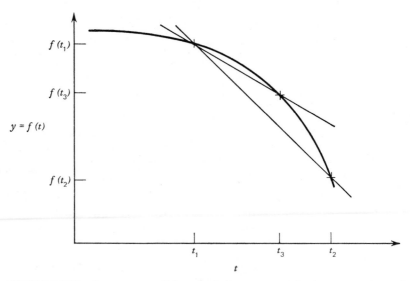

FIGURE 15.3   Average rate of change of y between t_1 and subsequent other points.

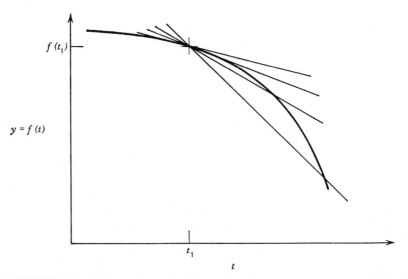

FIGURE 15.4   Average rate of change of y between t_1 and subsequent other points. As the later point approaches t_1 the average rate of change approaches the tangent.

chord approaches the tangent to $f(t)$ at $t = t_1$ more and more closely. We can write

$$\text{rate of change } at\ t_1 = \frac{\Delta f}{\Delta t} \text{ for vanishingly small } \Delta t.$$

Symbolically we denote the rate of change *at* t_1, the tangent, by df/dt and "$\Delta f/\Delta t$ for vanishingly small t" as $\lim\limits_{\Delta t \to 0} \Delta f/\Delta t$. Thus

$$\frac{df}{dt} = \lim_{\Delta t \to 0} \frac{\Delta f}{\Delta t} = \lim_{\Delta t \to 0} \frac{f(t_1 + \Delta t) - f(t_1)}{t_1 + \Delta t - t_1} \tag{15.1}$$

Example 15.1

$f(t) = t^2$. Compute df/dt.

$$\frac{df}{dt} = \lim_{\Delta t \to 0} \frac{f(t_1 + \Delta t) - f(t_1)}{\Delta t} = \lim_{\Delta t \to 0} \frac{[t_1 + \Delta t]^2 - [t_1]^2}{\Delta t}$$

$$= \lim_{\Delta t \to 0} \frac{2t_1 \Delta t + [\Delta t]^2}{\Delta t} = \lim_{\Delta t \to 0} 2t_1 + \Delta t = 2t_1$$

If we consider any value of t, then we can drop the subscript "1" and write

$$\frac{d[t^2]}{dt} = 2t$$

The slope or tangent, df/dt, is itself a function of the independent variable. We call this new function the *derivative of f* and frequently denote it by the symbol $f'(t)$. Its derivative is the second derivative of f, d^2f/dt^2 or $f''(t)$. The nth derivative is denoted d^nf/dt^n, or $f^{(n)}(t)$. The following useful relations concerning derivatives can all be obtained directly from the material presented here:

(a) $\qquad f(t) = t^k, \qquad \dfrac{df}{dt} = kt^{k-1}$ $\hfill (15.2)$

(b) $f(t)$ is the sum of N functions, $f_1(t), f_2(t), \ldots, f_N(t)$:

$$f(t) = \sum_{n=1}^{N} f_n(t), \qquad \frac{df}{dt} = \sum_{n=1}^{N} \frac{df_n}{dt} \tag{15.3}$$

(c) $f(t)$ is a constant multiple of a function $g(t)$:

$$f(t) = kg(t), \qquad \frac{df}{dt} = k\frac{dg}{dt} \tag{15.4}$$

(d) $f(t)$ is a product of two functions, $g(t)$ and $h(t)$:

$$f(t) = g(t)h(t), \qquad \frac{df}{dt} = g(t)\frac{dh}{dt} + \frac{dg}{dt}h(t) \tag{15.5}$$

(e) The function $f(t)$ gives a maximum value when $t = t_m$:

$$f(t_m) = \text{maximum}, \qquad \frac{df}{dt} = 0 \ \text{ and } \ \frac{d^2f}{dt^2} < 0 \ \text{ at } \ t = t_m \tag{15.6}$$

(f) The exponential function, written e^t or exp (t), can be defined by the series:

$$e^t = 1 + t + \frac{t^2}{2!} + \frac{t^3}{3!} + \frac{t^4}{4!} + \cdots$$

$$\frac{de^t}{dt} = e^t \tag{15.7}$$

(g) The functions sine and cosine can be defined by the series:

$$\sin t = t - \frac{t^3}{3!} + \frac{t^5}{5!} - \frac{t^7}{7!} + \cdots$$

$$\cos t = 1 - \frac{t^2}{2!} + \frac{t^4}{4!} - \frac{t^6}{6!} + \cdots$$

$$\frac{d \sin t}{dt} = \cos t, \qquad \frac{d \cos t}{dt} = -\sin t \tag{15.8}$$

15.4 CHAIN RULE

It often happens that the dependent variable in one relation is the independent variable in another. For example, pressure drop (independent variable) may determine a flow rate (dependent variable), but the flow rate (independent variable) will then determine a reactor conversion (dependent variable). We thus have

$$\text{function 1: } g(t) \qquad \text{function 2: } f(g)$$

or $f(g(t))$. Now, when t changes, this causes a change in g which in turn causes a change in f. We may wish to know the rate of change of f with t, df/dt.

From the definition, Equation 15.1,

$$\frac{df}{dt} = \lim_{\Delta t \to 0} \frac{f(g(t + \Delta t)) - f(g(t))}{\Delta t}$$

It is convenient to denote $g(t + \Delta t)$ as $g + \Delta g$ and to note that Δg vanishes as Δt goes to zero. Then

$$\frac{df}{dt} = \lim_{\Delta t \to 0} \frac{f(g + \Delta g) - f(g)}{\Delta t}$$

$$= \lim_{\Delta t \to 0} \frac{f(g + \Delta g) - f(g)}{\Delta g} \cdot \frac{\Delta g}{\Delta t} = \frac{df}{dg} \cdot \frac{dg}{dt} \tag{15.9}$$

This is known as the *chain rule of differentiation*.

Example 15.2

$$f(g) = e^{g(t)}, \qquad \frac{df}{dt} = e^{g(t)} \frac{dg}{dt}$$

If, say,

$$g(t) = kt^n, \qquad \frac{df}{dt} = e^{kt^n} \cdot nkt^{n-1}$$

The chain rule is sometimes useful in finding the derivative of a new function. For example, the natural logarithm, $\ln t$, can be defined by the relation

$$e^{\ln t} = t$$

The derivative of the logarithm can be found by setting $g(t) = \ln t, f(g) = e^g$. Then

$$\frac{df}{dt} = \frac{dt}{dt} = 1 = \frac{df}{dg}\frac{dg}{dt} = e^{\ln t} \cdot \frac{d \ln t}{dt} = t\frac{d \ln t}{dt}$$

or,

$$\frac{d \ln t}{dt} = \frac{1}{t} \qquad (15.10)$$

15.5 INTEGRATION

Thus far we have been concerned with differentiation, the procedure for computing the rate at which something occurs. Sometimes we are given the rate at which a process occurs and we wish to calculate the total extent of the process. (How many miles does an automobile travel in three hours at sixty miles an hour?) If the rate is constant this is easy:

$$\text{total} = \text{rate} \times \text{time}$$

This simple formula does not work, of course, if the rate is changing in time and a more precise effort is required.

Let $f(t)$ denote the total amount at any time t. The amount accumulated between times t_0 and t_N is then $f(t_N) - f(t_0)$. Notice that

$$
\begin{aligned}
f(t_N) - f(t_0) &= [f(t_N) - f(t_{N-1})] \\
&\quad + [f(t_{N-1}) - f(t_{N-2})] + \cdots + [f(t_1) - f(t_0)] \\
&= \sum_{n=1}^{N} [f(t_n) - f(t_{n-1})] \\
&= \sum_{n=1}^{N} \frac{[f(t_n) - f(t_{n-1})]}{t_n - t_{n-1}} [t_n - t_{n-1}] \\
&= \sum_{n=1}^{N} \frac{\Delta f}{\Delta t} \Delta t
\end{aligned}
$$

Now, if t_N represents some fixed value of t, then the value of the left-hand side of this equation does not depend on the number of increments into which we break up the interval from t_0 to t_N. As the number of increments gets large, Δt becomes small and the ratio $\Delta f / \Delta t$ approaches the derivative, df/dt. Thus, we can write

$$f(t_N) - f(t_0) = \lim_{\Delta t \to 0} \sum_{n=1}^{N} \frac{\Delta f}{\Delta t} \cdot \Delta t \qquad (15.11)$$

This limit, called the *integral*, is represented by the symbol $\int$, or script S for summation:

$$f(t_N) - f(t_0) = \int_{t_0}^{t_N} \frac{df}{dt} dt$$

For notational convenience we shall denote the derivative function, df/dt, by the name $\phi(t)$ and write

$$f(t_N) - f(t_0) = \int_0^{t_N} \phi(t)\, dt$$

The notation $\phi(t)\, dt$ refers to "all values of the independent variable between t_0 and t_N" and could just as easily be written $\phi(\tau)\, d\tau$, $\phi(s)\, ds$, etc. The argument is what is called a "dummy variable." If t_N refers to any time and we denote it simply by t then we have

$$f(t) = f(t_0) + \int_{t_0}^{t} \phi(\tau)\, d\tau; \qquad \phi \equiv \frac{df}{dt} \tag{15.12}$$

where, to avoid confusion between time t, the "upper limit of the integral," and "all times between t_0 and t," we have denoted the dummy variable by τ.

Most of the discussion here has really been one concerning nomenclature, but Equation 15.12 formally gives the relation between the rate and total amount. The *evaluation* of an integral is essentially one of antidifferentiation. Given $\phi(t)$ we find $f(t)$ by knowing what function $f(t)$ has the derivative $\phi(t)$. For example, suppose $\phi(t) = t^n$. We know

$$\frac{dt^n}{dt} = nt^{n-1}$$

Thus, if

$$f(t) = \frac{1}{n+1} t^{n+1}$$

$$\frac{df}{dt} = t^n$$

Then

$$\int_{t_0}^{t} \phi(\tau)\, d\tau = \int_{t_0}^{t} \tau^n\, d\tau = f(t) - f(t_0)$$

$$= \frac{1}{n+1} t^{n+1} - \frac{1}{n+1} t_0^{n+1}$$

[The notation $f(\tau)\Big|_{t_0}^{t}$ is sometimes used for $f(t) - f(t_0)$]. Similarly, since

$$\frac{de^{kt}}{dt} = ke^{kt}$$

$$\int_{t_0}^{t} e^{k\tau}\, d\tau = \frac{1}{k} e^{kt} - \frac{1}{k} e^{kt_0}$$

Tables of integrals are available which include large numbers of functions. To use a table it is important to distinguish between definite and indefinite integration. The *definite* integral is the relation that we have been using:

$$\int_{t_0}^{t} \phi(\tau)\, d\tau = f(t) - f(t_0)$$

The *indefinite* integral is written without limits as

$$\int \phi(\tau)\, d\tau = f(t) + \text{constant}$$

The constant, of course, is simply $-f(t_0)$. Tables list indefinite integrals and do not include the constant.

The following relations follow directly from the discussion here:

(a) $$\int_{t_0}^{t} [\phi(\tau) + \psi(\tau)]\, d\tau = \int_{t_0}^{t} \phi(\tau)\, d\tau + \int_{t_0}^{t} \psi(\tau)\, d\tau \qquad (15.13)$$

(b) $$\int_{t_0}^{t} a\phi(\tau)\, d\tau = a \int_{t_0}^{t} \phi(\tau)\, d\tau \qquad (15.14)$$

(c) $$\int_{t_0}^{t} \phi(\tau)\, d\tau = \int_{t_0}^{t_1} \phi(\tau)\, d\tau + \int_{t_1}^{t} \phi(\tau)\, d\tau \qquad (15.15)$$

(d) Functions $f(t)$ and $g(t)$ have derivatives $f'(t)$ and $g'(t)$, respectively. Then

$$\int_{t_0}^{t} f(\tau)g'(\tau)\, d\tau = f(t)g(t) - f(t_0)g(t_0) - \int_{t_0}^{t} f'(\tau)g(\tau)\, d\tau \qquad (15.16)$$

This last relation is known as *integration by parts*.

Example 15.3

$$\phi(t) = te^{kt}$$

Set

$$f(t) = t, \quad f'(t) = 1; \quad g(t) = \frac{1}{k} e^{kt}, \quad g'(t) = e^{kt}$$

$$\phi(t) = f(t)g'(t)$$

$$\int_{t_0}^{t} \phi(\tau)\, d\tau = \int_{t_0}^{t} \tau e^{k\tau}\, d\tau = \frac{1}{k} te^{kt} - \frac{1}{k} t_0 e^{kt_0} - \int_{t_0}^{t} \frac{1}{k} e^{k\tau}\, d\tau$$

$$= \frac{1}{k} te^{kt} - \frac{1}{k} t_0 e^{kt_0} - \frac{1}{k^2} e^{kt} + \frac{1}{k^2} e^{kt_0}$$

15.6   AREA AND NUMERICAL INTEGRATION

The integral can be given a simple geometrical meaning by returning to the definition in Equation 15.11. Consider the function $\phi(t)$ plotted on Figure 15.5. The difference quotient $\Delta f/\Delta t$ is a constant approximation to $\phi(t)$ over the interval Δt (Equation 15.1), so that we can plot on the same graph the lines $\Delta f/\Delta t$, which give us the "staircase" function shown in Figure 15.5. Now, $\Delta f/\Delta t$ (height) × Δt(base) is simply the area of a rectangle, so $\sum [\Delta f/\Delta t]\, \Delta t$ is the area under the staircase function. As Δt gets smaller, the number of rectangles increases and the staircase function approaches more and more closely to the function $\phi(t)$. *The integral*, the limit of this idea, *is then the area under the curve* $\phi(t)$. Though formalized by Leibniz, this principle dates in part to Archimedes.

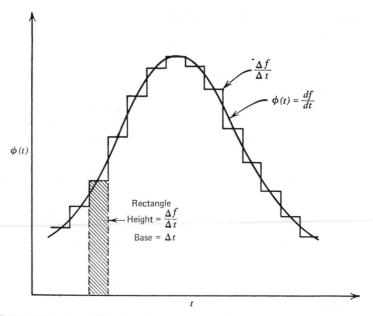

FIGURE 15.5 $\phi = df/dt$ plotted versus t. The function f, the integral of df/dt, is the area under the curve.

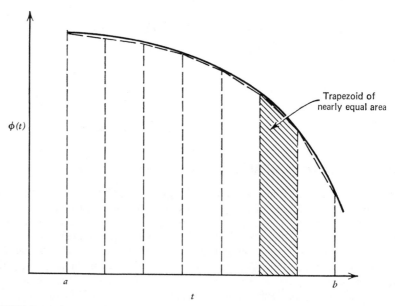

FIGURE 15.6 Approximation of the area under the curve by trapezoids of nearly equal area.

452

It is not always possible to find a simple function $f(t)$ for a given derivative $\phi(t)$. By exploiting the notion of the integral as an area we can, nevertheless, obtain numerical values for integrals. Consider Figure 15.6. The area under the curve $\phi(t)$ between a and b is close to the area under the trapezoids. Thus

$$\int_a^b \phi(t)\, dt \approx \sum_{N\ \text{trapezoids}} (\text{average height}) \times \Delta t$$

For the trapezoid between t_{n-1} and t_n, average height $= [\phi(t_{n-1}) + \phi(t_n)]/2$, while $\Delta t = [b - a]/N$. Thus

$$\int_a^b \phi(t)\, dt \approx \frac{b - a}{N} \{ \tfrac{1}{2}[\phi(a) + \phi(t_1)]$$

$$+ \tfrac{1}{2}[\phi(t_1) + \phi(t_2)] + \cdots + \tfrac{1}{2}[\phi(t_{N-1}) + \phi(b)]\}$$

$$= \frac{b - a}{N} \left[\tfrac{1}{2}\phi(a) + \sum_{n=1}^{N-1} \phi(t_n) + \tfrac{1}{2}\phi(b) \right] \quad (15.17)$$

Equation 15.17 is known as the *trapezoidal rule*. More accurate equations are available, but the trapezoidal rule is adequate for our purposes.

Example 15.4

Evaluate $\int_0^3 e^t\, dt$; $a = 0$, $b = 3$, $\phi(t) = e^t$

(a)   analytical: $\displaystyle\int_0^3 e^t\, dt = e^3 - e^0 = 19.086$

(b) trapezoidal rule:

(i) $N = 2$; three points, $a = 0$, $t_1 = 1.5$, $b = 3$

$$\int_0^3 e^t\, dt \approx \tfrac{3}{2}[\tfrac{1}{2}e^0 + e^{1.5} + \tfrac{1}{2}e^3]$$

$$= \tfrac{3}{2}[\tfrac{1}{2} \times 1 + 4.48 + \tfrac{1}{2} \times 20.09] = 22.54$$

(ii) $N = 4$; five points, $a = 0$, $t_1 = 0.75$, $t_2 = 1.5$, $t_3 = 2.25$, $b = 3.0$

$$\int_0^3 e^t\, dt \approx \tfrac{3}{2}[\tfrac{1}{2}e^0 + e^{0.75} + e^{1.50} + e^{2.25} + \tfrac{1}{2}e^3]$$

$$= \tfrac{3}{4}[\tfrac{1}{2} \times 1 + 2.12 + 4.48 + 9.49 + \tfrac{1}{2} \times 20.09] = 19.98$$

(iii) $N = 6$; seven points, $a = 0$, $t_1 = 0.5$, $t_2 = 1.0$, $t_3 = 1.5$, $t_4 = 2.0$, $t_5 = 2.5$, $b = 3$

$$\int_0^3 e^t\, dt \approx \tfrac{3}{6}[\tfrac{1}{2}e^0 + e^{0.5} + e^{1.0} + e^{1.5} + e^{2.0} + e^{2.5} + \tfrac{1}{2}e^{3.0}]$$

$$= \tfrac{3}{6}[\tfrac{1}{2} \times 1.0 + 1.65 + 2.72 + 4.48 + 7.39 + 12.18 + \tfrac{1}{2} \times 20.09]$$

$$= 19.48$$

Evidently, as the number of terms is increased, the numerical value estimated from the trapezoidal rule comes closer and closer to the exact value. Figure 15.7 is a

```
C        INTEGRATION USING TRAPEZOIDAL RULE.
C        THIS PROGRAM IS WRITTEN IN FORTRAN-IV LANGUAGE AND HAS BEEN
C        RUN ON A XDS9300 COMPUTER.
         READ (105,999) A,B,N
  999 FORMAT (2F10.0,I5)
         M = N-1
         X = 0.0
         SUM = 0.0
         DO 10  I = 1,M
         RN = N
         X = X+(B-A)/RN
   10 SUM = SUM+Y(X)
         AREA = (B-A)*(SUM+0.5*(Y(A)+Y(B)))/RN
         WRITE (108,998) AREA
  998 FORMAT (21H VALUE OF INTEGRAL = E10.4)
         CALL EXIT
         END
         FUNCTION Y(X)
C        INSERT FUNCTION OF INTEREST ON NEXT CARD.
         Y = EXP(X)
         RETURN
         END
C        DATA FOR EXAMPLE 15.4
         0.0         3.0      6
```

FIGURE 15.7   Fortran IV program for numerical integration using the trapezoidal rule.

Fortran IV program for integration using the trapezoidal rule. The data are for Example 15.4.

15.7 MEAN VALUE THEOREM

The interpretation of the integral as an area leads directly to a result known as the *mean value theorem*. The situation is shown in Figure 15.8. There must be a rectangle with base $b - a$ which has the same area as that under the curve of the function $\phi(t)$

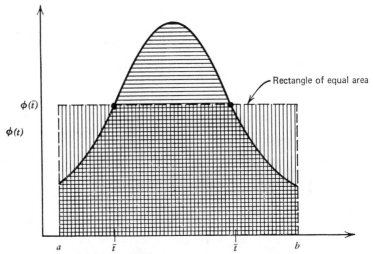

FIGURE 15.8   Geometrical interpretation of the mean value theorem. There is a rectangle with base *b-a* with the same area as the function $\phi(t)$.

between a and b. Furthermore, it is evident that the height of the rectangle is greater than the minimum value of $\phi(t)$ and less than the maximum. If $\phi(t)$ is continuous in the interval $a \leq t \leq b$ then $\phi(t)$ must take on all values between its maximum and minimum. Thus, the height of the rectangle must correspond to the value of $\phi(t)$ at some point, say $\bar{t}$, between a and b. That is

$$\phi \text{ continuous:} \int_a^b \phi(t)\, dt = \phi(\bar{t})[b - a], \qquad a \leq \bar{t} \leq b \qquad (15.18)$$

If ϕ represents mass and t position, then in mechanics the point $\bar{t}$ is known as the center of mass, or center of gravity.

15.8 TAYLOR'S THEOREM

Taylor's theorem is a direct consequence of the mean value theorem which enables us to estimate the value of a function at one point from its value and behavior at another. It is probably the most frequently used result in applications of the calculus. The theorem states that if we know the value of a function and its first n derivatives at $t = t_0$, then we can estimate the value of the function elsewhere with an error that goes like $(t - t_0)^{n+1}$.

In developing Taylor's theorem we shall use as an example the case $t_0 = 0$ and $n = 2$. The general case, which we will state and use, is obtained in the same way but at the expense of a bit more calculation. The function $f(t)$ will be expressed as a power series in t with an error term, the error representing the difference between the exact function at any value of t and the series:

$$f(t) = f_0 + f_1 t + e(t)$$

Here f_0 and f_1 are constants and $e(t)$ is a function which is zero at $t = 0$ and has a derivative, $e'(t)$, which is also zero at $t = 0$. Setting $t = 0$ we find that

$$f(0) = f_0$$

Taking the derivative of $f(t)$ yields

$$f'(t) = f_1 + e'(t)$$

and setting t to zero gives

$$f_1 = f'(0)$$

The second derivation is

$$f''(t) = e''(t)$$

and, since from the definition of an integral,

$$f''(t) = f''(0) + \int_0^t f'''(\tau)\, d\tau$$

we can apply the mean value theorem to obtain

$$e''(t) = f''(t) = f''(0) + f'''(\bar{t})t, \qquad 0 \leq \bar{t} \leq t$$

Notice that $f''(0)$ and $f'''(\bar{t})$ are constants. Then

$$e'(t) = \int_0^t e''(\tau)\,d\tau = f''(0)t + \tfrac{1}{2}f'''(\bar{t})t^2$$

$$e(t) = \int_0^t e'(\tau)\,d\tau = \tfrac{1}{2}f''(0)t^2 + \tfrac{1}{6}f'''(\bar{t})t^3$$

The function $f(t)$ can then be written

$$f(t) = f(0) + f'(0)t + \tfrac{1}{2}f''(0)t^2 + \tfrac{1}{6}f'''(\bar{t})t^3$$

That is, given the value of the function and its first two derivatives at $t = 0$, we can compute the function at other values of t within a term proportional to t^3. For small t, of course, t^3 is negligible.

The general relation for any n and any point t_0 is

$$f(t) = f(t_0) + f'(t_0)[t - t_0] + \frac{1}{2!}\,f''(t_0)[t - t_0]^2$$

$$+ \cdots + \frac{1}{n!}\,f^{(n)}(t_0)[t - t_0]^n + \frac{1}{(n+1)!}\,f^{(n+1)}(\bar{t})[t - t_0]^{n+1}$$

$$= \sum_{k=0}^{n} \frac{f^{(k)}(t_0)}{k!}\,[t - t_0]^k + \frac{f^{(k+1)}(\bar{t})}{(k+1)!}\,[t - t_0]^{k+1}, \qquad t_0 \le \bar{t} \le t$$

$$(15.19)$$

Geometrically this is simply a statement that in the neighborhood of a point t_0 a function can be approximated to various levels of accuracy by (Figure 15.9):

(a) a constant, $f(t) \approx f(t_0)$

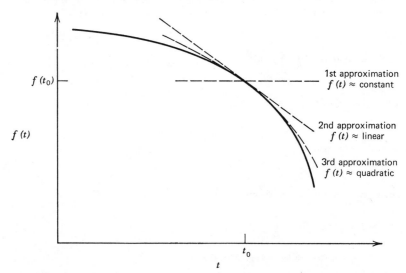

FIGURE 15.9   Various levels of approximation to the function $f(t)$ near $t = t_0$.

(b) a straight line, tangent at the point t_0:

$$f(t) \approx f(t_0) + f'(t_0)[t - t_0]$$

(c) a parabola, tangent at t_0 with the same curvature as the function:

$$f(t) \approx f(t_0) + f'(t_0)[t - t_0] + \tfrac{1}{2}f''(t_0)[t - t_0]^2$$

and so on. The term evaluated between t_0 and t gives an estimate of the error involved in the approximation.

Example 15.5

Compute the three term Taylor series for e^t about $t = 0$ and estimate the accuracy for $0 \le t \le 1$

$$f(t) = e^t, \quad f'(t) = e^t, \quad f''(t) = e^t, \quad f'''(t) = e^t$$

$$f(0) = 1, \quad f'(0) = 1, \quad f''(0) = 1$$

$$e^t = 1 + t + \tfrac{1}{2}t^2 + \tfrac{1}{6}e^t \cdot t^3$$

The error term, $\tfrac{1}{6}e^t \cdot t^3$, is always positive and is bounded in $0 \le t \le 1$ by $(e/6)t^3$. Thus, at any t, $0 \le t \le 1$

$$1 + t + \tfrac{1}{2}t^2 \le e^t \le 1 + t + \tfrac{1}{2}t^2 + \frac{e}{6}t^3$$

At $t = 0.5$, then, $1.625 \le e^{0.5} \le 1.681$, with the true value approximately 1.649.

15.9 SEPARABLE DIFFERENTIAL EQUATIONS

We sometimes have a situation in which we have an integral, $\int_{x_0}^{x} f(\xi)\, d\xi$, in which the upper limit, x, is a function of an independent variable, t, and we wish to compute the rate of change of the value of the integral as t changes. Then

$$\frac{d\left[\int_{x_0}^{x} f(\xi)\, d\xi\right]}{dt} = \lim_{\Delta t \to 0} \frac{\int_{x_0}^{x+\Delta x} f(\xi)\, d\xi - \int_{x_0}^{x} f(\xi)\, d\xi}{\Delta t}$$

$$= \text{(from Equation 15.15)} \lim_{\Delta t \to 0} \frac{\int_{x}^{x+\Delta x} f(\xi)\, d\xi}{\Delta t}$$

$$= \text{(from Equation 15.18)} \lim_{\Delta t \to 0} f(\bar{x}) \frac{\Delta x}{\Delta t} = f(x) \frac{dx}{dt} \qquad (15.20)$$

where the last step follows from the fact that if $x \le \bar{x} \le x + \Delta x$ then as $\Delta t \to 0$, $x + \Delta x \to x$ and so must any point in between.

A *differential equation* is an equation relating a function and its derivatives. A *first-order differential equation* involves only the first derivative. If x is a function of *t* then a first-order equation can be written

$$\frac{dx}{dt} = F(x(t), t) \tag{15.21}$$

That is, the rate of change of *x* depends on both *t* and the value of *x* itself at *t*. Notice that in order to solve for *x* as a function of *t* an integration is required, so *x* must be known at some value of *t* in order to compute the constant of integration.

A first-order equation is *separable* if it can be arranged into the form

$$f(x)\frac{dx}{dt} = g(t) \tag{15.22}$$

That is, separated into "*x*-terms" on the left and "*t*-terms" on the right. The solution then follows from Equation 15.20, for Equation 15.22 can then be rewritten

$$\frac{d}{dt}\left[\int_{x_0}^{x} f(\xi)\, d\xi\right] = g(t)$$

The left-hand side is a derivative, so if we integrate we obtain

$$\int_{x_0}^{x} f(\xi)\, d\xi = \int_{t_0}^{t} g(\tau)\, d\tau \tag{15.23}$$

Here t_0 is the time at which *x* has the value x_0.

Example 15.6

$$\frac{dx}{dt} = kx, \quad x = x_0 \text{ at } t = t_0$$

$$\frac{1}{x}\frac{dx}{dt} = k$$

$$\int_{x_0}^{x} \frac{d\xi}{\xi} = \int_{t_0}^{t} k\, d\tau = \ln\frac{x_0}{x} = k[t - t_0]$$

$$x = x_0 e^{k[t - t_0]}$$

Example 15.7

$$\frac{dx}{dt} = -kx^n, \qquad n \neq 1, \qquad x = x_0 \text{ at } t = t_0$$

$$\frac{1}{x^n}\frac{dx}{dt} = -k$$

$$\int_{x_0}^{x} \frac{d\xi}{\xi^n} = \int_{t_0}^{t} k\, d\tau = \frac{x^{1-n}}{1-n} - \frac{x_0^{1-n}}{1-n} = -k[t - t_0]$$

$$x = \{x_0^{1-n} - k[1 - n][t - t_0]\}^{1/[1-n]}$$

Example 15.8

$$\frac{dx}{dt} = -kx^n t^m, \qquad n \neq 1, \qquad m \neq 1, \qquad x = x_0 \text{ at } t = t_0$$

$$\frac{1}{x^n} \frac{dx}{dt} = -kt^m$$

$$\int_{x_0}^{x} \frac{d\xi}{\xi^n} = -\int_{t_0}^{t} k\tau^m \, d\tau = \frac{x^{1-n}}{1-n} - \frac{x_0^{1-n}}{1-n} = -\frac{kt^{m+1}}{m+1} + \frac{kt_0^{m+1}}{m+1}$$

$$x = \left\{ x_0^{1-n} - k \frac{1-n}{1+m} [t^{1+m} - t_0^{1+m}] \right\}^{1/[1-n]}$$

Equation 15.22 is sometimes separated into the shorthand form

$$f(x) \, dx = g(t) \, dt$$

This equation has no meaning, but if it is then integrated between the correct limits it clearly gives the correct answer.

15.10   LINEAR FIRST-ORDER DIFFERENTIAL EQUATION

$$\frac{dx}{dt} + a(t)x = b(t), \qquad x = x_0 \text{ at } t = t_0 \tag{15.24}$$

where $a(t)$ and $b(t)$ are functions of t, occurs often in applications. This equation is not separable but can be made separable by a trick. Rewrite the equation as

$$R(t) \frac{dx}{dt} = -R(t)a(t)x + R(t)b(t)$$

$$= \frac{d}{dt} Rx - \frac{dR}{dt} x$$

The relation on the right follows from the product rule, Equation 15.5. If $R(t)$ satisfies the separable equation

$$\frac{dR}{dt} = a(t)R$$

or

$$R(t) = R(t_0) \exp\left(\int_{t_0}^{t} a(\tau) \, d\tau \right)$$

then we are left with

$$\frac{d}{dt} Rx = R(t)b(t)$$

The left-hand side is a derivative, so we can integrate to obtain

$$R(t)x(t) - R(t_0)x(t_0) = \int_{t_0}^{t} R(\sigma)b(\sigma) \, d\sigma$$

Notice that as long as $R(t_0) \neq 0$ it cancels after substitution of $R(t)$ and we are left with

$$x = x_0 \exp\left(-\int_{t_0}^{t} a(\tau)\, d\tau\right)$$

$$+ \exp\left(-\int_{t_0}^{t} a(\tau)\, d\tau\right) \int_{t_0}^{t} \exp\left(+\int_{t_0}^{\sigma} a(\tau)\, d\tau\right) b(\sigma)\, d\sigma \qquad (15.25)$$

This is simplified slightly by noting that t is a constant insofar as the integration over σ is concerned, so $\exp\left[-\int_{t_0}^{t} a(\tau)\, d\tau\right]$ can be taken inside the integral. Then

$$\exp\left(-\int_{t_0}^{t} a(\tau)\, d\tau\right) \exp\left(+\int_{t_0}^{\sigma} a(\tau)\, d\tau\right)$$

$$= \exp\left(-\left[\int_{t_0}^{t} a(\tau)\, d\tau - \int_{t_0}^{\sigma} a(\tau)\, d\tau\right]\right) = \exp\left(-\int_{\sigma}^{t} a(\tau)\, d\tau\right)$$

Thus

$$x = x_0 \exp\left(-\int_{t_0}^{t} a(\tau)\, d\tau\right) + \int_{t_0}^{t} \exp\left(-\int_{\sigma}^{t} a(\tau)\, d\tau\right) b(\sigma)\, d\sigma \qquad (15.26)$$

Example 15.9

$$\frac{dx}{dt} + kx = b(t), \qquad k = \text{constant}, \qquad x = x_0 \text{ at } t = t_0$$

$$\int_{t_1}^{t_2} k\, d\tau = k[t_2 - t_1]$$

$$x = x_0 e^{-k[t-t_0]} + \int_{t_0}^{t} e^{-k[t-\sigma]} b(\sigma)\, d\sigma$$

$$x = x_0^{-k[t-t_0]} + e^{-kt} \int_{t_0}^{t} e^{k\sigma} b(\sigma)\, d\sigma$$

15.11   PARTIAL DERIVATIVE

A problem may have two independent variables, say space and time, or more than two. A function of two variables is denoted $f(x, y)$. We wish to compute how f changes as x and y change.

If one variable is fixed at a constant value, say $y = y^*$, then $f(x, y)$ really depends on only one variable, x, and we can compute the derivative as

$$\lim_{\Delta x \to 0} \frac{f(x + \Delta x, y^*) - f(x, y^*)}{\Delta x}$$

Similarly, if x is fixed at x^* the derivative is

$$\lim_{\Delta y \to 0} \frac{f(x^*, y + \Delta y) - f(x^*, y)}{\Delta y}$$

Such derivatives are called partial derivatives with respect to x (or y) holding y (or x) constant and denoted by a "curly" d:

$$\left.\frac{\partial f}{\partial x}\right)_{y=y*} = \lim_{\Delta x \to 0} \frac{f(x + \Delta x, y^*) - f(x, y^*)}{\Delta x} \tag{15.27a}$$

$$\left.\frac{\partial f}{\partial y}\right)_{x=x*} = \lim_{\Delta y \to 0} \frac{f(x^*, y + \Delta y) - f(x^*, y)}{\Delta y} \tag{15.27b}$$

Partial derivatives are computed just like ordinary derivatives.

Example 15.10

$$f(x, y) = x^2 e^{ky}$$

$$\left.\frac{\partial f}{\partial x}\right)_{y=\text{constant}} = 2xe^{ky} \qquad \left.\frac{\partial f}{\partial y}\right)_{x=\text{constant}} = kx^2 e^{ky}$$

In analogous fashion we compute *second partial derivatives:*

$$\frac{\partial^2 f}{\partial x^2} = \frac{\partial}{\partial x}\left(\frac{\partial f}{\partial x}\right) = 2e^{ky} \qquad\qquad \frac{\partial^2 f}{\partial y^2} = \frac{\partial}{\partial y}\left(\frac{\partial f}{\partial y}\right) = k^2 x^2 e^{ky}$$

$$\frac{\partial^2 f}{\partial x\, \partial y} = \frac{\partial}{\partial x}\left(\frac{\partial f}{\partial y}\right) = 2kxe^{ky} \qquad \frac{\partial^2 f}{\partial y\, \partial x} = \frac{\partial}{\partial y}\left(\frac{\partial f}{\partial x}\right) = 2kxe^{ky}$$

Notice that

$$\frac{\partial^2 f}{\partial x\, \partial y} = \frac{\partial^2 f}{\partial y\, \partial x}$$

15.12 CHAIN RULE

The chain rule can be extended to functions of more than one variable. For our purposes it suffices to consider the special case of a function of several variables— say, $f(x, y)$—where x and y each depend on a single variable, t. We want to compute the rate of change of f with respect to t.

Now, from the definition of a derivative,

$$\frac{df}{dt} = \lim_{\Delta t \to 0} \frac{f(x(t + \Delta t), y(t + \Delta t)) - f(x(t), y(t))}{\Delta t}$$

$$= \lim_{\Delta t \to 0} \left\{ \frac{f(x(t + \Delta t), y(t + \Delta t)) - f(x(t + \Delta t), y(t))}{\Delta t} \right.$$

$$\left. + \frac{f(x(t + \Delta t), y(t)) - f(x(t), y(t))}{\Delta t} \right\}$$

$$= \lim_{\Delta t \to 0} \left\{ \frac{f(x + \Delta x, y + \Delta y) - f(x + \Delta x, y)}{\Delta y}\frac{\Delta y}{\Delta t} \right.$$

$$\left. + \frac{f(x + \Delta x, y) - f(x, y)}{\Delta x}\frac{\Delta x}{\Delta t} \right\}$$

$$= \frac{\partial f}{\partial y}\frac{\partial y}{dt} + \frac{\partial f}{\partial x}\frac{dx}{dt}$$

The functions $\partial f/\partial x$ and $\partial f/\partial y$ are evaluated for each t at the values $x(t)$ and $y(t)$. The final step in the derivation uses the fact that $\lim\limits_{\Delta t \to 0} x + \Delta x = x$. The general result, when f depends on n functions of t, $x_1(t), x_2(t), \ldots, x_n(t)$ is

$$\frac{df}{dt} = \frac{\partial f}{\partial x_1}\frac{dx_1}{dt} + \frac{\partial f}{\partial x_2}\frac{dx_2}{dt} + \cdots + \frac{\partial f}{\partial x_n}\frac{dx_n}{dt} = \sum_{k=1}^{n}\frac{\partial f}{\partial x_k}\frac{dx_k}{dt} \qquad (15.28)$$

Example 15.11

$$f(x, y) = x^2 e^{ky}$$

$$\frac{df}{dt} = \frac{\partial f}{\partial x}\frac{dx}{dt} + \frac{\partial f}{\partial y}\frac{dy}{dt} = 2xe^{ky}\frac{dx}{dt} + kx^2 e^{ky}\frac{dy}{dt}$$

Let

$$x(t) = t^n, \qquad y(t) = \ln t; \qquad \frac{dx}{dt} = nt^{n-1}, \qquad \frac{dy}{dt} = \frac{1}{t}$$

$$\frac{df}{dt} = 2xe^{ky}nt^{n-1} + kx^2 e^{ky}t^{-1}$$

$$= 2t^n e^{k\ln t}nt^{n-1} + kt^{2n}e^{k\ln t}t^{-1}$$

$$= 2nt^{k+2n-1} + kt^{k+2n-1} = [k + 2n]t^{k+2n-1}$$

CHAPTER 16

Algebraic Equations

16.1 INTRODUCTION

In obtaining numerical solutions to problems we are often reduced to solving one or more algebraic equations. An algebraic equation is one involving some variable but not integrals or derivatives of that variable. The following are examples of algebraic equations:

$$-x^3 + 2x + 1 = 0 \tag{16.1}$$

$$5x - 6 = 0 \tag{16.2}$$

$$3x + y = 7 \tag{16.3a}$$

$$x + 3y = 5 \tag{16.3b}$$

Equations 16.1 and 16.2 involve a single equation for a single variable, x. Equations 16.3 are two equations that must be solved simultaneously for the two variables, x and y. Such systems of equations are sometimes referred to as *coupled* equations.

Algebraic equations are broadly classified as *linear* or *nonlinear*. Linear equations are those in which all variables appear only to the first power and, if there are two or more variables, there are no products. Equations 16.2 and 16.3 are linear. Equation 16.1 is nonlinear because x appears to the third power. Linear equations are significantly easier to solve.

16.2 LINEAR ALGEBRAIC EQUATIONS

Linear algebraic equations are solved by multiplying and dividing all terms by constants and adding and subtracting in order to manipulate the equations into a new form which gives the answer explicitly. For example, Equation 16.2 is solved by dividing each term by 5 to obtain

$$x - \tfrac{6}{5} = 0$$

or $x = 6/5$. Equations 16.3 are solved by multiplying each term of the second equation by 3 to yield

$$3x + y = 7$$

$$3x + 9y = 15$$

then subtracting the first from the second to give

$$8y = 8$$

or $y = 1$. Equation 16.3b can then be written

$$x + 3 = 5$$

or $x = 2$.

When the number of equations and unknowns is small—say, two or three—solution is quite elementary. With a large number of equations, however, a completely systematic approach that can be easily automated for computer solution is needed. Of the several such methods available *Gauss elimination* is probably the simplest. Suppose we have N equations for N variables which we denote by x_1, $x_2, \ldots, x_N$:

$$
\begin{aligned}
a_{11}x_1 &+ a_{12}x_2 + \cdots + a_{1N}x_N = b_1 \\
a_{21}x_1 &+ a_{22}x_2 + \cdots + a_{2N}x_N = b_2 \\
&\vdots \\
a_{N1}x_1 &+ a_{N2}x_2 + \cdots + a_{NN}x_N = b_N
\end{aligned}
\tag{16.4}
$$

Here, the a's and b's are given numbers. The system of equations is systematically manipulated into *triangular* form:

$$
\begin{aligned}
x_1 + c_{12}x_2 + c_{13}x_3 &+ \cdots + c_{1,N-1}x_{N-1} + c_{1N}x_N = d_1 \\
x_2 + c_{23}x_3 &+ \cdots + c_{2,N-1}x_{N-1} + c_{2N}x_N = d_2 \\
&\vdots \\
x_{N-1} &+ c_{N-1,N}x_N = d_{N-1} \\
&x_N = d_N
\end{aligned}
\tag{16.5}
$$

The solution is then obtained by solving in turn for $x_N, x_{N-1}, \ldots, x_1$.

The mechanics of Gauss elimination are illustrated by means of the following example:

$$+2x_1 + 3x_2 + 5x_3 = +9 \qquad \text{(a1)}$$

$$+4x_1 - x_2 + 3x_3 = +11 \qquad \text{(b1)}$$

$$-x_1 + 2x_2 + 2x_3 = -1 \qquad \text{(c1)}$$

$$x_1 + \tfrac{3}{2}x_2 + \tfrac{5}{2}x_3 = +\tfrac{9}{2} \qquad \text{(a2)} = \frac{+1}{+2}\,\text{(a1)}$$

$$-7x_2 - 7x_3 = -7 \qquad \text{(b2)} = \text{(b1)} - \frac{+4}{+2}\,\text{(a1)}$$

$$+\tfrac{7}{2}x_2 + \tfrac{9}{2}x_3 = \frac{+7}{2} \qquad \text{(c2)} = \text{(c1)} - \frac{-1}{+2}\,\text{(a1)}$$

$$x_1 + \tfrac{3}{2}x_2 + \tfrac{5}{2}x_3 = \tfrac{9}{2} \qquad \text{(a3)} = \text{(a2)}$$

$$x_2 + 1x_3 = +1 \qquad \text{(b3)} = \frac{+1}{-7}\,\text{(b2)}$$

$$1x_3 = 0 \qquad \text{(c3)} = \text{(c2)} - \frac{+\tfrac{7}{2}}{-7}\,\text{(b2)}$$

At this point the system is in the triangular form of Equation 16.5. One addition step leads to solution:

$$x_3 = 0 \qquad \text{(c4)} = \text{(c3)}$$

$$x_2 = +1 \qquad \text{(b4)} = \text{(b3)} - 1\text{(c4)}$$

$$x_1 = +3 \qquad \text{(a4)} = \text{(a3)} - \tfrac{3}{2}\text{(b4)} - \tfrac{5}{2}\text{(c4)}$$

The process of triangularization involves only dividing by the sequence of circled coefficients, called *pivots*, multiplication of each of the remaining equations by an appropriate constant, and subtraction. Figure 16.1 is a Fortran IV program for solution of N equations by this method, where the particular data shown are for the example used here.

In practice, two modifications are necessary. First, if a zero pivot should arise the equations must be rearranged. This modification is included in the program. Second, some prior rearrangement of terms is normally carried out to ensure that numbers of similar magnitude are subtracted from one another in order to maintain accuracy in computer calculations. We shall not go into those details here.

16.3 NONLINEAR ALGEBRAIC EQUATIONS

Nonlinear algebraic equations are called *nth order*, or *n*th order polynomial equations, when the unknown quantity appears to integer powers up to the *n*th

```
C       GAUSS ELIMINATION PROGRAM
C       THIS PROGRAM IS WRITTEN IN FORTRAN-IV LANGUAGE AND HAS BEEN
C       RUN ON A XDS9300 COMPUTER.
C       THIS PROGRAM SOLVES UP TO 99 SIMULTANEOUS ALGEBRAIC EQUATIONS.
C       *   *   *   *   *     NOTATION    *   *   *   *   *
C       N = NUMBER OF EQUATIONS
C       A = COEFFICIENTS
C       B = RIGHT HAND SIDE OF EQUATION
C
        DIMENSION A(99,99),B(99),X(99)
C       READ N, A, AND B.
C       COEFFICIENTS ARE READ IN BY COLUMNS.
        READ (105,999) N
999 FORMAT (I2)
        READ (105,998) ((A(I,J),I=1,N),J=1,N),(B(I),I=1,N)
998 FORMAT (8F10.0)
        M = N-1
        DO 10  I = 1,M
C       TEST FOR ZERO PIVOT AND REARRANGEMENT OF ROWS IF ZERO PIVOT OCCURS
        IF (A(I,I)) 98,97,98
97 DO 40  JJ = I,M
        JI = JJ+1
        IF (A(JI,I)) 96,40,96
40 CONTINUE
96 DO 50  KJ = I,N
        HOLD1 = A(I,KJ)
        A(I,KJ) = A(JI,KJ)
50 A(JI,KJ) = HOLD1
        HOLD2 = B(I)
        B(I) = B(JI)
        B(JI) = HOLD2
C       CONVERSION OF EQUATIONS TO TRIANGULAR FORM
98 KI = I+1
        DO 80  J== KI,N
        RATIO = A(J,I)/A(I,I)
        A(J,I) = 0.0
        DO 20  L = KI,N
20 A(J,L) = A(J,L)-RATIO*A(I,L)
80 B(J) = B(J)-RATIO*B(I)
        DO 70  IK = KI,N
70 A(I,IK) = A(I,IK)/A(I,I)
        B(I) = B(I)/A(I,I)
        A(I,I) = 1.
10 CONTINUE
C       SOLUTION FOR X(I) BY BACK-SUBSTITUTION
        X(N) = B(N)/A(N,N)
        J = M
99 II = J+1
        TOTAL = 0.0
        DO 30  L = II,N
30 TOTAL = TOTAL+A(J,L)*X(L)
        X(J) = B(J)-TOTAL
        J = J-1
        IF (J.GT.0) GO TO 99
C       WRITE RESULTS
        DO 60  J = 1,N
60 WRITE (108,997) J,X(J)
997 FORMAT (6H   X(,I2,3H)= ,E11.5)
        CALL EXIT
        END
C       DATA FOR EXAMPLE PROBLEM
 3
        2.0       4.0      -1.0       3.0      -1.0       2.0       5.0       3.0
        2.0       9.0      11.0      -1.0
```

FIGURE 16.1   Fortran IV program for solution of *N* linear algebraic equations using the Gauss elimination method.

power. Equation 16.1 is a third order, or cubic equation. An important property of nth order equations is that they have exactly n solutions, or roots. Equation 16.1, for example, is satisfied by each of the three values $x = -1.0$, -0.616, $+1.62$. Non-polynomial equations are often called transcendental and may have any number of roots. For example, the equation $\sin x = 0$ has an infinite number of roots, $x = 0$, $\pm\pi$, $\pm 2\pi$, ... In general, solutions can be obtained in terms of an exact formula for quadratic, cubic, and some quartic polynomial equations. In all other cases approximate procedures must be used.

Obviously a solution to a single nonlinear equation in one unknown can be obtained by simply graphing the function. This does not efficiently produce precise roots, however, and in particular does not generalize to the case of many equations. Thus we shall deal with analytical techniques.

Solutions to nonlinear equations are obtained by trial-and-error, or *iterative* methods. In an iterative solution we estimate the root of an equation and then use some systematic procedure to calculate a new (better) estimate from the amount by which the equation is not satisfied, continuing the process until the error in the solution is within some specified tolerance. The simplest iterative procedure is *direct substitution*. Here, we write the equation in the form

$$x = g(x) \tag{16.6}$$

For example, we would write Equation 16.1 as

$$x = \tfrac{1}{2}[x^3 - 1]$$

If the nth estimate of the solution is denoted x_n, then the $n + 1$st estimate is calculated from

$$x_{n+1} = g(x_n) \tag{16.7}$$

To solve Equation 16.1 we write

$$x_{n+1} = \tfrac{1}{2}[x_n{}^3 - 1]$$

If we take a first estimate as $x_0 = 0$ then we have

$$x_1 = \tfrac{1}{2}[0^3 - 1] = -0.500$$
$$x_2 = \tfrac{1}{2}[-0.500^3 - 1] = -0.562$$
$$x_3 = \tfrac{1}{2}[-0.562^3 - 1] = -0.589$$

Continuing the process we obtain the sequence of approximations x_4, x_5, x_6, ..., as -0.602, -0.609, -0.613, -0.615, -0.616, ... Convergence is evidently obtained to the root $x = -0.616$.

It is instructive to examine this procedure graphically. x and $g(x) = [x^3 - 1]/2$ are plotted versus x on the same coordinates in Figure 16.2. The solutions $x = -1$, $x = -0.616$, and $x = 1.62$ are at the intersections of the two curves. The iterative solution can be followed from the sequence of arrows. Initially, for $x_0 = 0$, $g(x_0) = -1/2$. Going to the line $x = x$ (the 45° line) we can read off the next value of $g(x)$ from the plot of $g(x)$ at the same x-value. Returning to the 45° line we obtain the next value, and so on, and it is clear how convergence occurs. Notice that starting to the left of the root convergence will occur in the same way.

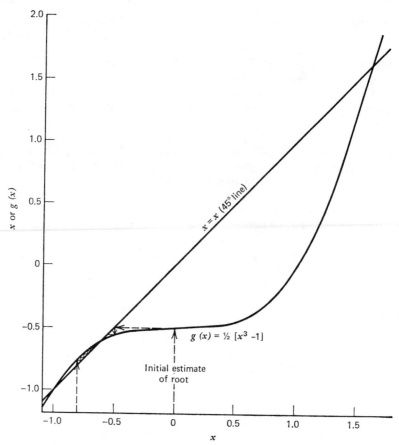

FIGURE 16.2 Graphical representation of the solution of Equation 16.1 using the method of direct substitution. Both initial estimates shown converge to the root of $x = -0.616$.

This latter point deserves further attention, for it is clear that no matter how close to the root at $x = -1$ we begin, we can never converge to it by this method, as shown graphically in Figure 16.3. A similar result holds at $x = +1.62$. Figure 16.4 shows other possible modes of convergence and divergence of the iterations for different functions $g(x)$. It can be shown that convergence to a root will occur only if $|dg/dx| < 1$ both at the root and in the region about the root including x_0. This criterion is violated in the example at the roots -1 and 1.62.

It is sometimes necessary to employ a trick to put the equation into the form of Equation 16.6. For example

$$x^2 - 2 = 0$$

is not of the required form, but the equivalent equation

$$x^2 - 2 - \beta x + \beta x = 0$$

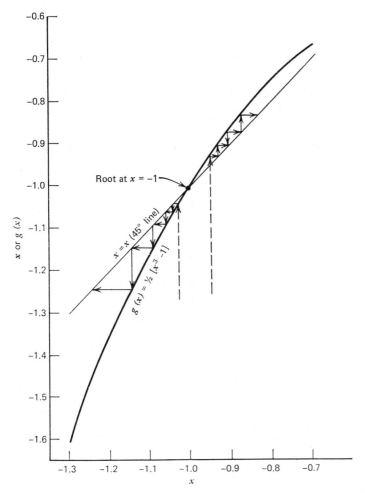

FIGURE 16.3   Graphical representation of the solution of Equation 16.1 in the neighborhood of the root at $x = -1$ using the method of direct substitution.

can be written

$$x_{n+1} = \frac{1}{\beta} \left[2 + \beta x_n - x_n^2 \right]$$

If we take $x_0 = 1$, then the convergence criterion requires $\beta > 1$. Taking $\beta = 2$ gives the sequence 1, 1.5, 1.375, 1.429, ... which is approaching the root 1.414 in an oscillatory manner.

There are more elegant modifications of direct substitution with better convergence properties, but we shall not examine them here. The method clearly requires little manipulation and calculation, but the strong possibility that a root will not be found, even with a good estimate of the solution, is the price paid for simplicity.

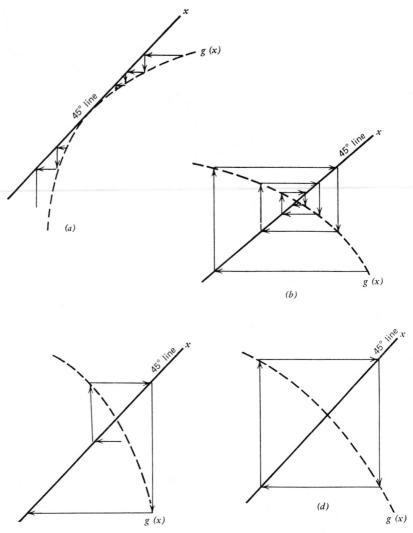

FIGURE 16.4 Modes of convergence for the method of direct substitution. (a) Convergence from one side, divergence from the other. (b) Oscillatory convergence. (c) Oscillatory divergence. (d) Cycling.

The extension to several nonlinear equations is obvious. If we have two equations in the variables x and y we write them in the form

$$x = g(x, y) \qquad y = h(x, y) \tag{16.8}$$

The iterations are then carried out with the equations

$$x_{n+1} = g(x_n, y_n) \qquad y_{n+1} = h(x_n, y_n) \tag{16.9}$$

More rapid convergence might be obtained using the alternate formulation employing the new value of x to calculate y:

$$x_{n+1} = g(x_n, y_n) \qquad y_{n+1} = h(x_{n+1}, y_n) \tag{16.10}$$

Example 16.1

The equations for effluent temperature and concentration with a first-order reaction in an adiabatic stirred tank reactor are derived in Section 12.9 as

$$qc_{Af} - qc_A - Vk_0 e^{-E/RT} c_A = 0$$
$$\rho q \underline{c}_p T_f - \rho q \underline{c}_p T + Vk_0 e^{-E/RT}[-\Delta \underline{H}_R]c_A = 0$$

The following numerical values are used:

$$q = 100 \text{ ft}^3/\text{hr} \qquad c_{Af} = 0.5 \text{ moles/ft}^3 \qquad V = 30 \text{ ft}^3$$
$$k_0 = 10^9 \text{ min}^{-1} \qquad E = 1.85 \times 10^4 \text{ cal/g-mole}$$
$$R = 1.987 \text{ cal/mole }^\circ\text{K} \qquad \rho = 62.4 \text{ lb}_m/\text{ft}^3$$
$$\underline{c}_p = 1.0 \text{ BTU/lb}_m \text{ }^\circ\text{F} = 1.8 \text{ BTU/lb}_m^\circ\text{k} \qquad T_f = 100^\circ\text{F} = 311^\circ\text{K}$$
$$\Delta \underline{H}_R = -3.8 \times 10^4 \text{ BTU/lb-mole}$$

Denoting c_A by c the equations then become

$$1.667c + 3 \times 10^{10} e^{-9310/T} c - 0.8335 = 0 \tag{16.11a}$$
$$1.872 \times 10^2 T - 11.4 \times 10^{14} e^{-9310/T} c - 5.823 \times 10^4 = 0 \tag{16.11b}$$

or, in the form of Equation 16.8,

$$c = 0.5 - 1.8 \times 10^{10} e^{-9310/T} c$$
$$T = 3.11 \times 10^2 + 6.09 \times 10^{12} e^{-9310/T} c$$

The iteration formula, Equation 16.10, is then

$$c_{n+1} = 0.5 - 1.8 \times 10^{10} e^{-9310/T_n} c_n$$
$$T_{n+1} = 3.11 \times 10^2 + 6.09 \times 10^{12} e^{-9310/T_n} c_{n+1}$$

From initial estimates $c_0 = 0.20$, $T_0 = 200^\circ\text{F} = 366.7^\circ\text{K}$ the following sequence results:

$n =$	0	1	2	3	4
$c_n =$	0.20	0.4661	0.4991	0.4988	0.4991
$T_n = $	366.7	337.7	314.2	311.4	311.3

The results $c = 0.499$, $T = 311.4$, which correspond to essentially no reaction, are obtained from all starting estimates. Yet these nonlinear equations also have roots at $c = 0.2568$, $T = 393.4$ and $c = 0.0080$, $T = 477.5$. These roots cannot be found by direct substitution. In fact, it can be shown that convergence can be obtained only to roots satisfying $T < 316.23$.

In passing it should be noted that although this problem is an excellent example for demonstrating the solution of two nonlinear equations in two unknowns, its particular structure is such that it can be converted into a single equation for T and solved as one complicated equation in one unknown.

16.4 NEWTON–RAPHSON ITERATION

The *Newton–Raphson* method is a calculus-based iterative method for solving nonlinear algebraic equations that will always converge from a good starting estimate and that converges much more rapidly than direct substitution. More effort is required to obtain the solution, however.

We shall represent the algebraic equation in the form

$$f(x) = 0 \tag{16.12}$$

For example, in Equation 16.1 $f(x) = -x^3 + 2x + 1$. We have an estimate x_n of the root, and we assume that x_{n+1} is the true root. Then

$$f(x_{n+1}) = 0$$

But if x_n and x_{n+1} are close together we may expand $f(x_{n+1})$ in a Taylor series about x_n and write

$$f(x_{n+1}) = f(x_n) + f'(x_n)[x_{n+1} - x_n] + \cdots = 0$$

and, neglecting the quadratic terms in the Taylor series, we can solve for x_{n+1}:

$$x_{n+1} = x_n - \frac{f(x_n)}{f'(x_n)} \tag{16.13}$$

Example 16.2

$$f(x) = -x^3 + 2x + 1 = 0$$

$$f'(x) = -3x^2 + 2$$

The iteration formula is then

$$x_{n+1} = x_n - \frac{-x_n{}^3 + 2x_n + 1}{-3x_n{}^2 + 2} = \frac{2x_n{}^3 + 1}{3x_n{}^3 - 2}$$

From various starting values the sequence of calculations is then as follows:

x_0	x_1	x_2	x_3
0	−0.50	−0.60	−0.616
−0.80	−0.50	−0.60	−0.616
−0.85	−1.105	−1.02	−1.00
+3.0	+2.20	+1.78	+1.62

Clearly, depending on the starting value, rapid convergence can be obtained to each of the three roots.

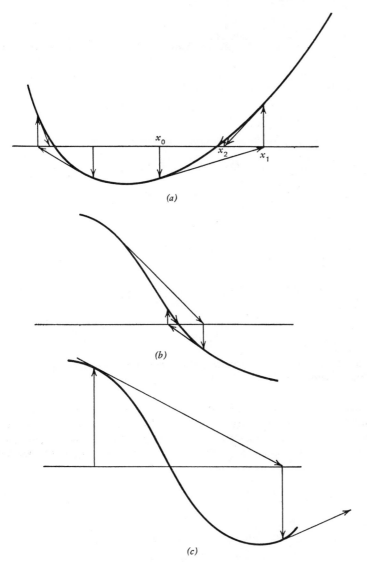

FIGURE 16.5   Modes of convergence of the Newton–Raphson method. (*a*) Convergence to different roots. (*b*) Oscillatory convergence. (*c*) Divergence.

Newton–Raphson iteration is equivalent to approximating the function by its tangent line at the estimate of the root. The types of behavior to be expected are shown in Figure 16.5. Convergence is seen to occur in general provided no maximum or minimum of the function lies between the iteration and the root. Figure 16.6 is a Fortran IV program for Newton–Raphson iteration with the particular function being the one used here.

```
C      NEWTON-RAPHSON PROGRAM
C      THIS PROGRAM IS WRITTEN IN FORTRAN-IV LANGUAGE AND HAS BEEN
C      RUN ON A XDS9300 COMPUTER.
C      READ INITIAL VALUE OF X, CONVERGENCE CRITERION, AND MAXIMUM NUMBER
C      OF ITERATIONS.
       READ (105,999) X,EPS,N
  999 FORMAT (2F10.0,I5)
C      INITIALIZE ITERATION COUNTER
       I = 1
C      VALUES OF THE FUNCTION AND ITS FIRST DERIVATIVE ARE OBTAINED FROM
C      FUNCTION SUBPROGRAMS.
   20 XN = X-Y(X)/DYDX(X)
       WRITE (108,998) XN
  998 FORMAT (E12.5)
C      TEST FOR CONVERGENCE
       IF (ABS(XN-X).LE.EPS.OR.I.GT.N)  GO TO 10
       I = I+1
C      UPDATE X VALUE
       X = XN
       GO TO 20
   10 CALL EXIT
       END
       FUNCTION Y(X)
C      INSERT FUNCTION OF INTEREST ON NEXT CARD.
       Y = -X**3+2.*X+1.
       RETURN
       END
       FUNCTION DYDX(X)
C      INSERT DERIVATIVE OF FUNCTION ON NEXT CARD.
       DYDX = -3.*X*X+2.
       RETURN
       END
C      DATA FOR EXAMPLE PROBLEM
       3.0     .00001    50
```

FIGURE 16.6   Fortran IV program for solution of a nonlinear algebraic equation using the Newton–Raphson method.

The extension of Newton–Raphson iteration to more than one nonlinear equation in more than one unknown requires the use of the multivariate version of Taylor's theorem, which we have not developed. It suffices to note that the procedure is equivalent to approximating an N-dimensional surface by its tangent hyperplane in $N-1$ dimensions. For two equations in two unknowns

$$f(x, y) = 0 \qquad g(x, y) = 0 \tag{16.14}$$

The expansion for iterates x_{n+1}, y_{n+1} about x_n, y_n is

$$f(x_{n+1}, y_{n+1}) = f(x_n, y_n) + \frac{\partial f}{\partial x}(x_n, y_n)[x_{n+1} - x_n]$$

$$+ \frac{\partial f}{\partial y}(x_n, y_n)[y_{n+1} - y_n] + \cdots$$

$$g(x_{n+1}, y_{n+1}) = g(x_n, y_n) + \frac{\partial g}{\partial x}(x_n, y_n)[x_{n+1} - x_n]$$

$$+ \frac{\partial g}{\partial y}(x_n, y_n)[y_{n+1} - y_n] + \cdots$$

and the new estimates x_{n+1}, y_{n+1} are found as solutions of the coupled *linear* equations

$$\left[\frac{\partial f}{\partial x}(x_n, y_n)\right]x_{n+1} + \left[\frac{\partial f}{\partial y}(x_n, y_n)\right]y_{n+1}$$

$$= -f(x_n, y_n) + \frac{\partial f}{\partial x}(x_n, y_n)x_n + \frac{\partial f}{\partial y}(x_n, y_n)y_n \qquad (16.15a)$$

$$\left[\frac{\partial g}{\partial x}(x_n, y_n)\right]x_{n+1} + \left[\frac{\partial g}{\partial y}(x_n, y_n)\right]y_{n+1}$$

$$= -g(x_n, y_n) + \frac{\partial g}{\partial x}(x_n, y_n)x_n + \frac{\partial g}{\partial y}(x_n, y_n)y_n \qquad (16.15b)$$

Example 16.3

Equations 16.11 are written

$$f(c, T) = 1.667c + 3 \times 10^{10}e^{-9310/T}c - 0.8335 = 0$$

$$g(c, T) = 1.872 \times 10^{2}T - 11.4 \times 10^{14}e^{-9310/T}c$$

$$-5.823 \times 10^{4} = 0$$

The required partial derivatives are then

$$\frac{\partial f}{\partial c}(c_n, T_n) = 1.667 + 3 \times 10^{10}e^{-9310/T_n}$$

$$\frac{\partial f}{\partial T}(c_n, T_n) = \frac{9310}{T_n^{2}} \times 3 \times 10^{10}e^{-9310/T_n}c_n$$

$$\frac{\partial g}{\partial c}(c_n, T_n) = -11.4 \times 10^{14}e^{-9310/T_n}$$

$$\frac{\partial g}{\partial T}(c_n, T_n) = 1.872 \times 10^{2} - \frac{9310}{T_n^{2}} \times 11.4 \times 10^{14}e^{9310/T_n}c_n$$

Equations 16.15 are then solved with x and y replaced by c and T, respectively. Convergence of the iterations from various starting estimates is shown in Table 16.1. It is evident that the three solutions can be obtained rapidly, the particular solution depending on the initial estimate.

TABLE 16.1 Convergence of the iterations for concentration and temperature in a CFSTR from various starting values using the Newton–Raphson method

$n =$	0	1	2	3	4	5	6	7
$c_n =$	0.3	0.5296	0.4993	0.4991				
$T_n =$	350.0	301.0	311.3	311.4				
$c_n =$	0.3	0.2568						
$T_n =$	388.0	393.4						
$c_n =$	0.260	0.2518	0.2566	0.2568				
$T_n =$	400.0	395.1	393.4	393.4				
$c_n =$	0.4	−0.0738	−0.0336	−0.0069	0.0054	0.0079	0.0080	
$T_n =$	500.00	505.2	491.6	482.6	478.4	477.6	477.5	
$c_n =$	0.001	0.0042	0.0078	0.0080				
$T_n =$	500.0	478.9	477.6	477.5				
$c_n =$	0.05	−0.1548	−0.0974	−0.0513	−0.0180	0.0011	0.0074	0.0080
$T_n =$	450.0	532.6	513.2	497.6	486.3	479.9	477.8	477.5
$c_n =$	0.023	0.0074	0.0053	0.0079	0.0080			
$T_n =$	450.0	482.8	478.5	477.6	477.5			

CHAPTER 17

Ordinary Differential Equations

17.1 INTRODUCTION

Equations involving a function and its derivative are called *differential equations*. The following are examples of differential equations:

$$2\frac{d^2x}{dt^2} + \frac{dx}{dt} - x = 0 \tag{17.1}$$

$$\frac{d^3x}{dt^3} + 6\frac{d^2x}{dt^2} + 11\frac{dx}{dt} + 6x = 0 \tag{17.2}$$

$$-\tfrac{1}{5}t^3\frac{d^3x}{dt^3} + \tfrac{1}{2}t^2\frac{d^2x}{dt^2} - t\frac{dx}{dt} + x = 0 \tag{17.3}$$

$$\frac{d^2x}{dt^2} - \frac{dx}{dt}x = 0 \tag{17.4}$$

The *order* of a differential equation refers to the highest derivative. Equations 17.1 and 17.4 are second order, for the highest derivative of x is the second, while Equations 17.2 and 17.3 are third order. Equations are also classified as linear or nonlinear. An equation is *linear* if it contains no products or powers of x and its derivatives. Equations 17.1 to 17.3 are all linear, while Equation 17.4 is nonlinear. Notice that by this definition the coefficient of x or its derivatives may depend on the independent variable, as in Equation 17.3. Linear equations are substantially easier to deal with than nonlinear equations. We have already dealt with linear first-order equations in Section 15.10 and those nonlinear first-order equations that are separable in Section 15.9.

17.2 GENERAL PROPERTIES

Differential equations of the form of Equations 17.1 to 17.4 in which each term involves the dependent variable, x, or its derivative, are denoted as *homogeneous* (this usage is accurate only for the linear equations, but we will use it generally). There is an intimate relationship between nth-order homogeneous differential equations and nth-order algebraic equations. For example, like the algebraic equation, the nth order homogeneous differential equation has n solutions. We shall establish this in some generality subsequently, but for now we simply present solutions to the example equations and suggest that the reader substitute them into the equations and verify that they are indeed satisfied:

Equation 17.1 $x = e^{t/2}$ or e^{-t}

Equation 17.2 $x = e^{-t}, e^{-2t}$, or e^{-3t}

Equation 17.3 $x = t, t^2$, or $t^{5/2}$

Equation 17.4 $x = 1$ or $-2t^{-1}$

The linear equations have some further properties. If we denote a solution by x_h, then Cx_h, where C is any constant, is also a solution. Consider Equation 17.1, for example. If e^{-t} is a solution then so is Ce^{-t}, for

$$\frac{dx}{dt} = C\frac{dx_h}{dt} = -Ce^{-t}$$

$$\frac{d^2x}{dt^2} = C\frac{d^2x_h}{dt^2} = Ce^{-t}$$

Thus

$$2\frac{d^2x}{dt^2} + \frac{dx}{dt} - x = 2Ce^{-t} - Ce^{-t} - Ce^{-t} = 0$$

We write the general linear nth order homogeneous equation as

$$a_n(t)\frac{d^nx}{dt^n} + a_{n-1}(t)\frac{d^{n-1}x}{dt^{n-1}} + \cdots + a_1(t)\frac{dx}{dt} + a_0(t)x = 0 \tag{17.5}$$

If x_h denotes a solution, then letting $x = Cx_h$

$$a_n(t)\frac{d^nx}{dt^n} = Ca_n(t)\frac{d^nx_h}{dt^n}, \qquad a_{n-1}(t)\frac{d^{n-1}x}{dt^{n-1}} = Ca_{n-1}(t)\frac{d^{n-1}x_h}{dt^{n-1}}, \ldots$$

and Equation 17.5 can be written

$$Ca_n(t)\frac{d^nx_h}{dt^n} + Ca_{n-1}(t)\frac{d^{n-1}x_h}{dt^{n-1}} + \cdots + Ca_1(t)\frac{dx_h}{dt} + Ca_0(t)x_h$$

$$= C\left[a_n(t)\frac{d^nx_h}{dt^n} + a_{n-1}(t)\frac{d^{n-1}x_h}{dt^{n-1}} + \cdots + a_1(t)\frac{dx_h}{dt} + a_0(t)x_h\right] = 0$$

since the term in brackets is simply the left-hand side of Equation 17.5, to which x_h is a solution. This establishes the result that *any solution of a linear homogeneous equation can be multiplied by an arbitrary constant and remain a solution.*

This result does not extend generally to nonlinear equations. Consider Equation 17.4, which has a solution $x_h = -2t^{-1}$. An arbitrary multiple would then be $x = Ct^{-1}$, in which case

$$\frac{dx}{dt} = -Ct^{-2}, \qquad \frac{d^2x}{dt^2} = 2Ct^{-3}, \qquad x\frac{dx}{dt} = -C^2t^{-3}$$

so that

$$\frac{d^2x}{dt^2} - \frac{dx}{dt}x = 2Ct^{-3} + C^2t^{-3}$$

which equals zero only for $C = -2$, the value given, and the trivial case $C = 0$.

There is a second significant property of linear homogeneous equations. If we have two solutions, say x_{h1} and x_{h2}, then any linear combination is also a solution. We will establish this for the general equation 17.5. Let

$$x = C_1 x_{h1} + C_2 x_{h2}$$

in which case

$$\frac{dx}{dt} = C_1 \frac{dx_{h1}}{dt} + C_2 \frac{dx_{h2}}{dt}$$

$$\frac{d^2x}{dt^2} = C_1 \frac{d^2x_{h1}}{dt^2} + C_2 \frac{d^2x_{h2}}{dt^2}$$

$$\cdot$$
$$\cdot$$
$$\cdot$$

$$\frac{d^n x}{dt^n} = C_1 \frac{d^n x_{h1}}{dt^n} + C_2 \frac{d^n x_{h2}}{dt^n}$$

Substituting into Equation 17.5 the left-hand side becomes

$$a_n(t)\left[C_1 \frac{d^n x_{h1}}{dt^n} + C_2 \frac{d^n x_{h2}}{dt^n} \right]$$

$$+ a_{n-1}(t)\left[C_1 \frac{d^{n-1}x_{h1}}{dt^{n-1}} + C_2 \frac{d^{n-1}x_{h2}}{dt^{n-1}} \right] + \cdots$$

$$+ a_1(t)\left[C_1 \frac{dx_{h1}}{dt} + C_2 \frac{dx_{h2}}{dt} \right] + a_0(t)[C_1 x_{h1} + C_2 x_{h2}]$$

$$= C_1\left[a_n(t)\frac{d^n x_{h1}}{dt^n} + a_{n-1}(t)\frac{d^{n-1}x_{h1}}{dt^{n-1}} + \cdots \right.$$

$$\left. + a_1(t)\frac{dx_{h1}}{dt} + a_0(t)x_{h1} \right]$$

$$+ C_2\left[a_n(t)\frac{d^n x_{h2}}{dt^n} + a_{n-1}(t)\frac{d^{n-1}x_{h2}}{dt^{n-1}} + \cdots \right.$$

$$\left. + a_1(t)\frac{dx_{h2}}{dt} + a_0(t)x_{h2} \right] = 0$$

The equation is satisfied because the coefficients of C_1 and C_2 are each solutions and hence the brackets equal zero. *If we have n solutions to the nth order linear homogeneous equation, denoted* $x_{h1}, x_{h2}, \ldots, x_{hn}$, *then a general solution may be written*

$$x = C_1 x_{h1} + C_2 x_{h2} + \cdots + C_n x_{hn}$$

The constants $C_1, C_2, \ldots, C_n$ represent the n constants of integration obtained by integrating the nth-order equation. For example, a general solution to Equation 17.3 would be

$$x(t) = C_1 t + C_2 t^2 + C_3 t^{5/2}$$

Notice that solutions to Equation 17.4 cannot be added because of the nonlinearity.

17.3 LINEAR HOMOGENEOUS EQUATIONS WITH CONSTANT COEFFICIENTS

When the coefficients $a_0, a_1, a_2, \ldots, a_n$ are *constants*, the n solutions to the homogeneous linear equation can be found easily. We have

$$a_n \frac{d^n x}{dt^n} + a_{n-1} \frac{d^{n-1} x}{dt^{n-1}} + \cdots + a_1 \frac{dx}{dt} + a_0 x = 0 \tag{17.6}$$

Thus, at all values of t the same linear combination of x and its derivatives must sum to zero. This proportionality between a function and its derivatives suggests exponential behavior. We therefore seek a solution of the form

$$x = e^{mt}$$

Then

$$\frac{dx}{dt} = m e^{mt}, \qquad \frac{d^2 x}{dt^2} = m^2 e^{mt}, \ldots, \frac{d^n x}{dt^n} = m^n e^{mt}$$

and the left-hand side of Equation 17.6 becomes

$$a_n m^n e^{mt} + a_{n-1} m^{n-1} e^{mt} + \cdots + a_1 m e^{mt} + a_0 e^{mt} = 0$$

or, factoring out e^{mt},

$$e^{mt}[a_n m^n + a_{n-1} m^{n-1} + \cdots + a_1 m + a_0] = 0$$

Since e^{mt} cannot vanish, the bracket must be equal to zero. *Thus, e^{mt} is a solution of Equation* 17.6 *if and only if m is a solution of the algebraic equation*

$$a_n m^n + a_{n-1} m^{n-1} + \cdots + a_1 m + a_0 = 0 \tag{17.7}$$

Equation 17.7 is called the *characteristic equation*, and its roots are called *characteristic values*, or sometimes, *eigenvalues*. Notice that the characteristic equations can be written down by inspection from the differential equation by substituting m^k for $d^k x/dt^k$. Equation 17.7 is an nth-order algebraic equation with n roots. If we denote these roots by $m_1, m_2, \ldots, m_n$ then

$$x_{h1} = e^{m_1 t}, \ x_{h2} = e^{m_2 t}, \cdots, x_{hn} = e^{m_n t}$$

and the general solution to Equation 17.7 is

$$x(t) = C_1 e^{m_1 t} + C_2 e^{m_2 t} + \cdots + C_n e^{m_n t}$$

Example 17.1

The characteristic equation corresponding to Equation 17.1 is

$$2m^2 + m - 1 = 0$$

with roots $m = 1/2, -1$. Thus

$$x(t) = C_1 e^{t/2} + C_2 e^{-t}$$

Example 17.2

The characteristic equation corresponding to Equation 17.2 is

$$m^3 + 6m^2 + 11m + 6 = 0$$

with roots $m = -1, -2, -3$. Thus

$$x(t) = C_1 e^{-t} + C_2 e^{-2t} + C_3 e^{-3t}$$

Example 17.3

$$\frac{dx}{dt} - kx = 0$$

This linear equation is a separable first-order equation which was solved in Section 15.9. The technique developed here produces a characteristic equation

$$m - k = 0$$
$$m = +k$$
$$x(t) = C e^{kt}$$

17.4 COMPLEX CHARACTERISTIC ROOTS

The characteristic equation might have complex or imaginary roots. If a root is complex then its complex conjugate must also be a root. We will break a complex root into real and imaginary parts and write

$$m = m_R + i m_I$$

where m_R and m_I are real numbers and $i = \sqrt{-1}$. Then

$$m = m_R - i m_I$$

is also a root and the solution includes terms

$$x(t) = C_1 e^{[m_R + i m_I]t} + C_2 e^{[m_R - i m_I]t}$$

The imaginary exponential can be expressed in terms of sine and cosine using relations defined in Section 15.3:

$$e^{i m_I t} = 1 + im_I t + \frac{[im_I t]^2}{2!} + \frac{[im_I t]^3}{3!} + \frac{[im_I t]^4}{4!} + \frac{[im_I t]^5}{5!} + \cdots$$

$$= 1 - \frac{[m_I t]^2}{2!} + \frac{[m_I t]^4}{4!} + \cdots$$

$$+ i \left\{ m_I t - \frac{[m_I t]^3}{3!} + \frac{[m_I t]^5}{5!} + \cdots \right\}$$

$$= \cos m_I t + i \sin m_I t$$

Thus

$$x(t) = C_1 e^{m_R t} [\cos m_I t + i \sin m_I t] + C_2 e^{m_R t} [\cos m_I t - i \sin m_I t]$$

$$= [C_1 + C_2] e^{m_R t} \cos m_I t + i[C_1 - C_2] e^{m_R t} \sin m_I t$$

Since $C_1 + C_2$ and $i[C_1 - C_2]$ are just constants the terms $e^{m_R t} \cos m_I t$ and $e^{m_R t} \sin m_I t$ must be solutions to the homogeneous equation. Thus, when we have complex roots the corresponding homogeneous solutions are

$$x_h = e^{m_R t} \cos m_I t, \, e^{m_R t} \sin m_I t$$

Example 17.4

$$\frac{d^2 x}{dt^2} + \lambda^2 x = 0$$

The characteristic equation is

$$m^2 + \lambda^2 = 0$$

$$m = i\lambda, \, -i\lambda$$

$$m_R = 0, \quad m_I = \lambda$$

Thus the solution may be written either as

$$x(t) = C_1 e^{i\lambda t} + C_2 e^{-i\lambda t}$$

or

$$x(t) = K_1 \cos \lambda t + K_2 \sin \lambda t$$

where, of course, the C's and K's are different constants.

Example 17.5

$$\frac{d^3 x}{dt^3} - 3\frac{d^2 x}{dt^2} + 4\frac{dx}{dt} - 2x = 0$$

$$m^3 - 3m^2 + 4m - 2 = 0$$

$$m = 1, 1 + i, 1 - i$$

$$m_R = 1, \quad m_I = 1$$

$$x(t) = C_1 e^t + C_2 e^{[1+i]t} + C_3 e^{[1-i]t}$$

or

$$x(t) = K_1 e^t + K_2 e^t \cos t + K_3 e^t \sin t$$

17.5   CONSTANTS OF INTEGRATION

The constants of integration are evaluated by information specified about the system at *particular* times.

Example 17.6 (Equation 17.1)

$$2\frac{d^2x}{dt^2} + \frac{dx}{dt} - x = 0 \qquad \begin{aligned} & x = 1 \quad \text{at} \quad t = 0 \\ & \frac{dx}{dt} = -2 \quad \text{at} \quad t = 0 \end{aligned}$$

$$x(t) = C_1 e^{t/2} + C_2 e^{-t}$$

$$\frac{dx}{dt} = \tfrac{1}{2}C_1 e^{t/2} - C_2 e^{-t}$$

at $t = 0$ $\qquad x = C_1 + C_2 = 1$

$$\frac{dx}{dt} = \tfrac{1}{2}C_1 - C_2 = -2$$

$$C_1 = -\tfrac{2}{3}, \qquad C_2 = +\tfrac{5}{3}$$
$$x(t) = -\tfrac{2}{3}e^{t/2} + \tfrac{5}{3}e^{-t}$$

Example 17.7 (Equation 17.1)

$$x = 1 \quad \text{at} \quad t = 0$$
$$x = 2 \quad \text{at} \quad t = 1$$
$$x(t) = C_1 e^{t/2} + C_2 e^{-t}$$
$$x(t) = C_1 e^{t/2} + C_2 e^{-t}$$

at $t = 0$ $\quad x = C_1 + C_2 = 1$

at $t = 1$ $\quad x = C_1 e^{1/2} + C_2 e^{-1} = 2$

$$C_1 = \frac{2 - e^{-1}}{e^{1/2} - e^{-1}}, \qquad C_2 = \frac{e^{1/2} - 2}{e^{1/2} - e^{-1}}$$

$$x(t) = \frac{2 - e^{-1}}{e^{1/2} - e^{-1}} e^{t/2} + \frac{e^{1/2} - 2}{e^{1/2} - e^{-1}} e^{-t}$$

Example 17.8

$$\frac{d^3x}{dt^3} - 3\frac{d^2x}{dt^2} + 4\frac{dx}{dt} - 2x = 0 \qquad \begin{aligned} & \text{at } t = 0 \qquad x = 1 \\ & \frac{dx}{dt} = 0 \\ & \frac{d^2x}{dt^2} = 3 \end{aligned}$$

From Example 17.5

$$x(t) = K_1 e^t + K_2 e^t \cos t + K_3 e^t \sin t$$

$$\frac{dx}{dt} = K_1 e^t + K_2 e^t \cos t - K_2 e^t \sin t + K_3 e^t \sin t + K_3 e^t \cos t$$

$$\frac{d^2 x}{dt^2} = K_1 e^t - 2K_2 e^t \sin t + 2K_3 e^t \cos t$$

at $t = 0$ $\quad x = K_1 + K_2 = 1$

$$\frac{dx}{dt} = K_1 + K_2 + K_3 = 0$$

$$\frac{d^2 x}{dt^2} = K_1 + 2K_3 = 3$$

$$K_1 = 5, \qquad K_2 = -4, \qquad K_3 = -1$$

$$x(t) = 5e^t - 4e^t \cos t - e^t \sin t$$

When all of the information is given at a particular time—say, $t = 0$—the problem is called an *initial value problem*. When more than one time is involved we have a *boundary value problem*. There are some differences in behavior, but most are beyond our brief treatment here.

17.6 REPEATED ROOTS

The characteristic equations might have repeated roots, in which case we obtain only $n - 1$ solutions to the homogeneous differential equation. Since we need n homogeneous solutions to obtain a general solution with n constants of integration, we must obtain an additional homogeneous solution. The following procedure is not at all rigorous but does suggest the correct form, which can then be verified by substitution into the equation.

Consider the case when two roots are almost identical, and call these m_1 and $m_2 = m_1 + \varepsilon$. The solution then contains terms

$$x(t) = C_1 e^{m_1 t} + C_2 e^{[m_1 + \varepsilon]t} + \cdots$$

Since any linear combination of homogeneous solutions is also a homogeneous solution we can write the solution instead in terms of different constants K_1 and K_2 as

$$x(t) = K_1 e^{m_1 t} + K_2 \left\{ \frac{e^{[m_1 + \varepsilon]t} - e^{m_1 t}}{\varepsilon} \right\} + \cdots$$

The second term is a difference quotient and goes to the derivative in the limit as $m_2 \to m_1$ and $\varepsilon \to 0$. Thus we have

$$\varepsilon \to 0: \quad x(t) = K_1 e^{m_1 t} + K_2 \frac{d}{dm_1} \{e^{m_1 t}\} + \cdots$$

or

$$x(t) = K_1 e^{m_1 t} + K_2 t e^{m_1 t} + \cdots$$

That is, if m_1 is a repeated root then $te^{m_1 t}$ is also a solution to the homogeneous equation. In general, *if a root m is repeated k times, then e^{mt}, te^{mt}, $t^2 e^{mt}$, . . . , $t^{k-1} e^{mt}$ are k independent solutions to the homogeneous equation.*

Example 17.9

$$\frac{d^3 x}{dt^3} - 4\frac{d^2 x}{dt^2} + 5\frac{dx}{dt} - 2x = 0$$

$$m^3 - 4m^2 + 5m - 2 = 0$$

$$m = +1, +1, +2$$

$$x(t) = C_1 e^t + C_2 te^t + C_3 e^{2t}$$

Example 17.10

$$\frac{d^3 x}{dt^3} - 6\frac{d^2 x}{dt^2} + 12\frac{dx}{dt} - 8x = 0$$

$$m^3 - 6m^2 + 12m - 8 = 0$$

$$m = +2, +2, +2$$

$$x(t) = C_1 e^{2t} + C_2 te^{2t} + C_3 t^2 e^{2t}$$

17.7 LINEAR NONHOMOGENEOUS EQUATION

The general linear nth-order differential equation is written

$$a_n(t)\frac{d^n x}{dt^n} + a_{n-1}(t)\frac{d^{n-1} x}{dt^{n-1}} + \cdots + a_1(t)\frac{dx}{dt} + a_0(t)x = f(t) \qquad (17.8)$$

where $f(t)$ is often referred to as the *forcing function*. The homogeneous case, Equation 17.5, corresponds to $f(t) = 0$. A solution to Equation 17.8 which satisfies the n conditions specified at particular times is found by the indirect approach of obtaining the n homogeneous solutions, $x_{h1}(t)$, $x_{h2}(t)$, . . . , $x_{hn}(t)$, and an *arbitrary* solution of Equation 17.8, denoted $x_p(t)$ for *particular solution*. An arbitrary solution will be easier to find than a solution that satisfies n additional conditions. The solution is then

$$x(t) = C_1 x_{h1}(t) + C_2 x_{h2}(t) + \cdots + C_n x_{hn}(t) + x_p(t) \qquad (17.9)$$

It is readily verified that Equation 17.9 is a solution to Equation 17.8 because of the linearity of the equation. The constants of integration $C_1, C_2, \ldots, C_n$ are then chosen to satisfy the n extra conditions.

Example 17.11

$$-\tfrac{1}{5}t^3\frac{d^3 x}{dt^3} + \tfrac{1}{2}t^2\frac{d^2 x}{dt^2} - t\frac{dx}{dt} + x = t^3$$

$$\text{at } t = 1, \quad x = \frac{dx}{dt} = \frac{d^2 x}{dt^2} = 0$$

When $f(t) = 0$ this is Equation 17.3, for which the three homogeneous solutions are t, t^2, and $t^{5/2}$. A particular solution for $f(t) = t^3$ is readily verified to be $-5t^3$. Notice that this particular solution does not satisfy any of the conditions imposed at $t = 1$. The general solution is then

$$x(t) = C_1 t + C_2 t^2 + C_3 t^{5/2} - 5t^3$$

as may be verified by substitution back into the equation. The constants of integration are evaluated from the conditions at $t = 1$:

$$\text{at } t = 1, \quad x = C_1 + C_2 + C_3 - 5 = 0$$

$$\frac{dx}{dt} = C_1 + 2C_2 + \tfrac{5}{2}C_3 - 15 = 0$$

$$\frac{d^2x}{dt^2} = 2C_2 + \tfrac{15}{4}C_3 - 30 = 0$$

which can be solved to obtain $C_1 = 5/3$, $C_2 = -10$, $C_3 = 40/3$. The complete solution satisfying the differential equation and initial conditions is then

$$x(t) = \tfrac{5}{3}t - 10t^2 + \tfrac{40}{3}t^{5/2} - 5t^3$$

An important property that follows from linearity and should be noted is that if we have two different forcing functions, say $f_1(t)$ and $f_2(t)$, and $x_{p1}(t)$ and $x_{p2}(t)$ are particular solutions corresponding to $f_1(t)$ and $f_2(t)$, respectively, then a particular solution corresponding to $f(t) = f_1(t) + f_2(t)$ is $x_p(t) = x_{p1}(t) + x_{p2}(t)$.

17.8 METHOD OF UNDETERMINED COEFFICIENTS

The method of undetermined coefficients is a procedure for obtaining particular solutions under two restrictions:

1. The coefficients $a_0, a_1, a_2, \ldots, a_n$ in Equation 17.8 are *constants*.
2. The function $f(t)$ consists only of terms that are each products of powers of t, exponentials, and sines and cosines.

The latter limits the situations for which solutions can be obtained, but the method is so simple that it is an important one to know.

The method for computing $x_p(t)$ is as follows:

1. Let $\phi(t)$ be a typical term in $f(t)$. Set

$$x_p(t) = \beta_0 \phi + \beta_1 \frac{d\phi}{dt} + \beta_2 \frac{d^2\phi}{dt^2} + \beta_3 \frac{d^3\phi}{dt^3} + \cdots \qquad (17.10)$$

(For the types of functions to which the method applies the series will have a finite number of terms!)

2. Substitute $x_p(t)$ into Equation 17.8 and compute $\beta_0, \beta_1, \beta_2, \ldots$ by equating coefficients of like terms.

We shall demonstrate by example, and for algebraic simplicity we will restrict attention to the case $n = 2$ with $a_2 = 1$

$$\frac{d^2x}{dt^2} + a_1 \frac{dx}{dt} + a_0 x = f(t)$$

though it will be apparent that the method applies to any n.

Example 17.12

$$f(t) = \text{constant} = A_0$$

The derivatives of A_0 are all zero, so

$$x_p(t) = \beta_0, \qquad \frac{dx_p}{dt} = \frac{d^2x}{dt^2} = 0$$

$$a_0 \beta_0 = A_0$$

$$\beta_0 = \frac{A_0}{a_0}$$

$$x(t) = C_1 e^{m_1 t} + C_2 e^{m_2 t} + \frac{A_0}{a_0}$$

Example 17.13

$$f(t) = \text{polynomial of order } N = \sum_{k=0}^{N} A_k t^k$$

Each term in x_p is thus a power of t with derivatives that are also powers of t up to t^N. Thus

$$x_p = \sum_{k=0}^{N} \beta_k t^k$$

$$\frac{dx_p}{dt} = \sum_{k=0}^{N} k \beta_k t^{k-1}$$

$$\frac{d^2x_p}{dt^2} = \sum_{k=0}^{N} k[k-1] \beta_k t^{k-2}$$

Then substituting into the differential equation

$$\sum_{k=0}^{N} k[k-1]\beta_k t^{k-2} + a_1 \sum_{k=0}^{N} k\beta_k t^{k-1} + a_0 \sum_{k=0}^{N} \beta_k t^k = \sum_{k=0}^{N} A_k t^k$$

or equating coefficients of each power of t

$$k = N: \qquad\qquad\qquad\qquad\qquad a_0 \beta_N = A_N$$

$$k = N-1: \qquad\qquad\qquad a_0 \beta_{N-1} + a_1 N \beta_N = A_{N-1}$$

$$k = N-2: \quad a_0 \beta_{N-2} + a_1[N-1]\beta_{N-1} + N[N-1]\beta_N = A_{N-2}$$

and for all other k, $0 \le k < N - 2$

$$a_0 \beta_k + a_1[k + 1]\beta_{k+1} + [k + 1][k + 2]\beta_k = A_k$$

Thus, starting from β_N we may solve sequentially for each coefficient in turn. Notice that this set of linear equations for the β's is already in triangular form.

A specific numerical example would be helpful here. Consider

$$\frac{d^2 x}{dt^2} + 5 \frac{dx}{dt} + 6x = -1 + 4t + 3t^2 + 6t^3$$

$$x = 0, \qquad \frac{dx}{dt} = 1 \quad \text{at} \quad t = 0$$

The characteristic equation is

$$m^2 + 5m + 6 = 0$$

with roots $m = -2, -3$. Thus

$$x(t) = C_1 e^{-2t} + C_2 e^{-3t} + x_p(t)$$

Now

$$x_p(t) = \beta_0 + \beta_1 t + \beta_2 t^2 + \beta_3 t^3$$

$$\frac{dx_p}{dt} = \qquad \beta_1 + 2\beta_2 t + 3\beta_3 t^2$$

$$\frac{d^2 x_p}{dt^2} = \qquad\qquad 2\beta_2 + 6\beta_3 t$$

Substituting into the differential equation gives

$$[2\beta_2 + 6\beta_3 t] + [5\beta_1 + 10\beta_2 t + 15\beta_3 t^2]$$
$$+ [6\beta_0 + 6\beta_1 t + 6\beta_2 t^2 + 6\beta_3 t^3] = -1 + 4t + 3t^2 + 6t^3$$

Equating coefficients of like powers of t gives

$$
\begin{array}{llll}
t^0: & 6\beta_0 + 5\beta_1 + 2\beta_2 & & = -1 \\
t^1: & \qquad 6\beta_1 + 10\beta_2 + 6\beta_3 & = 4 \\
t^2: & \qquad\qquad 6\beta_2 + 15\beta_3 & = 3 \\
t^3: & \qquad\qquad\qquad 6\beta_3 & = 6
\end{array}
$$

The solution is $\beta_3 = 1$, $\beta_2 = -2$, $\beta_1 = 3$, $\beta_0 = -2$. Thus

$$x = C_1 e^{-2t} + C_2 e^{-3t} - 2 + 3t - 2t^2 + t^3$$

Now, at $t = 0$

$$x(0) = 0 = C_1 + C_2 - 2$$

$$\frac{dx}{dt}(0) = 1 = -2C_1 - 3C_2 + 3$$

or $C_1 = 4$, $C_2 = -2$. Thus, finally, the complete solution to the example is

$$x(t) = 4e^{-2t} - 2e^{-3t} - 2 + 3t - 2t^2 + t^3$$

Example 17.14

$$f(t) = A_0 e^{\alpha t}$$

All derivatives of $e^{\alpha t}$ are also proportional to $e^{\alpha t}$. Thus

$$x_p = \beta_0 e^{\alpha t}, \qquad \frac{dx_p}{dt} = \alpha \beta_0 e^{\alpha t}, \qquad \frac{d^2 x_p}{dt^2} = \alpha^2 \beta_0 e^{\alpha t}$$

The equation is then

$$\alpha^2 \beta_0 e^{\alpha t} + a_1 \alpha \beta_0 e^{\alpha t} + a_0 \beta_0 e^{\alpha t} = A_0 e^{\alpha t}$$

or

$$\beta_0 = \frac{A_0}{\alpha^2 + a_1 \alpha + a_0}$$

$$x(t) = C_1 e^{m_1 t} + C_2 e^{m_2 t} + \frac{A_0}{\alpha^2 + a_1 \alpha + a_0} e^{\alpha t}$$

Example 17.15

$$f(t) = A_0 + A_1 \sin \omega t + A_2 \cos \omega t$$

The derivatives of sines and cosines are themselves sinces and cosines so the particular solution will have the form

$$x_p = \beta_0 + \beta_1 \sin \omega t + \beta_2 \cos \omega t$$

$$\frac{dx_p}{dt} = \omega \beta_1 \cos \omega t - \omega \beta_2 \sin \omega t$$

$$\frac{d^2 x_p}{dt^2} = -\omega^2 \beta_1 \sin \omega^2 t - \omega^2 \beta_2 \cos \omega t$$

Substituting into the equation gives

$$[-\omega^2 \beta_1 \sin \omega t - \omega^2 \beta_2 \cos \omega t] + a_1[\omega \beta_1 \cos \omega t - \omega \beta_2 \sin \omega t]$$
$$+ a_0[\beta_0 + \beta_1 \sin \omega t + \beta_2 \cos \omega t] = A_0 + A_1 \sin \omega t + A_2 \cos \omega t$$

Equating the coefficients of t^0, $\sin \omega t$, and $\cos \omega t$ gives

$$t^0: \qquad\qquad a_0 \beta_0 = A_0$$
$$\sin \omega t: \quad \beta_1[a_0 - \omega^2] - a_1 \omega \beta_2 = A_1$$
$$\cos \omega t: \quad a_1 \omega \beta_1 + [a_0 - \omega^2]\beta_2 = A_2$$

or

$$\beta_0 = A_0/a_0$$

$$\beta_1 = \frac{A_1[a_0 - \omega^2] + a_1 \omega A_2}{[a_0 - \omega^2]^2 + a_1^2 \omega^2}$$

$$\beta_2 = \frac{A_2[a_0 - \omega^2] + a_1 \omega A_1}{[a_0 - \omega^2]^2 + a_1^2 \omega^2}$$

The complete solution is then

$$x(t) = C_1 e^{m_1 t} + C_2 e^{m_2 t} + A_0/a_0 + \frac{1}{[a_0 - \omega^2]^2 + a_1^2 \omega^2}$$

$$\times \{[A_1 \langle a_0 - \omega^2 \rangle + a_1 \omega A_2] \sin \omega t + [A_2 \langle a_0 - \omega^2 \rangle + a_1 \omega A_1] \cos \omega t\}$$

It is a useful aside here to note that if a_0 and a_1 are positive, then $e^{m_1 t}$ and $e^{m_2 t}$ go to zero, so that for long times the system response is a sinusoidal oscillation about a mean value A_0/a_0, independent of the initial conditions, which enter only through C_1 and C_2.

17.9 CONVOLUTION INTEGRAL

For the equation with constant coefficients

$$a_n \frac{d^n x}{dt^n} + a_{n-1} \frac{d^{n-1} x}{dt^{n-1}} + \cdots + a_1 \frac{dx}{dt} + a_0 x = f(t)$$

a particular solution can be written for arbitrary $f(t)$. This solution has the property that $x_p(t)$ and its first $n - 1$ derivatives are all zero at $t = 0$. We shall not derive the result, which is obtained as follows:

1. Let $w(t) = K_1 e^{m_1 t} + K_2 e^{m_2 t} + \cdots + K_n e^{m_n t}$ be a homogeneous solution. Evaluate $K_1, K_2, \ldots, K_n$ from the following conditions at $t = 0$.

$$w = \frac{dw}{dt} = \frac{d^2 w}{dt^2} = \cdots = \frac{d^{n-2} w}{dt^{n-2}} = 0$$

$$\frac{d^{n-1} w}{dt^{n-1}} = \frac{1}{a_n}$$

(17.11)

2. Then

$$x_p(t) = \int_0^t \{K_1 e^{m_1 [t-\tau]} + K_2 e^{m_2 [t-\tau]} + \cdots + K_n e^{m_n [t-\tau]}\} f(\tau) \, d\tau$$

(17.12)

$$x_p = \frac{dx_p}{dt} = \frac{d^2 x_p}{dt^2} = \cdots = \frac{d^{n-1} x_p}{dt^{n-1}} = 0 \quad \text{at} \quad t = 0$$

Example 15.9 is a special case of this result when $n = 1$.

Example 17.16

$$\frac{d^2 x}{dt^2} + 5 \frac{dx}{dt} + 6x = f(t)$$

$$x = 0, \quad \frac{dx}{dt} = 1 \quad \text{at} \quad t = 0$$

Then

$$w(t) = K_1 e^{-2t} + K_2 e^{-3t}$$

$$w(0) = K_1 + K_2 = 0$$

$$\frac{dw}{dt}(0) = -2K_1 - 3K_2 = \frac{1}{a_2} = 1$$

$$K_1 = 1, \qquad K_2 = -1$$

$$x_p = \int_0^t \{e^{-2[t-\tau]} - e^{-3[t-\tau]}\} f(\tau)\, d\tau$$

$$x(t) = C_1 e^{-2t} + C_2 e^{-3t} + \int_0^t \{e^{-2[t-\tau]} - e^{-3[t-\tau]}\} f(\tau)\, d\tau$$

$$x(0) = 0 = C_1 + C_2 + 0$$

$$\frac{dx}{dt}(0) = 1 = -2C_1 - 3C_2 + 0$$

$$C_1 = 1, \qquad C_2 = -1$$

$$x(t) = e^{-2t} - e^{-3t} + \int_0^t \{e^{-2[t-\tau]} - e^{-3[t-\tau]}\} f(\tau)\, d\tau$$

If we now set $f(t) = -1 + 4t + 3t^2 + 6t^3$ and carry out the integration we obtain the result found by the method of undetermined coefficients in Example 17.13.

The function $w(t)$ is known as the *weighting function*. An integral of the form of Equation 17.12 is known as a *convolution*, or *Faltüng integral*.

17.10 COUPLED EQUATIONS

In applications it often happens that we obtain not a single differential equation of nth order, but rather n coupled equations of first order. These situations are equivalent, as we shall demonstrate for the second-order system.

Suppose we have two equations for the functions $x(t)$ and $y(t)$:

$$\frac{dx}{dt} = -a_{11}x - a_{12}y + f_1(t) \tag{17.13a}$$

$$\frac{dy}{dt} = -a_{21}x - a_{22}y + f_2(t) \tag{17.13b}$$

For convenience we take a_{11}, a_{12}, a_{21}, and a_{22} as constants. We can eliminate y from these equations by first differentiating Equation 17.13a to obtain

$$\frac{d^2x}{dt^2} = -a_{11}\frac{dx}{dt} - a_{12}\frac{dy}{dt} + \frac{df_1}{dt}$$

We substitute dy/dt from Equation 17.13b

$$\frac{d^2x}{dt^2} = a_{11}\frac{dx}{dt} - a_{12}[-a_{21}x - a_{22}y + f_2(t)] + \frac{df_1}{dt}$$

and finally we obtain $a_{12}y$ from Equation 17.13a:

$$\frac{d^2x}{dt^2} = -a_{11}\frac{dx}{dt} + a_{12}a_{21}x - a_{22}\left[\frac{dx}{dt} + a_{11}x - f_1(t)\right] - a_{12}f_2(t) + \frac{df_1}{dt}$$

Rearranging we have

$$\frac{d^2x}{dt^2} + a_1\frac{dx}{dt} + a_0x = f(t) \tag{17.14}$$

where

$$a_1 = a_{11} + a_{22}, \qquad a_0 = a_{11}a_{22} - a_{12}a_{21},$$

$$f(t) = a_{22}f_1(t) + \frac{df_1}{dt} - a_{12}f_2(t) \tag{17.15}$$

If we are given initial conditions $x = x_0$, $y = y_0$ at $t = t_0$ then we have equivalent conditions from Equation 17.13a:

$$x = x_0, \qquad \frac{dx}{dt} = -a_{11}x_0 - a_{12}y_0 + f_1(t_0) \quad \text{at} \quad t = t_0$$

It easily follows that $y(t)$ satisfies the second-order equation

$$\frac{d^2y}{dt^2} + a_1\frac{dy}{dt} + a_0y = a_{11}f_2(t) + \frac{df_2}{dt} - a_{21}f_1(t)$$

$$y = y_0, \qquad \frac{dy}{dt} = -a_{21}x_0 - a_{22}y_0 + f_2(t_0) \quad \text{at} \quad t = t_0$$

That is, the homogeneous equations (and thus the characteristic equations) are the same.

It is also true that one second-order equation is equivalent to two coupled first-order equations, a result which is sometimes helpful in numerical computation. Here the result is not unique, as long as Equation 17.15 is satisfied. For example, Equation 17.14 can be represented by the equivalent equations

$$\frac{dx}{dt} = -a_1x + y, \qquad \frac{dy}{dt} = -a_0x + f(t)$$

or

$$\frac{dx}{dt} = y, \qquad \frac{dy}{dt} = -a_0x - a_1y + f(t)$$

17.11 NONLINEAR EQUATIONS

We have developed some rather general analytical procedures for the solution of linear ordinary differential equations. Such procedures do not exist in general for nonlinear equations. In some very special cases analytical solutions for nonlinear equations can be obtained. For example, equations of the form

$$\frac{d^2x}{dt^2} + \frac{dF(x)}{dx} = 0$$

where $F(x)$ is some nonlinear function of x, can be solved by multiplying the equation by dx/dt to obtain

$$\frac{dx}{dt}\frac{d^2x}{dt^2} + \frac{dF}{dx}\frac{dx}{dt} = 0$$

This is equivalent to

$$\frac{d}{dt}\left\{\frac{1}{2}\left[\frac{dx}{dt}\right]^2 + F(x)\right\} = 0$$

The derivative is then integrated to

$$\frac{1}{2}\left[\frac{dx}{dt}\right]^2 + F(x) = C_1$$

where

$$C_1 = \frac{1}{2}\left[\frac{dx}{dt}\right]^2 + F(x) \quad \text{at} \quad t = 0$$

Thus, we obtain the separable equation

$$[C_1 - F(x)]^{-1/2}\frac{dx}{dt} = 2^{1/2}$$

which has a solution

$$t = \int_{x_0}^{x} \frac{d\xi}{2^{1/2}[C_1 - F(\xi)]^{1/2}}$$

Even in this special case it is likely that the final integration will require a numerical solution using the trapezoid rule or some other method. In most cases nonlinear equations must simply be solved numerically from the start.

17.12   NUMERICAL SOLUTION

In developing a numerical solution of a differential equation we compute the value of the function at a sequence of discrete values of the independent variable. Suppose we seek the function $x(t)$ at points $t_0, t_1, t_2, \ldots$, where $t_{n+1} - t_n = \Delta t$ is a fixed number. For simplicity we denote $x(t_n)$ by x_n. If $x(t)$ is the solution of the (usually nonlinear) equation

$$\frac{dx}{dt} = F(x, t), \qquad x_0 \text{ given} \tag{17.16}$$

then we may formally integrate between t_n and t_{n+1} to write

$$x_{n+1} = x_n + \int_{t_n}^{t_{n+1}} F(x(t), t)\, dt \tag{17.17}$$

The problem is how to evaluate the integral, since, of course, we do not know the function $x(t)$.

The simplest approximation is to take $F(x(t), t)$ as constant between t_n and t_{n+1}. Then

$$\int_{t_n}^{t_{n+1}} F(x(t), t) \, dt \simeq F(x_n, t_n) \, \Delta t$$

and

$$x_{n+1} \simeq x_n + F(x_n, t_n) \, \Delta t \tag{17.18}$$

Since x_0 is known we can compute, in turn, $x_1, x_2, \ldots$. This simplest approximation is known as *Euler's method*.

Example 17.7

$$\frac{dx}{dt} = F(x, t) = x, \qquad x_0 = 1$$

We consider this linear equation, which has an exact solution $x = e^t$, so that we can compare the numerical and exact solution. Equation 17.18 becomes

$$x_{n+1} = x_n + x_n t = x_n[1 + t]$$

Take $\Delta t = 0.10$. Then we obtain the following results.

t	0	0.1	0.2	0.3	0.4
Euler's method	1.00	1.10	1.21	1.33	1.46
Exact solution	1.00	1.11	1.22	1.35	1.49

There is, clearly, an error in the calculation which is cumulative as t gets larger. For this case we can compute the error exactly, for x_n can be shown to be equal to

$$x_n = [1 + \Delta t]^n = \left[1 + \frac{t}{n}\right]^n$$

As $n \to \infty$ for fixed t this is, of course, e^t, so it is clear that the smaller Δt, the better the approximation. The relative error is

$$\text{relative error} = \frac{|e^t - x_{n, \text{Euler}}|}{e^t} = \frac{|e^t - \{1 + t/n\}^n|}{e^t}$$

At $t = 1$, for example, for $\Delta t = 0.2, 0.1,$ and 0.05 ($n = 5, 10, 20$), respectively, we obtain a relative error of $0.08, 0.04,$ and 0.03.

A better approximation to the integral in Equation 17.17 is obtained from the trapezoidal rule, Equation 15.17:

$$\int_{t_n}^{t_{n+1}} F(x(t), t) \, dt \simeq \tfrac{1}{2}[F(x_n, t_n) + F(x_{n+1}, t_{n+1})] \, \Delta t$$

Thus

$$x_{n+1} \simeq x_n + \tfrac{1}{2}F(x_n, t_n) \, \Delta t + \tfrac{1}{2}F(x_{n+1}, t_{n+1}) \, \Delta t$$

Since we do not know x_{n+1} in order to evaluate $F(x_{n+1}, t_{n+1})$ we can approximate it by, say, Equation 17.18 and write, finally,

$$x_{n+1} \simeq x_n + \tfrac{1}{2}F(x_n, t_n) \, \Delta t + \tfrac{1}{2}F(x_n + F(x_n, t_n) \, \Delta t, t_{n+1}) \, \Delta t \tag{17.19}$$

Equation 17.19 is the simplest of a class of numerical schemes called *Runge-Kutta methods*. It is a second-order Runge-Kutta equation.

Example 17.18

$$F(x, t) = x.$$

Then

$$F(x_n, t_n) = x_n$$

$$F(x_n + F(x_n, t_n) \, \Delta t, t_n) = x_n + F(x_n, t_n) \, \Delta t = x_n + x_n \, \Delta t$$

Thus, Equation 17.19 becomes

$$x_{n+1} \simeq x_n + \tfrac{1}{2} x_n \, \Delta t + \tfrac{1}{2} [x_n + x_n \, \Delta t] \, \Delta t$$

$$= x_n [1 + \Delta t + \tfrac{1}{2} \Delta t^2]$$

For $\Delta t = 0.10$ and $x_0 = 1.00$ we obtain $x_1 = 1.11, x_2 = 1.22, x_3 = 1.35, x_4 = 1.49$, and so on, which agrees to three significant figures with the exact solution.

The error can be computed exactly again for this linear example by noting that

$$x_n = [1 + \Delta t + \tfrac{1}{2}\Delta t^2]^n = \left[1 + \frac{t}{n} + \frac{1}{2}\frac{t^2}{n^2} \right]^n$$

Thus, at $t = 1$ the relative error is less than 0.01 for $n = 5$ ($\Delta t = 0.20$), a considerable improvement over the simple Euler method for little extra computational effort. Figure 17.1 is a Fortran IV program for the second-order Runge–Kutta method, with the data shown from Example 17.18.

The most commonly used Runge–Kutta method for digital computation is the fourth order.

```
C       SECOND ORDER RUNGE-KUTTA METHOD
C       THIS PROGRAM IS WRITTEN IN FORTRAN-IV LANGUAGE AND HAS BEEN
C       RUN ON A XDS9300 COMPUTER.
C       READ INITIAL VALUES OF T AND X, INTERVAL BETWEEN SUCCESSIVE VALUES
C       OF T, AND MAXIMUM VALUE OF T.
        READ (105,999) T,X,DELTAT,TMAX
999 FORMAT (4F10.0)
C       WRITE HEADINGS
        WRITE (108,998)
998 FORMAT (20H      T                X//)
        WRITE (108,997) T,X
997 FORMAT (E10.4,E15.4)
20 XNEW = X+0.5*F(X,T)*DELTAT+0.5*F(X+F(X,T)*DELTAT,T+DELTAT)*DELTAT
        T = T+DELTAT
        WRITE (108,997) T,XNEW
        IF (T.GE.TMAX) GO TO 10
        X = XNEW
        GO TO 20
10 CALL EXIT
        END
        FUNCTION F(X,T)
C       INSERT FUNCTION OF INTEREST ON NEXT CARD.
        F = X
        RETURN
        END
C       DATA FOR EXAMPLE PROBLEM
        0.0        1.0        0.1        1.0
```

FIGURE 17.1   Fortran IV program for the numerical solution of a differential equation using the second order Runge–Kutta method.

Index